CRISPR-Cas:
Applications in Gene Editing & Beyond

Roohi Bansal, PhD

ISBN: 978-93-5891-217-3

More books from the author

Antibodies and their role in therapeutics

CRISPR-Cas system based diagnosis of infectious diseases

About the book

Whether you're a curious learner intrigued by the promise of CRISPR, a researcher planning to use it in your experiment, or someone concerned about the ethical dilemmas surrounding this cutting-edge technology, this book " CRISPR-Cas: Applications in Gene Editing & Beyond" is the perfect resource for you.

Written in an engaging style, this comprehensive guide covers everything you need to know about CRISPR, from its core mechanism, evolutionary roots, components, and regulation to its delivery methods. The book delves deep into the details of gene editing, how CRISPR functions as a gene-editing tool, and what distinguishes it from other gene editing techniques like ZFNs and TALENs. Additionally, this book explores the wide-ranging applications of CRISPR and discusses how this revolutionary tool is reshaping agriculture, animal science, gene drives, therapeutics, and much more. Alongside these breakthroughs, the book thoughtfully examines the potential risks and unknown consequences of CRISPR technology, providing a balanced perspective on its pros and cons. Furthermore, it addresses the ethical concerns surrounding this technology, emphasizing the significance of safety, human dignity, and ethical principles while also taking into account its potential long-term impact on future generations.

With insights into the engineering of Cas9 variants, Cas9 orthologs, and CRISPR-mediated precise gene editing techniques such as prime editing, and base editing, you'll learn how to enhance the specificity and efficiency of CRISPR while minimizing unintended consequences.

But the wonders of CRISPR don't end with gene editing – this book also illuminates its potential for gene regulation, infectious disease diagnostics, chromatin immunoprecipitation (ChIP), live cell imaging, and more. The possibilities are truly limitless!

Perfect for curious minds from all backgrounds, "CRISPR-Cas: Applications in Gene Editing & Beyond" is an invaluable resource for medical students, doctors, researchers, and anyone interested in biotechnology, biology, immunology, genetics, molecular biology, cell biology, or bioinformatics. With every turn of the page, you'll uncover eye-opening insights and get inspired by the boundless potential of CRISPR technology, transforming the way you view the future of science and medicine.

So are you ready to unlock the possibilities of one of the most revolutionary discoveries of our time?

Contents

Chapter 1: Basic Molecular Biology concepts

1.1. What is a genome and DNA?

The 2020 Nobel Prize in Chemistry was awarded to two scientists Jennifer Doudna and Emmanuelle Charpentier for the discovery of the CRISPR- gene editing mechanism, also known as CRISPR/Cas9.

In this book, we will understand the CRISPR/Cas system and how it can be used for gene editing and other purposes.

But before doing that, it is important to understand what a genome is, what gene editing is, and why gene editing has drawn significant attention as a promising strategy for the treatment of numerous genetic diseases like sickle cell disease, cystic fibrosis, etc., to the generation of engineered crops, wiping out of the malaria-causing mosquitoes and much more.

The term "**genome**" refers to the entire set of genetic instructions that make up a living organism. Your genome is the code that directs your cells how to behave. Cells interacting together lead to the formation of tissues, and the cooperation of tissues makes organs. And finally, the cooperation of organs ultimately creates an organism, you! So you are who you are in large part because of your genome. The same is the case with other organisms. For instance, the genome of a fish guides the growth of fins and gills that enable it to swim and breathe underwater. Similarly, the genome of a plant directs the production of leaves and chlorophyll to harvest energy from the sun. Even in the case of humans, the information stored in our genomes shapes our eyesight, skin tone, hair, height, and susceptibility to illnesses, among other characteristics.

The genome is made up of a molecule called Deoxyribonucleic acid or simply written as DNA. DNA serves as the hereditary material in humans and most organisms except for some viruses that possess RNA as their genetic material (*Fig 1-1*).

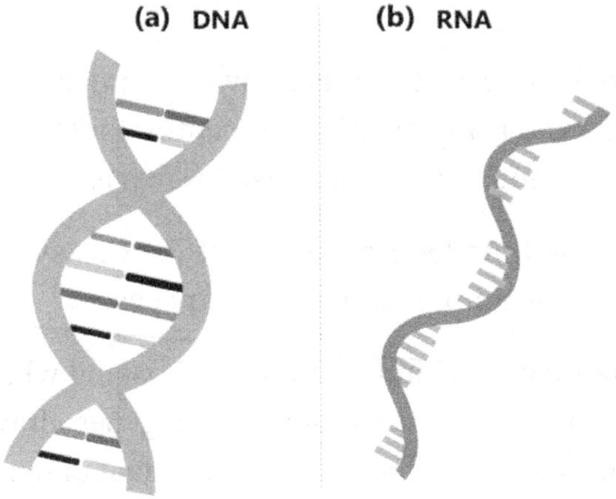

Fig 1-1: (a) DNA and (b) RNA

Like in the case of computers where the data is stored in 0 and 1 binary digits, in human DNA, the information of life is stored as a code comprised of four nitrogenous bases: A, G, C, and T, shorthand for the adenine, guanine, cytosine, and thymine (*Fig 1-2*).

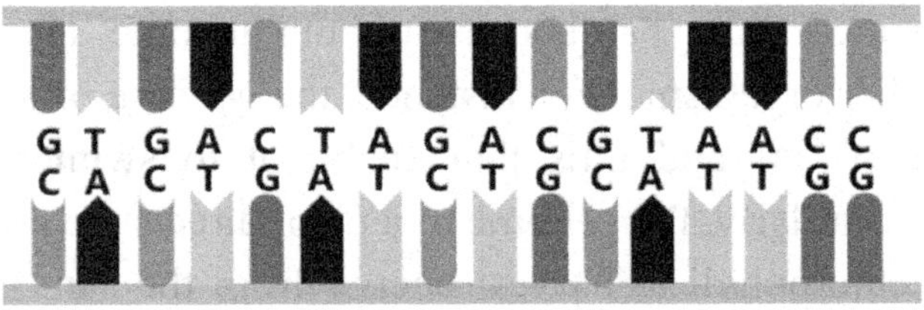

Fig 1-2: Code for the information of life is made up of bases: A, G, C, and T

The arrangement, or sequence of these bases determines the information required for building and maintaining an organism, similar to how letters of the alphabet appear in a certain order to form meaningful words and sentences.

Each nitrogenous base is attached to a sugar molecule and a phosphate molecule. Together, a base, sugar, and phosphate are called a nucleotide (*Fig 1-3*). The nitrogenous bases are derivatives of two-parent compounds, either purine or pyrimidine. Adenine and Guanine bases are two purine bases, and Cytosine and Thymine bases are two pyrimidine bases.

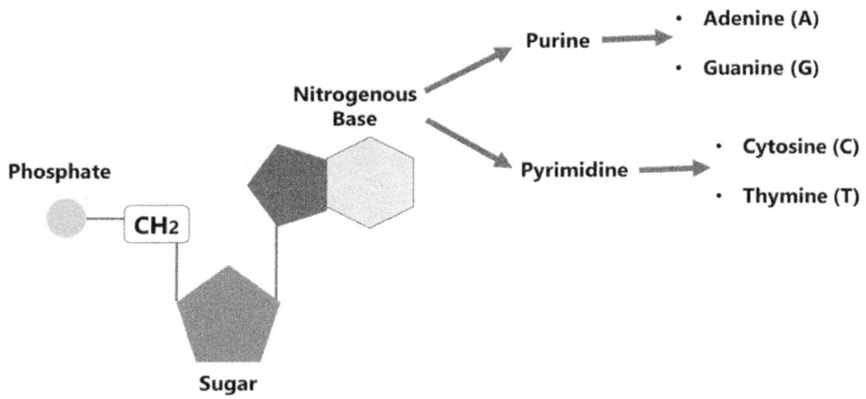

Fig 1-3: Structure of the nucleotide

Many nucleotides are joined together to form a DNA strand. Two of these strands come together to form the double helix structure of DNA.

The backbone of each strand is identical. It is a chain of alternating sugar and phosphate groups. There is a 3' carbon with a hydroxyl group at one end of the strand, this is called the 3'end of the strand (*Fig 1-4*). And there is a 5' carbon attached to a phosphate group at the other end, this is called the 5'end of the DNA strand. In a DNA molecule, one strand runs from 3' to 5', and the other strand runs from 5' to 3'. The two strands are said to be **antiparallel**, which means that the two strands of DNA run in opposite directions.

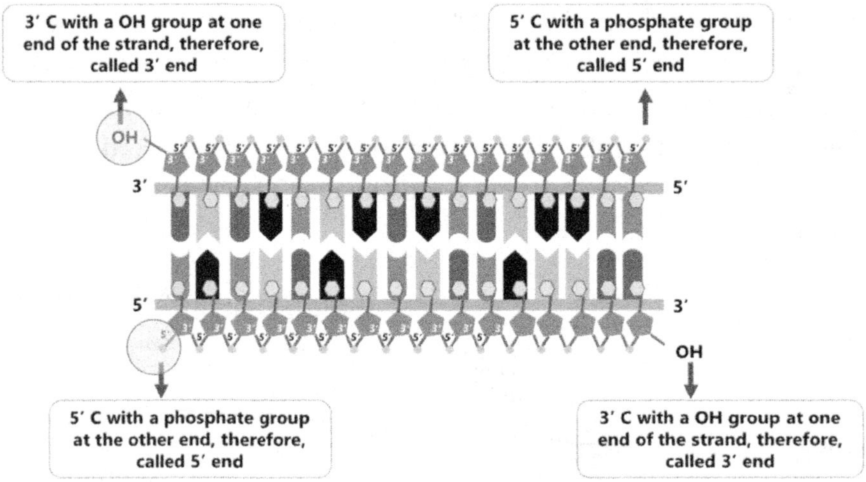

Fig 1-4: 3′ and 5′ ends of DNA strands

Think of the double helix structure of the DNA as a ladder that is twisted into a long, spiraling coil in which the two strands of DNA wrap along a central axis.

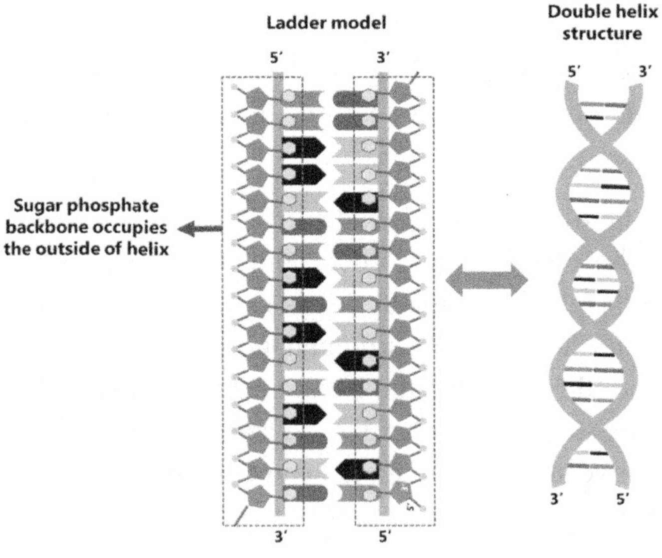

Fig 1-5: Ladder model and double helix structure of DNA

The sugar-phosphate backbone of each strand occupies the outside of the helix and the four different bases are positioned inside (*Fig 1-5*). The position where the nucleotides of both strands meet forms the rungs of the ladder, and the sugar and phosphate molecules form the sides of the ladder.

One strand is held to another by hydrogen bonds between the nucleotide bases. Or in other words, the bases on one strand pair with the bases on another strand, i.e., the base A from one strand always pairs with the base T on the other strand by 2 Hydrogen bonds (*Fig 1-6*). On the other hand, the base G from one strand always pairs with the base C on the other strand by 3 Hydrogen bonds. After pairing, these are known as base pairs and the pairing is called complementary base pairing.

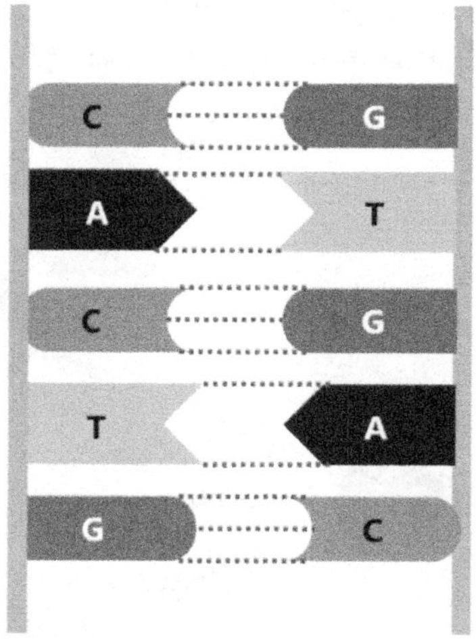

Fig 1-6: A pairs with the T by 2 H bonds and C pairs with G by 3 H bonds

Simply put, A pairs with T and C pairs with G. Or in other words, if there is A on one strand, then there will be T on the other strand. And if there is G on one strand, then definitely there will be C on the opposite strand. And the two strands are then said to be complementary to each other. Thus if we know the sequence of one strand, then using the complementary base pairing rule, we can know the sequence of the other strand.

To sum up, alternating sugar and phosphate groups are linked together to form a backbone, and complementary bases are attached to them. Due to the bond angle of the sugar-phosphate molecules, the linkages will eventually form a double helix structure.

1.2. Concept of directionality

The double helix of DNA and the complementary base pairing rule suggests the molecular basis of heredity and how DNA can transmit genetic information to the two daughter cells during cell division through a process called DNA replication.

DNA replication is a biological process of producing two identical replicas of the DNA from one original DNA (*Fig 1-7*). As a result, at the end of the cell division, we have two completely new cells, each with its own DNA set.

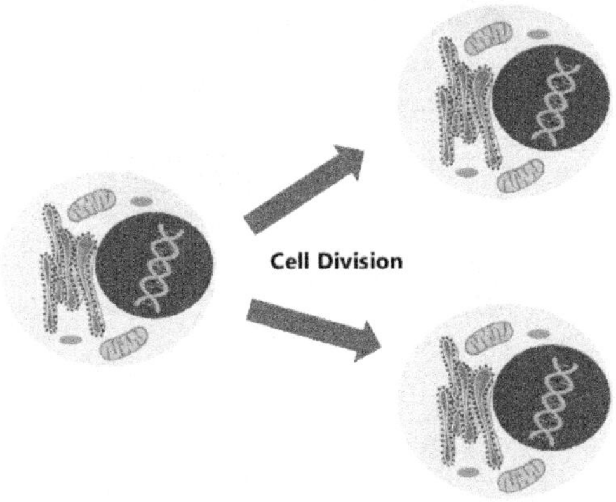

Fig 1-7: During DNA replication, two completely new cells are formed, each with its own DNA set

We have already discussed about DNA structure and complementary base pairing rules. Now let's start with an overview of the process of DNA replication. If you untwist the double helix structure, you will get a linear and ladder model of the DNA, and it is also a simple depiction of the structure of DNA.

In this model, the periphery light grey lines represent the sugar-phosphate backbone of the DNA, and the dark grey lines in the center, represent the nitrogenous base pairs of the DNA (*Fig 1-8*). In the process of DNA replication, the two strands of DNA separate from each other. The original DNA which separates to form these two strands is called parent DNA, and the two separated strands act as templates upon which new strands of DNA are formed to generate two molecules of double-stranded (also written as **ds**) DNA.

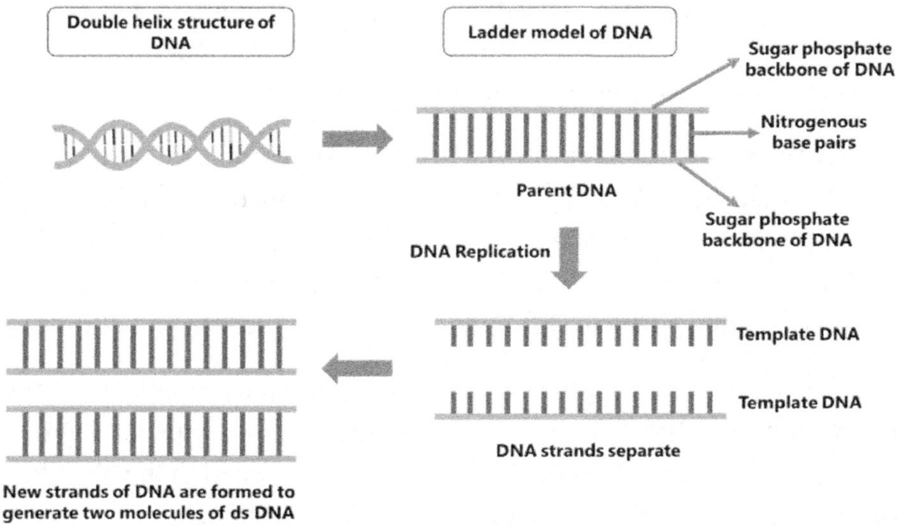

Fig 1-8: Overview of DNA replication process

This process is made possible by means of complementary base pairing. In a nutshell, base Adenine on one strand always pairs with base Thymine on the other strand, and base Cytosine on one strand pairs with base Guanine on the other strand and vice versa.

So when the parent DNA separates, each strand acts as a template for the synthesis of a new strand; this leads to the formation of two completely new sets of DNA which are called daughter DNA.

Now, if you take a closer look at the ladder model of DNA, you can see that the nitrogenous bases of the two strands are held together in the center by hydrogen bonds. The other important feature of DNA is that the two strands have direction. And these two strands run in opposite directions to each other, which means they are antiparallel. In DNA, the directions are determined by 5'end and 3'end. In a DNA molecule, one strand runs from 3' to 5', and the other strand runs from 5' to 3'. To better understand the concept of direction in DNA, let's take a close look at the structure of DNA. We know that the different nucleotides join together to form a DNA strand.

Structure of one nucleotide: As already discussed, each nitrogenous base is attached to a sugar molecule and a phosphate molecule. In DNA, the 5-carbon sugar is deoxyribose; thus, the full name of DNA is deoxyribonucleic acid. The carbons in a deoxyribose sugar are numbered with primes to differentiate between them (*Fig 1-9*).

The 1' carbon is the carbon that is bonded to the nitrogenous base, and the 5' carbon connects to the phosphate group. This phosphate group at the 5' end then binds to the hydroxyl group attached to the 3' carbon of the next nucleotide through a phosphodiester bond.

Thus all the nucleotides in a single strand (also written as **ss**) of DNA are joined by a phosphodiester bond between their 3' and 5' carbons.

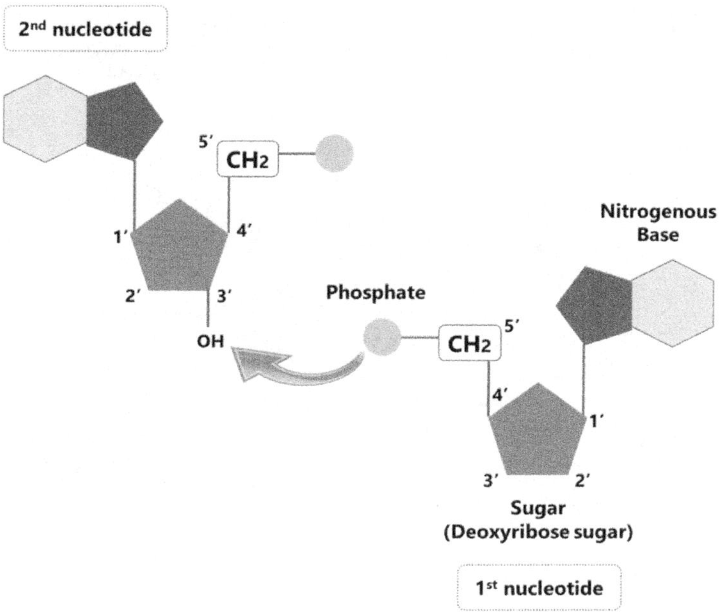

Fig 1-9: Nucleotides are joined by a phosphodiester bond between their 3' and 5' carbons

This means that the respective 5' and 3' carbons are exposed at either end of the DNA strand, which are therefore called the 5' end and the 3' end. 5' is the end where the phosphoryl group is exposed, which is also called the phosphoryl end.

On the other hand, the 3' end is the end where the hydroxyl group is exposed; therefore, also called the hydroxyl end (*Fig 1-10*).

Now, if you see the opposite strand, you will see that it runs in the opposite direction, 3'-5'. It is because of the different orientations of the sugar molecules in the two strands.

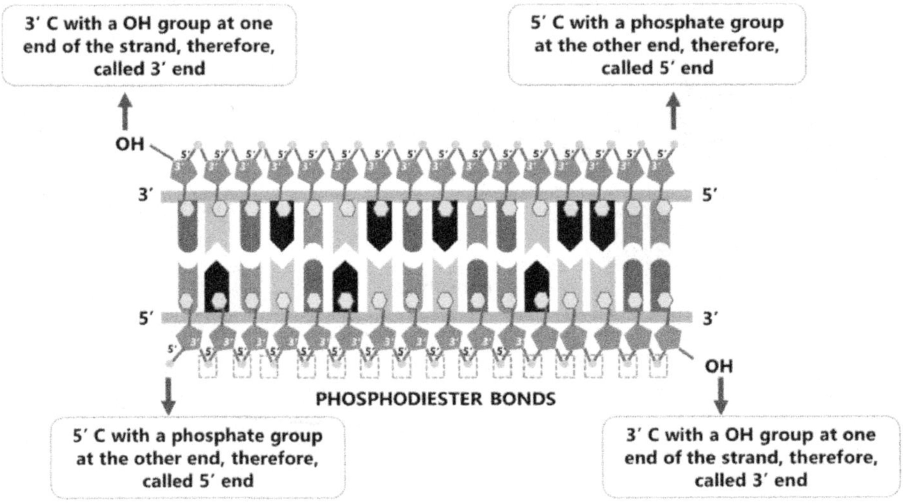

Fig 1-10: 3' and 5' ends of DNA strands

1.3. DNA Replication

The concept of directionality is important in the DNA replication process. It is because of the enzyme called DNA polymerase, which is very important in undergoing DNA replication. This enzyme can add nucleotides only at the 3' end of the DNA. Or in other words, it can synthesize new strands only in the 5'-3' direction. DNA polymerase also requires a primer to synthesize new strands. Primers are single-stranded sequences that are complementary to the template strands but their length is much shorter. They bind only to the very short sequences present at the 3' end of each strand.

The primers have a free 3' hydroxyl group to which new nucleotides can be added (*Fig 1-11*).

In other words, part of the new strand must always be in place because all DNA polymerases can only add nucleotides to a preexisting strand. Therefore, it cannot synthesize new DNA from scratch.

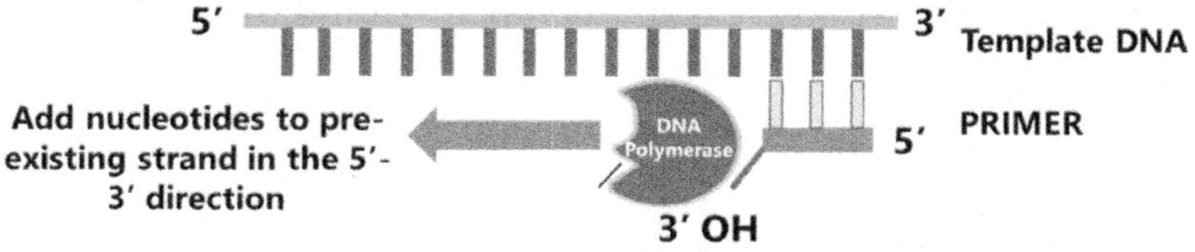

Fig 1-11: Primers have a free 3'OH group to which new nucleotides are added in the 5'-3' direction

(i) For the process of DNA replication to begin, the two strands of DNA need to open up. This job is done by an enzyme called DNA helicase, also known as the unzipping enzyme. The enzyme helicase unzips DNA by acting on the hydrogen bonds present between the base pairs of opposite strands (*Fig 1-12a*). Separating the two single strands of DNA creates a 'Y' shape structure called a replication 'fork.'

(ii) To prevent the two separated strands of DNA from joining, the special type of proteins known as single-stranded binding proteins hold the two strands of DNA apart.

(iii) The two separated strands now act as templates for making the new strands of DNA. Let's suppose the upper strand is 3'-5' and the lower strand is 5'-3'. The enzyme called primase synthesizes a short piece of RNA called a primer complementary to the upper strand (*Fig 1-12b*), and then this primer acts as the starting point for DNA synthesis. And the DNA polymerases can only add nucleotides to a preexisting nucleotide strand (*Fig 1-12c*).

(iv) DNA polymerase, which has the job of forming new strands of DNA, can only form a strand in the direction 5' to 3'. It synthesizes the strand by adding new nucleotides to the 3' end of the growing chain.

(v) Now, since the upper daughter strand of the parent DNA is running from 5'-3' direction, the DNA polymerase can easily synthesize it. Since the helicase is also unzipping the DNA in the same direction, the DNA polymerase can form this upper daughter strand continuously. Due to this reason, the upper strand of DNA is known as the leading strand.

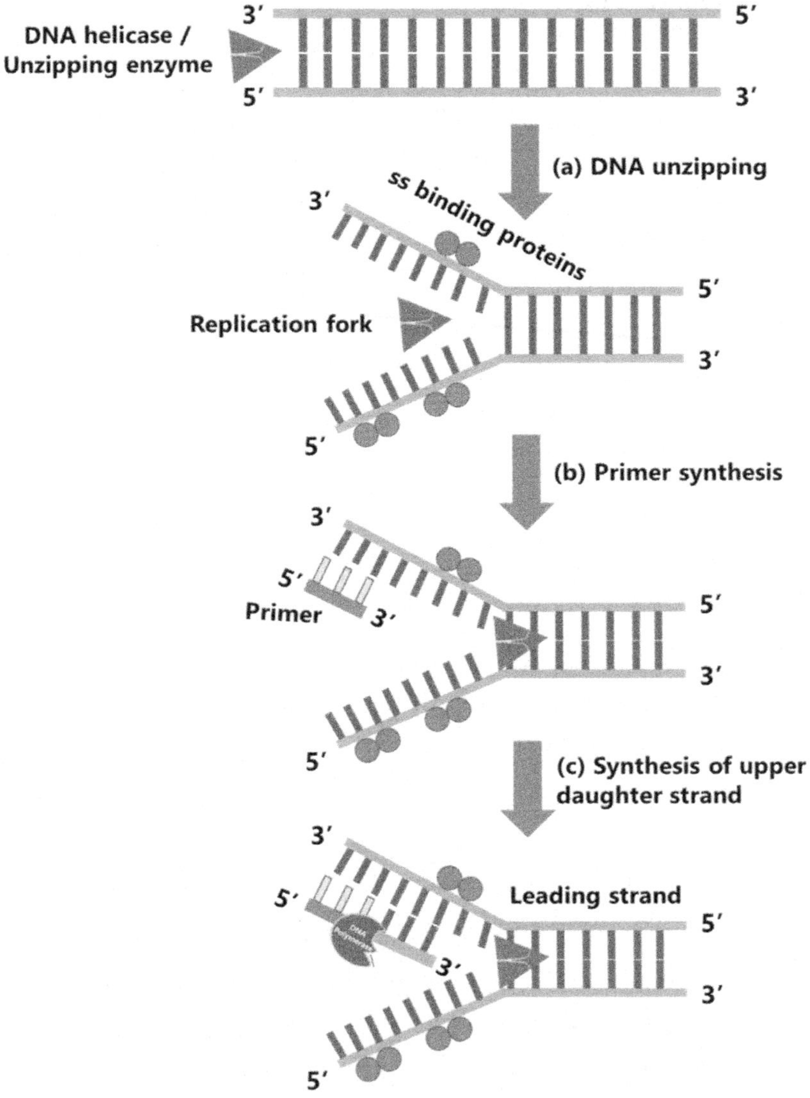

Fig 1-12: During replication, DNA polymerase can synthesize the upper daughter strand continuously.

(vi) But things are different when we look at the opposite strand of DNA, which is running in the opposite direction, 5'-3'. Since the DNA polymerase has to synthesize the daughter complementary strand in the 5'-3' direction and DNA helicase unzips DNA in the opposition direction, the DNA polymerase cannot continuously synthesize the new strand.

Well, this problem is solved by forming the new strand in short segments known as the Okazaki fragments (*Fig 1-13*). For this, the primase enzyme synthesizes a short piece of primer, and then DNA polymerase synthesizes a short fragment of DNA.

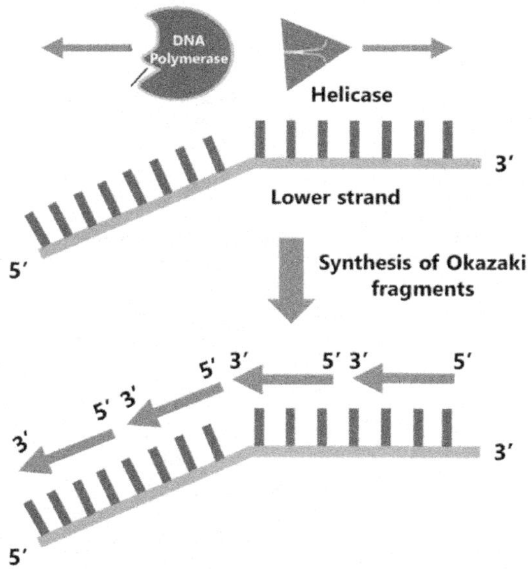

Fig 1-13: New strand is formed in short segments known as Okazaki fragments

When enzyme helicase has opened more of the DNA helix, then again a primase enzyme synthesizes a new primer which is followed by synthesizing one more short fragment of the daughter strand by DNA polymerase (*Fig 1-14*). This process is repeated multiple times; thus, this sort of replication is known as discontinuous DNA synthesis. Due to this reason, the opposite strand of DNA is known as the lagging strand.

(vii) Once the synthesis of the daughter strand complementary to the lagging strand is done, it appears as short segments of DNA with interrupting segments consisting of RNA primers.

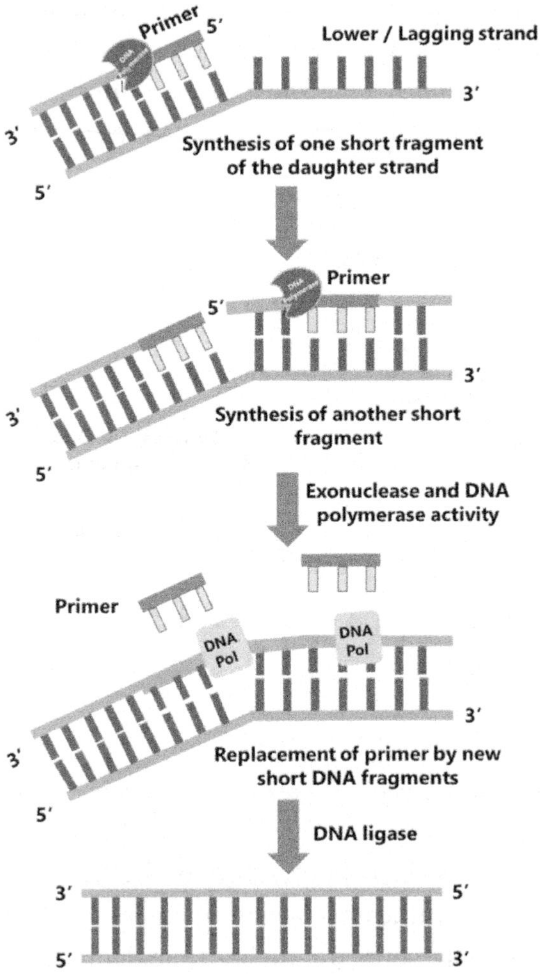

Fig 1-14: Synthesis of daughter strand complementary to the lagging strand

In prokaryotes, elongation of both the leading and lagging strands during DNA replication is primarily carried out by **DNA polymerase III.** While in eukaryotes, DNA polymerase δ (Pol δ) is the primary polymerase enzyme responsible for the elongation of the lagging strand during DNA replication, while DNA polymerase ε (Pol ε) is responsible for the elongation of the leading strand.

(viii) Then an enzyme called exonuclease comes into play which strips away the primers, and the gaps are then replaced by new short segments of DNA by another type of DNA polymerase, i.e., DNA Polymerase I in prokaryotes.

(ix) Finally, an enzyme called DNA ligase seals up or joins the DNA fragments to form a continuous DNA strand.

(x) Lastly, checking for errors in the newly synthesized DNA is essential. This is done by DNA polymerases I or II in prokaryotes (and polymerases δ, ε or α in eukaryotes), which proofread the newly synthesized DNA for any wrong bases that may have been inserted, and if they find a wrong base, they remove the base and replace it with the correct base.

(xi) After the replication process is completed, two double-stranded DNA molecules are formed. Each newly formed DNA consists of one new strand and one original strand. This is why DNA replication is described as **semi-conservative**, because half of the DNA is part of the original DNA molecule, and the other half is brand new.

(xii) Following the replication, the newly synthesized DNA molecules coil up to form a double helix structure.

So, this was a brief overview of DNA replication.

1.4. Genes

You can think of a cell's total DNA content as a cookbook containing detailed recipes for every aspect of the organism, from physical appearance to metabolism, reproduction, etc. The recipes are the genes. A gene is a fragment of DNA that accounts for a specific characteristic, such as hair color, skin color, height, or any other information pertaining to the organism's appearance or functioning. You can also describe genes as the working subunits of DNA.

Each gene is a long stretch of DNA containing a specific sequence of As, Gs, Cs, and Ts that code for a particular set of instructions required for synthesizing a functional biological product, whether a protein or RNA. The human genome contains 3 billion base pairs, and there are about 30,000 protein-coding genes. The size of genes ranges from a few hundred DNA bases to over 2 million bases.

Although thousands of genes code for certain proteins, large stretches of DNA in the human genome do not code for proteins. These regions are known as **introns**, making up about 95% of the genome. Some of these non-coding DNA have a purely regulatory function. For example, they may provide signals that denote the beginning or end of genes, influence the transcription of genes or function as initiation points of replication, etc.

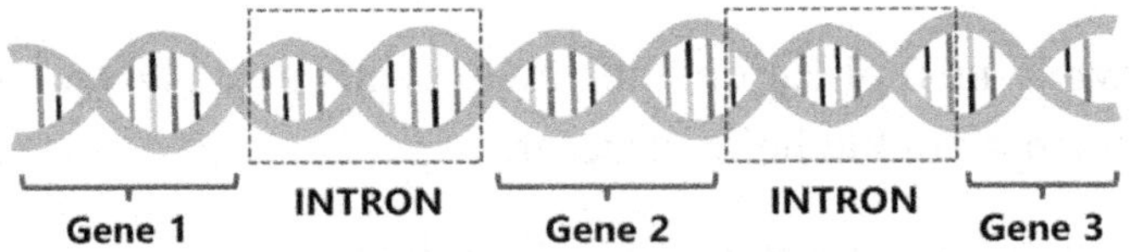

Fig 1-15: Introns make up about 95% of the genome

All the human genome information is coded in the DNA in genes in the form of nucleotide base pairs. Thus, the specific sequence of nucleotides on genes provides instructions to produce physical traits like hair color, skin color, height, eye color, and instructions for breaking down food, movement, vision, synthesis of enzymes, etc. And the final product of these instructions is proteins.

For instance, pigments responsible for providing hair, eye, or skin color are proteins. Likewise, hemoglobin is a protein that transports oxygen to the cells. Enzymes are the proteins that help digest food, DNA replication, and break down toxins in the body. Antibodies are also proteins that are produced by the immune system to help remove foreign substances and fight infections.

In short, proteins form the basis of living cells.

1.5. DNA transcription

To transform the instructions contained in DNA into proteins, the cells use an intermediary molecule called ribonucleic acid or RNA (*Fig 1-16*). RNA is produced from the DNA template via a process called transcription.

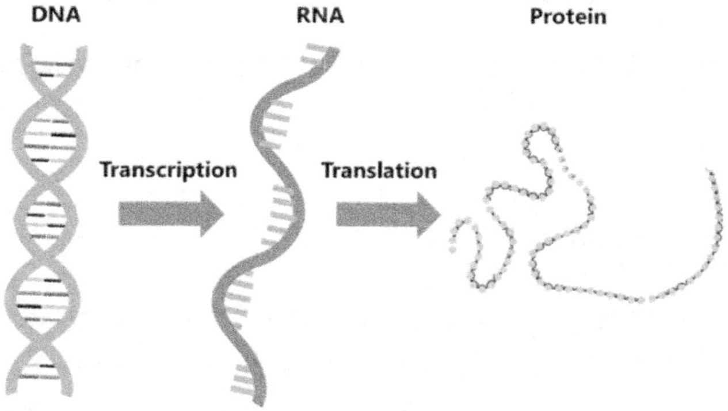

Fig 1-16: RNA is produced from DNA via transcription

The sugar that makes up the backbone of RNA has additional oxygen than the sugar backbone in DNA (*Fig 1-17*). Hence DNA stands for deoxyribonucleic acid, and RNA stands for ribonucleic acid.

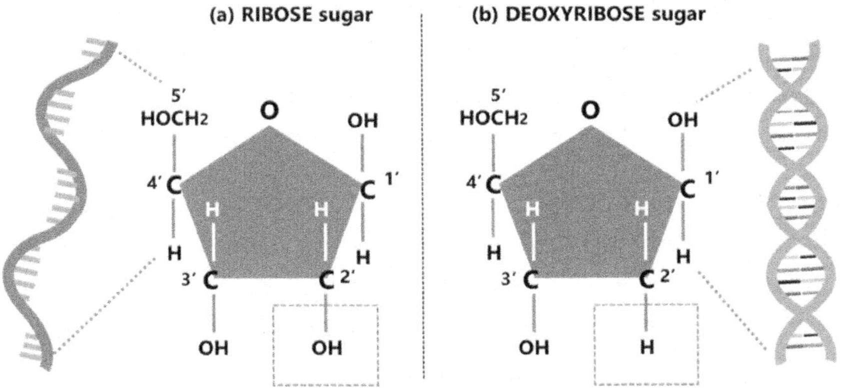

Fig 1-17: (a) Ribose sugar in RNA and (b) deoxyribose sugar in DNA

The process of transcription occurs in 2 broad steps:

1. First, the enzyme RNA polymerase forms a pre-messenger RNA molecule. This happens in the nucleus by adding the base pairs complementary to the base sequence of the DNA template strand.

2. In the second step, pre-mRNA is edited to produce the desired mRNA, which is the final product of the transcription.

The first step of the transcription takes place in 3 phases:

- initiation,
- elongation
- and termination

The main enzyme involved in the transcription process is RNA polymerase. The RNA polymerase has the main job of copying the information present in the DNA to a new RNA molecule. RNA polymerase synthesizes the RNA strand from 5'-3' direction adding new nucleotides to the 3' end.

(i) During initiation, the RNA polymerase binds to the promoter region of the DNA to initiate transcription. Let's assume the gene, shown in black color, needs to be transcribed (*Fig 1-18*). At the beginning of the gene, there is a sequence of base pairs known as the promoter region of the gene. The promoter region serves as the guide, which tells the RNA polymerase where to attach.

Then the RNA polymerase binds to the promoter region with the help of specific proteins called transcription factors (denoted as TF). Once the RNA polymerase binds to the promoter region, it separates the two strands of DNA.

One of the strands serves as the template strand or antisense strand, which is used to generate mRNA. And the other strand is the non-template strand or sense strand. Note that the template strand always runs in the 3'-5' direction. It is because RNA polymerase can only synthesize RNA in the 5'-3' direction, so it can only use the DNA strand as a template that runs from the 3' to 5' direction.

(ii) After the RNA polymerase binds to the promoter region of the DNA, the second step begins, which is elongation. During this process, the information present on the template DNA in the form of nucleotides is copied to form a new complementary RNA molecule.

Unlike DNA polymerase, RNA polymerase does not require a primer for mRNA synthesis. But like DNA synthesis, the synthesis of RNA molecules also utilizes complementary base pairing rules.

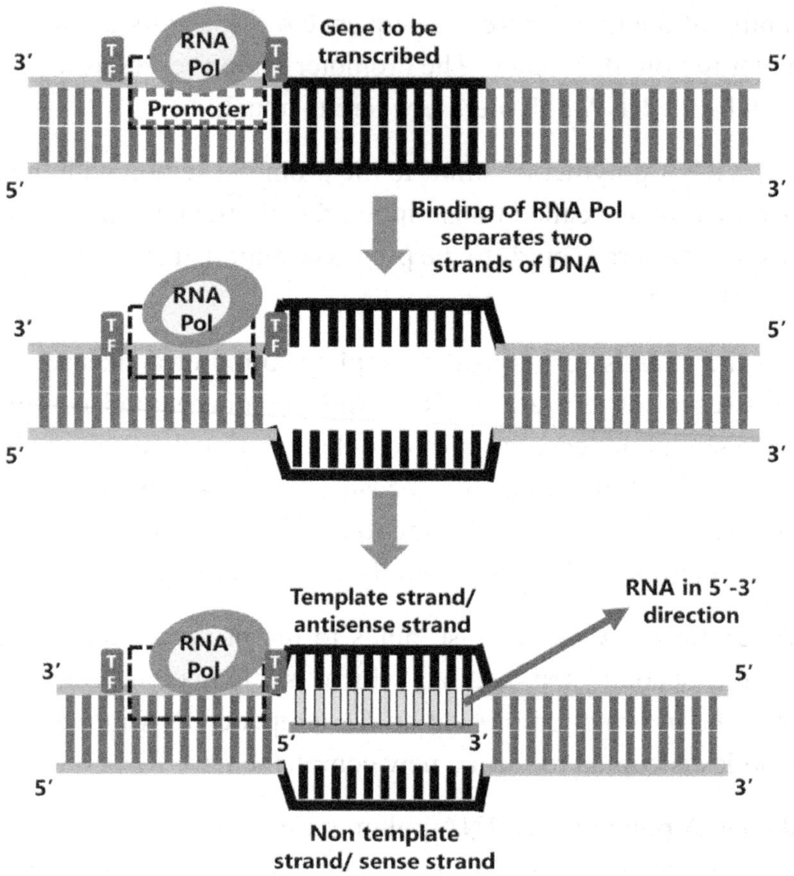

Fig 1-18: The initiation phase of transcription

As already discussed in DNA replication, the DNA polymerase adds the base adenine (A) on the new strand when there is base thymine (T) on the template strand and adds base cytosine (C) on the new strand when there is base guanine (G) on the template strand.

Similarly, in the case of RNA as well, there are four bases: The three bases A, G, and C are the same as found in the case of DNA, but the fourth base, thymine is replaced with the base uracil. So there are no T's in the nucleotide sequence of RNA.

Therefore, whenever the RNA polymerase enzyme detects the base adenine in the template strand, it adds uracil in the RNA strand (*Fig 1-19*). This is how the RNA polymerase adds new base pairs at the 3' end of the growing RNA strand.

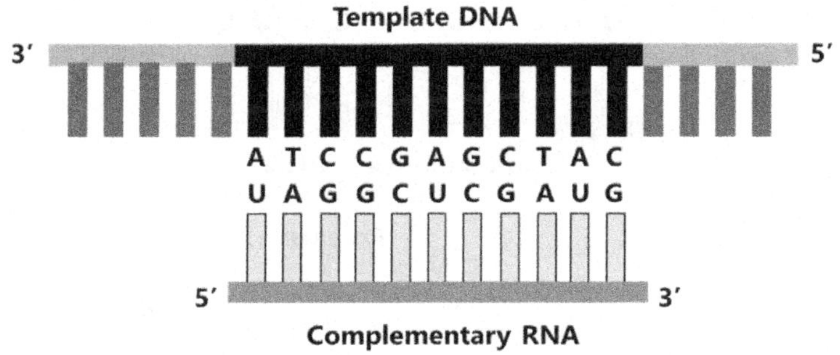

Fig 1-19: Complementary base pairing between template DNA and mRNA

(iii) The next step is termination. Just like the promoter region, which signals the RNA polymerase where to start, the gene also has a region that signals the RNA polymerase enzyme to stop the transcription. This signal is known as a terminator. It signals RNA polymerase that the RNA transcript is now complete.

Once the RNA polymerase reads and transcribes the terminator region, the newly formed RNA transcript is released from the transcription complex (*Fig 1-20*). The RNA formed is pre-messenger RNA represented as pre-mRNA.

Then in the next step, the pre-mRNA is converted to form a mature mRNA molecule.

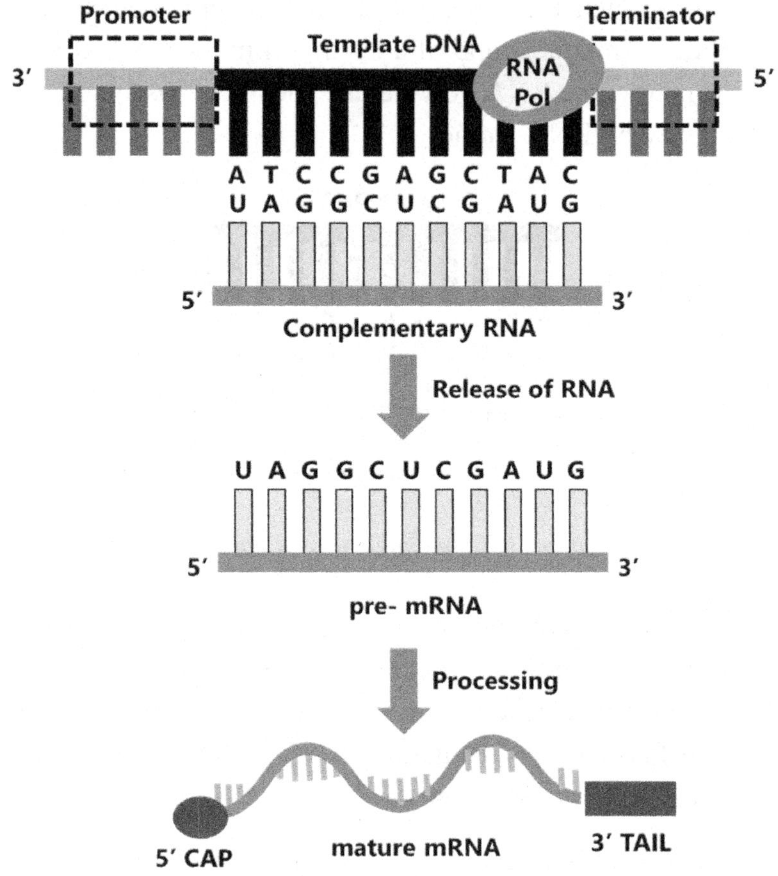

Fig 1-20: pre-mRNA is formed in the termination phase

Imagine that you want to publish a new book and you have printed all the pages. But you cannot sell the book yet because this new book needs both front and back covers. So, before selling it, it is also important to check any blank or wrong pages that may have been inserted.

These two processes essentially represent how the pre-mRNA is processed.

(i) The first step involves the addition of a cap and tail to the ends of this pre-mRNA molecule. This cap and tail are short amino acid sequences that protect the pre-mRNA molecule because the free ends of pre-mRNA are vulnerable to getting attacked by many enzymes (*Fig 1-21*).

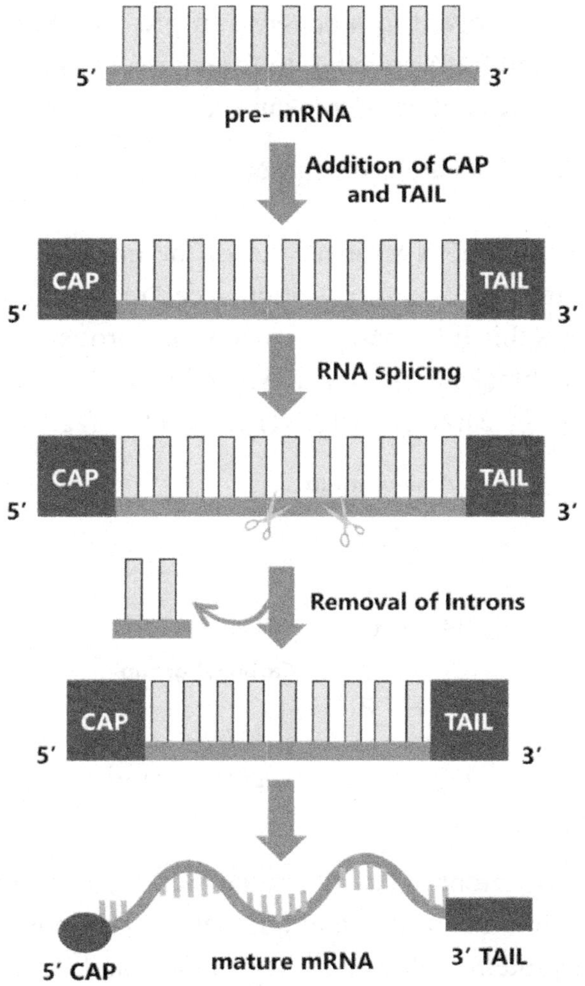

Fig 1-21: pre-mRNA processing to generate mature mRNA

(ii) The second step of pre-mRNA processing involves the removal of any junk sequences present that do not code for proteins. These non-coding sequences are known as **introns**. The process of removing introns is known as RNA splicing, in which the RNA-protein complex known as spliceosome removes these non-coding sequences or introns. The rest of the RNA joins together to form a mature mRNA molecule.

Note: The steps involved in processing of pre-mRNA (i.e., addition of cap and tail; and RNA splicing) to form a mature mRNA molecule occurs in eukaryotes but not in prokaryotes.

1.6. Translation or Protein synthesis

After transcription, mRNA moves out of the nucleus and enters the cytoplasm. In the process of translation, a cell reads information from mRNA and uses this information to synthesize proteins. Amino acids are the building blocks for proteins. All amino acids contain an amino (NH2) group and a carboxyl (COOH) group (*Fig 1-22*).

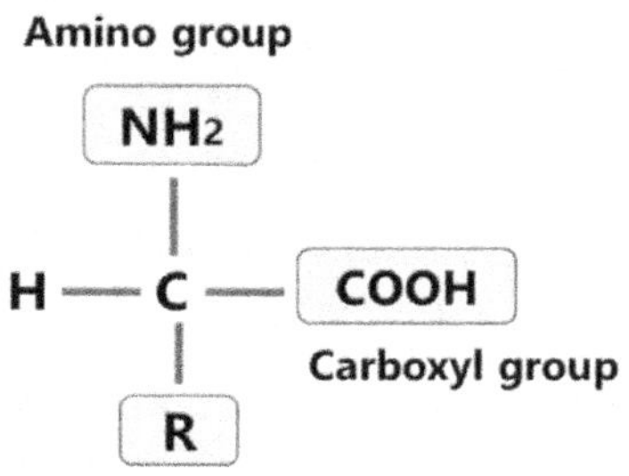

Fig 1-22: Amino acid structure

There are 20 different types of amino acids commonly found in proteins. And in terms of the total number, 300 or more amino acids are found per protein molecule. Before understanding the translation process, it is important to get familiar with polypeptide and protein.

A polypeptide is formed when amino acids are joined together through peptide bonds. A peptide bond results from a condensation reaction between two amino acids: the carboxyl group of one amino acid reacts with the amino group of adjacent amino acid, releasing a molecule of water (*Fig 1-23*).

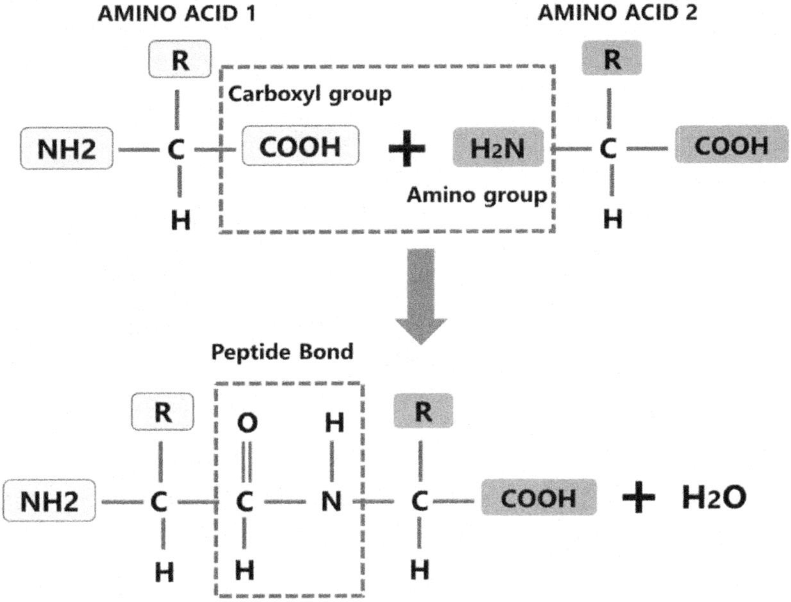

Fig 1-23: Peptide bond between two amino acids

And the term protein is used to describe the three-dimensional structure formed by folding one or more polypeptides. Simply put, a polypeptide is the primary structure of more complex proteins.

mRNA formed from DNA contains the instructions for forming a polypeptide. Within an mRNA, the instructions required for building a polypeptide occur in groups of three nucleotides called codons, where each codon corresponds to a specific amino acid (*Fig 1-24*). Translation occurs in such a way that these nucleotide triplets or codons are read in a non-overlapping fashion. There is no punctuation between codons for successive amino acids of a polypeptide. Thus insertion or deletion of a base pair can alter the sequence of codons, eventually changing the amino acids coded by the mRNA.

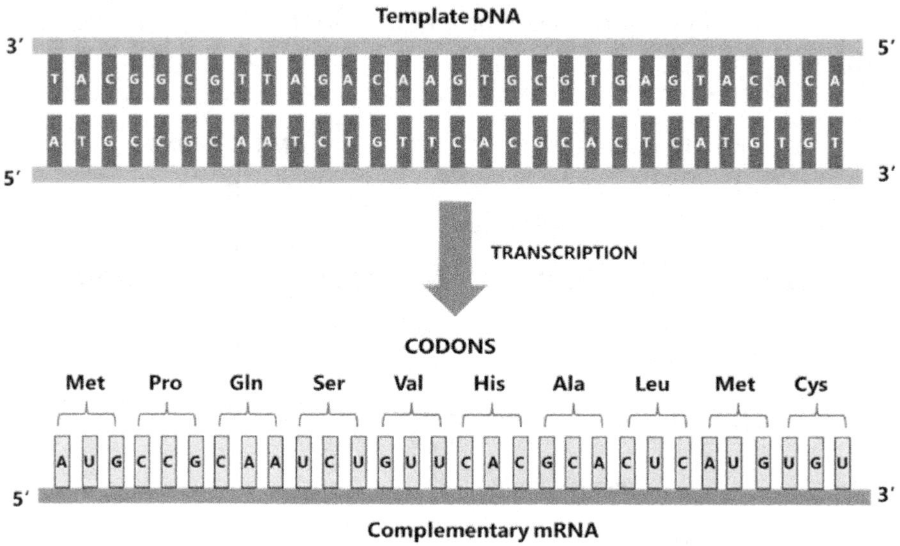

Fig 1-24: In an mRNA, instructions for building a polypeptide are in groups of three nucleotides called codons; each codon codes for a specific amino acid

Here are some key characteristics of codons:

- Since there are four nucleotides and a set of 3 nucleotides is called a codon; therefore 4^3 gives 64 different possible codons. Out of 64 codons, 61 codons code for 20 amino acids. For example, if there are three uracil bases in sequence on mRNA, it means the codon will be UUU and this codon codes for phenylalanine amino acid. Similarly, the codon CCU codes for amino acid proline and so on.
- There are start codons and stop codons that signal to start and stop the translation process. For example, the codon AUG is called the start codon and codes for methionine. This codon signals the translation machinery to start synthesizing the protein.
- Similarly, there are three unique codons called stop codons, which signal the end of the polypeptide synthesis and do not code for any amino acids. These are: UAG, UGA and UAA.

- Another striking feature of genetic code is that one amino acid may be coded by more than one codon, so the genetic code is also described as degenerative. Though there are 61 codons in number, only 20 amino acids exist. Therefore, it is a possibility that an amino acid may have two or more codons. For instance, the alanine amino acid is coded by four codons: GCU, GCC, GCA, and GCG. But remember that each codon codes for a specific amino acid only.

During translation, the codons of an mRNA are read in a sequential order from the 5' end to the 3' end by the molecules called transfer RNAs, or simply written as tRNAs. Each tRNA has an anticodon, which is a set of three nucleotides that binds to a corresponding mRNA codon through base pairing. While the other end of the tRNA carries the amino acid specified by the codon. A minimum of 32 tRNAs are required to translate all 61 codons—31 tRNAs to encode the amino acids and 1 tRNA for initiation.

Apart from tRNAs, ribosomes play a central role in translation. These complex molecular structures inside all living cells act as a site for protein synthesis. The ribosome consists of 2 major components: a smaller subunit and a larger subunit. The unit of measurement used to describe the ribosomal subunits is the Svedberg unit which is a measure of the rate of sedimentation in centrifugation.

In Eukaryotic ribosomes, the smaller subunit is 40S, and the larger subunit is 60S (*Fig 1-25*). When these subunits are joined, they are denoted as 80S ribosomes. These subunits lie separately in the cytoplasm until they are required to come together for translation.

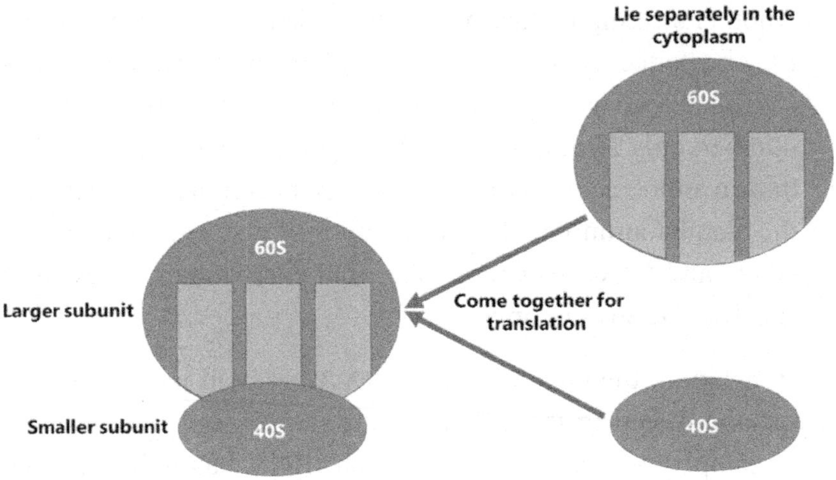

Fig 1-25: Subunits of the ribosome: 60S (larger) and 40S (smaller)

Now let's get into the details of the translation mechanism:

1. As discussed, amino acids are joined together by peptide bonds to form a polypeptide. But for synthesizing a polypeptide, first, all the 20 amino acids must be attached to their corresponding tRNAs. This is done by enzymes called aminoacyl transfer RNA synthetases (*Fig 1-26*). When tRNAs attach to their corresponding amino acid, they are said to be charged.

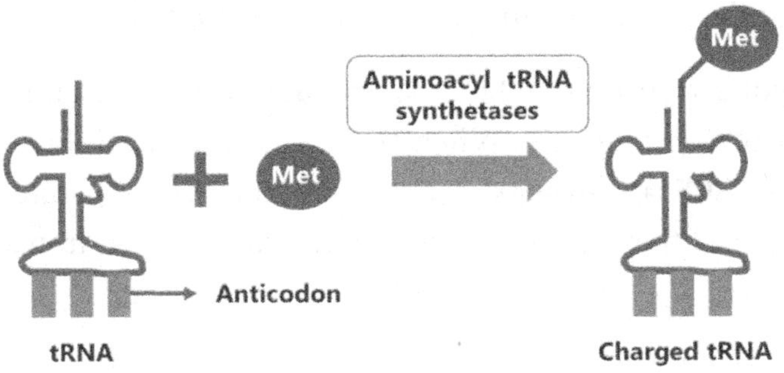

Fig 1-26: tRNAs attach to their corresponding amino acids

2. After charging of tRNAs, the translation process begins. In the initiation step, the subunits of ribosomes lie separately in the cytoplasm. The 40S subunit binds to the initiation factors to prevent the 40S and 60S subunits from combining prematurely. Then the 40S subunit moves along the mRNA in the 5' to 3' direction until it recognizes the AUG start codon, which specifies the methionine residue, and signals the beginning of the protein synthesis (*Fig 1-27a*). This initiation step requires the tRNA attached to methionine amino acid residue to begin translation (*Fig 1-27b*).

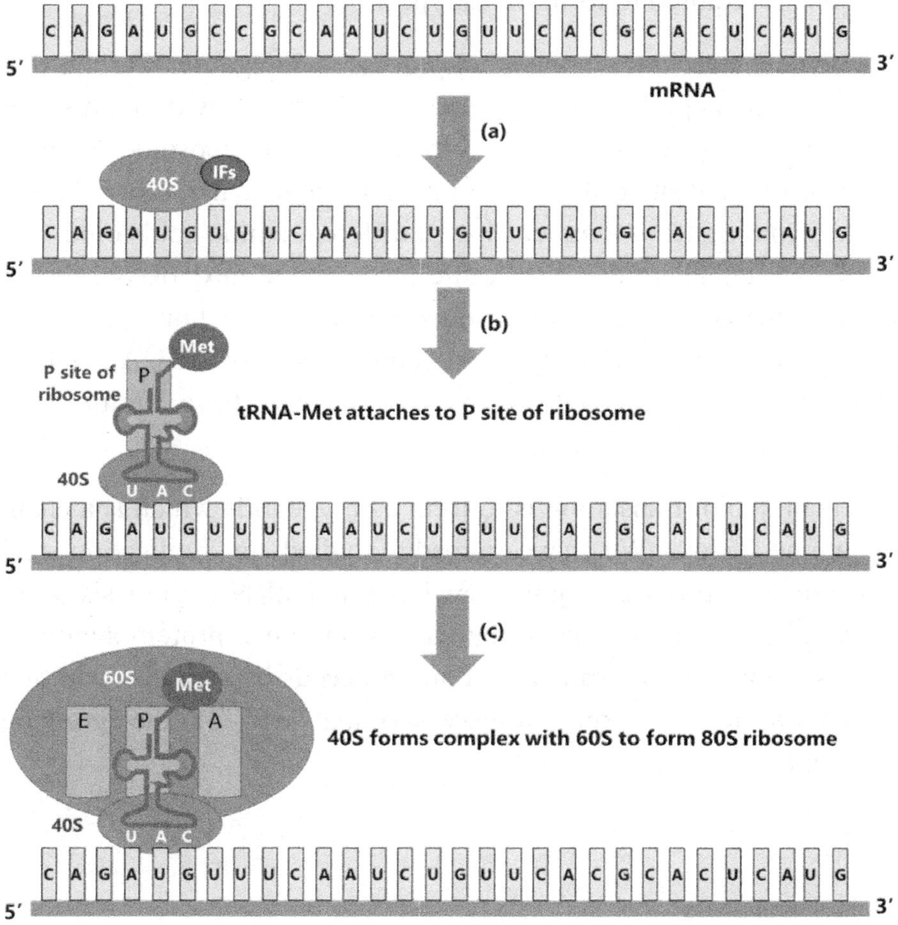

Fig 1-27: Initiation step of translation

The ribosomes have 3 sites: E, P, and A which are present in both the subunits. The start codon AUG is positioned at the P site of the 40S subunit. For initiation of protein synthesis, the charged transfer RNA, which has an amino acid methionine bound to it, attaches to the P site of the smaller subunit 40S of the ribosome. At this site, the anticodon of this tRNA (UAC) pairs with the AUG codon of the mRNA. As soon as the tRNA finds the start codon AUG, the 40S subunit with attached tRNA immediately forms a complex with the larger subunit 60S to form the complete 80S ribosome (*Fig 1-27c*). Assembly of ribosome subunits is required to form the polypeptide.

3. The next step in the translation process is elongation. During this step, the polypeptide chain gets longer. The tRNA with amino acid methionine attached to it occupies the P site of the ribosome. Now on one side of the P site is the E site and on the other is the A site. All the other incoming tRNAs attached with different amino acids bind first to the A site. Therefore, the A site receives the next tRNA, which matches the codon next in the sequence in mRNA. For example, if after the first codon AUG, the next codon present in mRNA is UUU, then the tRNA attached to the A site will have amino acid phenylalanine (*Fig 1-28a*).

The main job of translation is to form a polypeptide which is a chain of amino acids. We have already studied in replication and transcription processes that the new DNA and mRNA synthesis occur in the 5’-3' direction. But in the case of proteins, protein synthesis begins at the amino-terminal end and proceeds by stepwise addition of amino acids to the carboxyl-terminal end of the growing polypeptide.

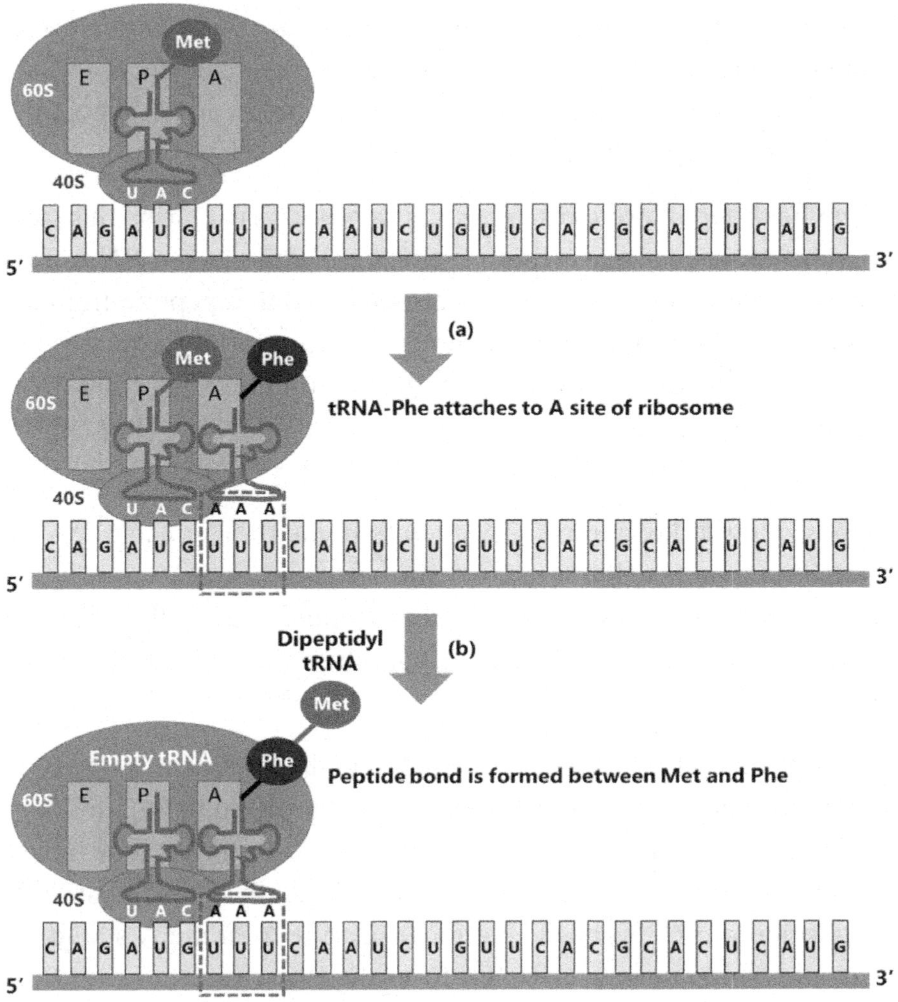

Fig 1-28: Elongation step of translation

The methionine residue coded by the AUG start codon determines the amino terminal of the polypeptide. Therefore the next step is to form a peptide bond between the two amino acids bound to their tRNAs at the P and A sites of the ribosome, and the following amino acid is added to the carboxyl-terminal of the growing polypeptide. This occurs by the transfer of the initiating methionine amino acid attached to the tRNA at the P site to the amino group of the second amino acid, phenylalanine, in the A site (*Fig 1-28b*). After this reaction, the tRNA in the P site loses its amino acid, and on the other hand, dipeptidyl tRNA is produced in the A site.

4. The final step of protein synthesis is translocation. During this step, the ribosome moves one codon towards the 3' end of the mRNA. This movement shifts the position of tRNA attached to dipeptide from the A site to the P site and shifts the empty tRNA or tRNA which is not attached to any amino acid from the P site, to the E site. From the E site, the uncharged tRNAs are released into the cytoplasm (*Fig 1-29a*). Therefore, now the third codon lies in the A site.

Suppose the third codon is CCA; therefore, the tRNA attached to the glutamine comes and binds to the A site because the CCA codon codes for glutamine amino acid. Again, the peptide bond will be formed between the amino acid in the P site and the amino acid in the A site (*Fig 1-29b*).

Therefore dipeptide attached to the tRNA at the P site is transferred to the amino group of the third amino acid glutamine in the A site. Thus leaving tRNA empty in the P site (*Fig 1-29c*). As the ribosome moves one more codon towards the 3' end of the mRNA, this movement shifts the position of tRNA attached to the tripeptide from the A site to the P site and shifts the empty tRNA from the P site to the E site, from where the uncharged tRNAs are released into the cytoplasm. This is how the elongation and translocation steps repeat until the ribosome reads the termination codon which signals the termination of polypeptide synthesis. Termination codons can be UAA, UAG, or UGA; these termination codons do not code for any amino acid. Once the termination codon occupies the A site of the ribosome (*Fig 1-30a*), the polypeptide is released from the ribosome (*Fig 1-30b*).

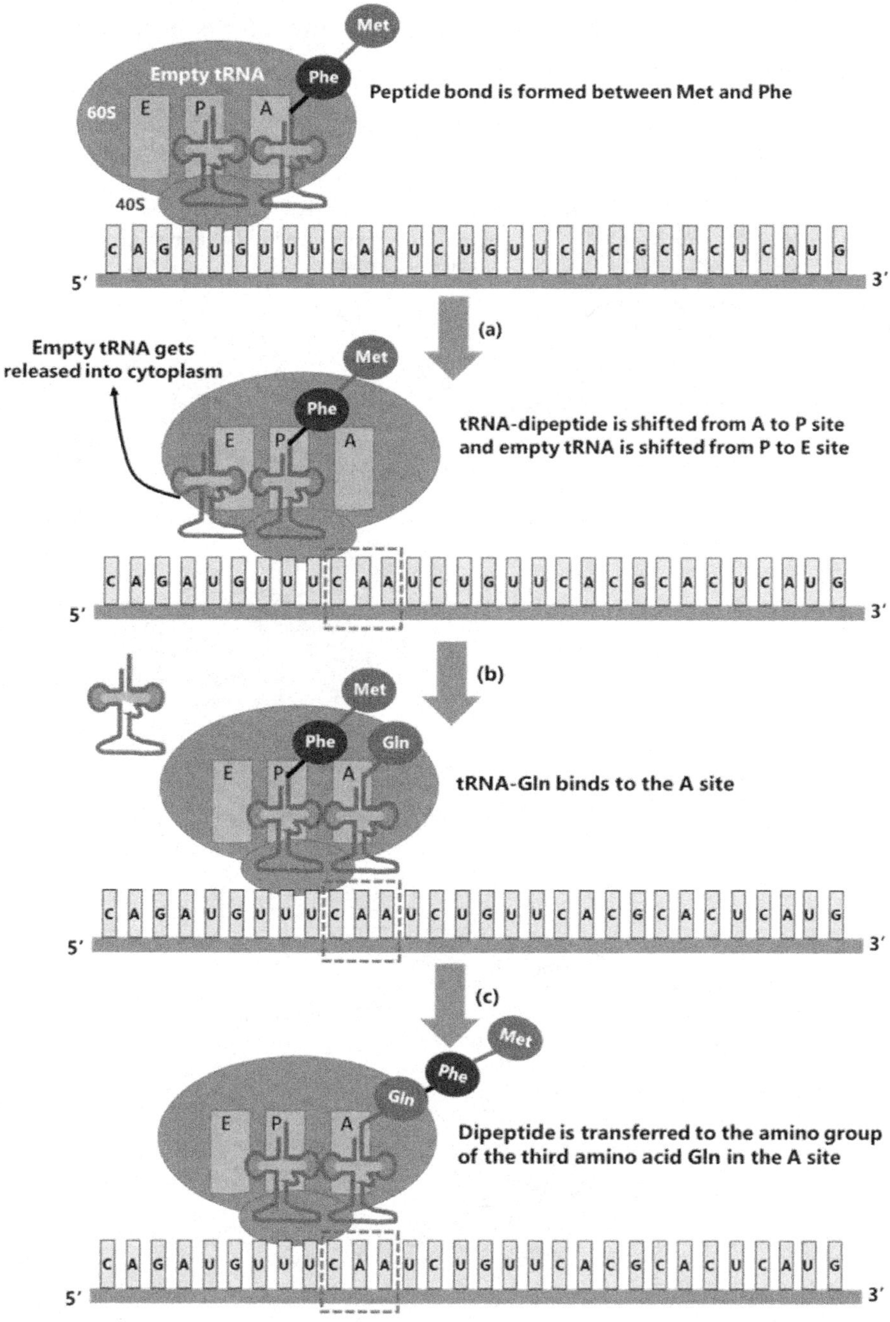

Fig 1-29: Translocation step of translation

5. In the final stage of protein synthesis, the polypeptide chain is folded and processed into its biologically active form (*Fig 1-30c*).

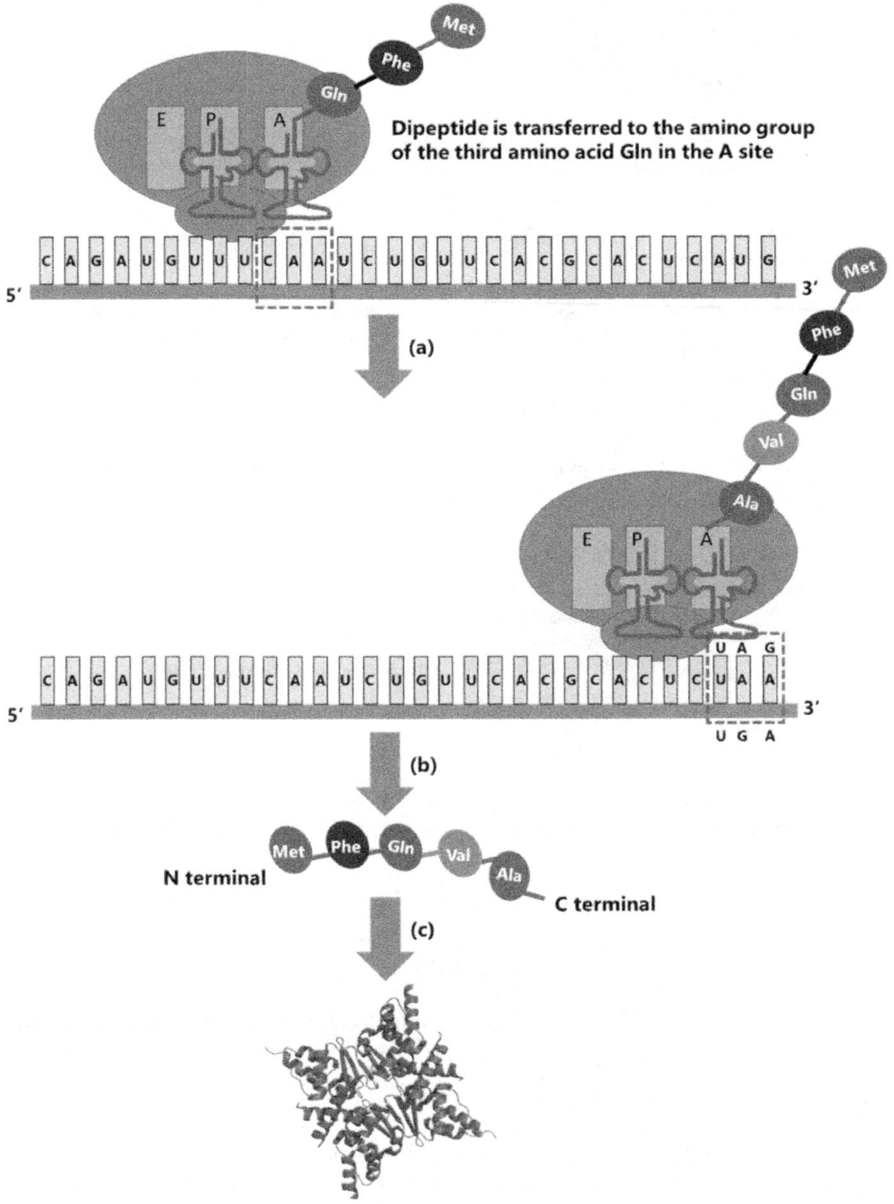

Fig 1-30: Final step of protein synthesis

So in the two-step process of transcription and translation, DNA is transcribed into an mRNA, and mRNA is translated into a protein. Since every gene codes for a specific protein and proteins make up most of what you are, from your muscle tissue and organ tissue to all your receptors and enzymes, this is how DNA carries the code for a living organism.

In a nutshell, this overall flow of genetic information from DNA to RNA to protein is known as **central dogma**, and it is the language used to communicate and express life.

Note: Translation, the protein synthesis process, occurs in the cytoplasm of both eukaryotic and prokaryotic cells. However, there are some differences in the details of translation between the two types of cells. In eukaryotes, after transcription occurs in the nucleus, the mature mRNA molecule is exported to the cytoplasm, where it undergoes translation by ribosomes. While in prokaryotes, transcription and translation occur simultaneously in the cytoplasm. Additionally, prokaryotic ribosomes are smaller than eukaryotic ribosomes, with a sedimentation coefficient of the 70S (composed of a 30S small subunit and a 50S large subunit) in contrast to eukaryotic ribosomes, which are larger with a sedimentation coefficient of 80S (composed of a 40S small subunit and a 60S large subunit).

1.7. DNA Packaging and chromosomes

The DNA within the human genome is composed of approximately 3 billion base pairs. Since human cells are diploid, each cell has 2 sets of the human genome, compared to haploid cells which contain 1 set of the genome. All cells are diploid in humans except for the egg and sperm cells. That means each cell has 6 billion base pairs of DNA or 12 billion bases per cell. A single base pair measures approximately 0.34 nanometers in length, with a nanometer representing one-billionth of a meter. Therefore, each diploid cell contains approximately 2 meters of DNA, as calculated by multiplying the length of a single base pair (0.34×10^{-9}) by the total number of base pairs in a cell (6×10^9).

Moreover, it is estimated that the human body contains about 50 trillion cells, meaning that the total DNA length in the human body will be around 100 trillion meters or 10^{11} km of DNA per human. Compare this with the circumference of the Earth, which is $4*10^4$ km, or the distance between the Earth and the Sun, which is $1.5 * 10^8$ km. In simple terms, each of us has enough DNA to go from the Earth to the Sun and back over 300 times, or around Earth's equator 2.5 million times! You must be wondering how is it possible that so much DNA can reside in the nucleus of the cells that are typically only 10 to 20 microns in diameter?

In order to fit into the nucleus, DNA must be packaged into a highly compacted structure. In the first step of DNA packaging, DNA is condensed into an 11 nm fiber called chromatin that represents an approximate 6-fold level of compaction (*Fig 1-31*). This compaction of DNA is done by certain proteins called histones.

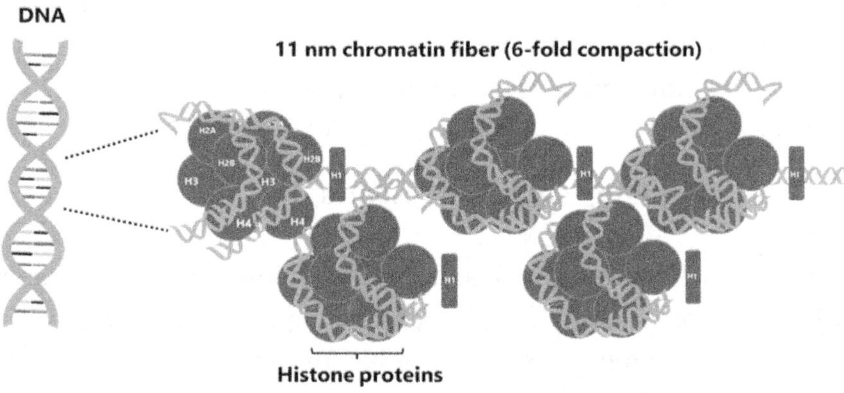

Fig 1-31: Histones condense DNA into an 11 nm fiber, 6-fold compaction

Histones are a group of small, positively charged proteins that include H1, H2A, H2B, H3, and H4 (*Fig 1-32*). Since DNA is negatively charged due to the phosphate groups in its phosphate-sugar backbone, histones tightly bind with DNA. One repeating unit of chromatin is called a nucleosome, and multiple nucleosomes formed in a linear fashion along the DNA molecule produce the 11 nm fiber called chromatin.

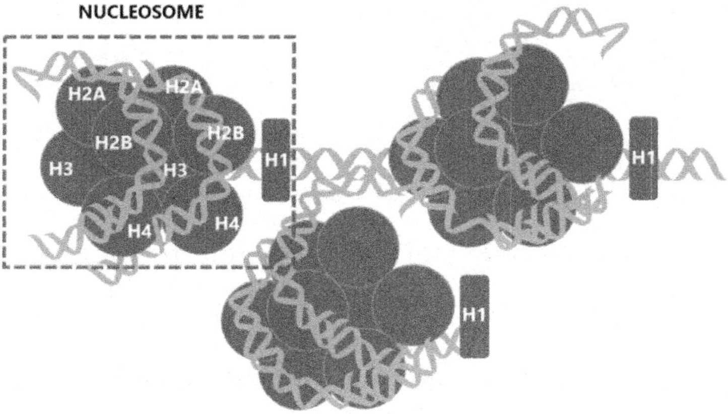

Fig 1-32: One repeating unit of chromatin is called a nucleosome.

One nucleosome is formed when 147 base pairs of DNA wrap around the histone octamer 1.75 times. Histone octamer is formed from the 2 copies of each of the H2A, H2B, H3, and H4 histone proteins. Adjacent nucleosomes are then linked together by a linker DNA, which is typically around 20-80 base pairs in length, and is usually bound to the H1 histone protein. This arrangement of nucleosomes in chromatin gives the appearance of chromatin similar to beads on a string (*Fig 1-33*). The packaging of DNA into nucleosomes results in a reduction of its length by approximately 6-7 fold. However, the level of compaction achieved through the formation of the 11 nm nucleosome fiber is inadequate to package the entire genome within the nucleus, which is only 10 to 20 microns in diameter.

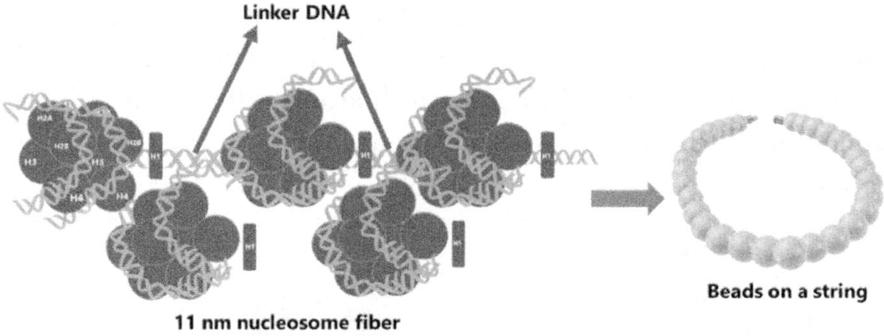

Fig 1-33: Adjacent nucleosomes are connected via linker DNA which is between 20-80 bps in length.

Therefore, a beaded string of 11 nm chromatin fiber undergoes additional folding and coiling to form a 30nm fiber. In this process, multiple nucleosomes are coiled together, and these are then stacked on top of each other. The end result is 30 nm chromatin fiber which exists in the form of a solenoid and is condensed to a thickness of approximately 30 nanometers in diameter (*Fig 1-34*). This 30 nm fiber is the second level of compaction, which provides an approximately 100-fold compaction of DNA. Still, this packaging level is insufficient to fit DNA into a nucleus. Therefore, another group of proteins helps in the further packaging of DNA. These proteins are called scaffold proteins. The scaffold protein helps in further coiling of 30nm fiber and, as a result, forms the chromatin loop of 300 nm in diameter. Further condensation of the 300 nm chromatin loop forms the chromatid, which is 700 nm in diameter.

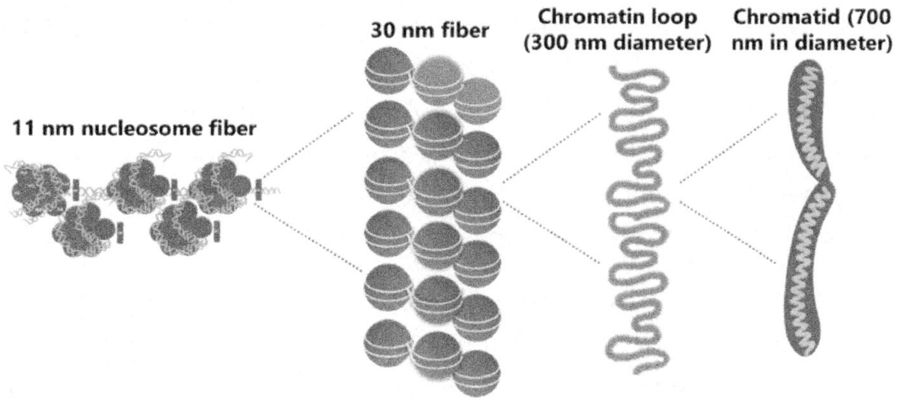

Fig 1-34: Levels of DNA compaction to form chromatid

The condensation of a 300 nm chromatin loop to form the chromatid of 700 nm in diameter is the final level of DNA compaction, which creates chromatids of a chromosome (*Fig 1-35*). Each chromosome has two sister chromatids, and each chromatid is attached to one centromere. The centromere of the chromosome divides the chromosome into two arms. The short arm of the chromosome is labeled as the "p arm." While the long arm of the chromosome is labeled as the "q arm."

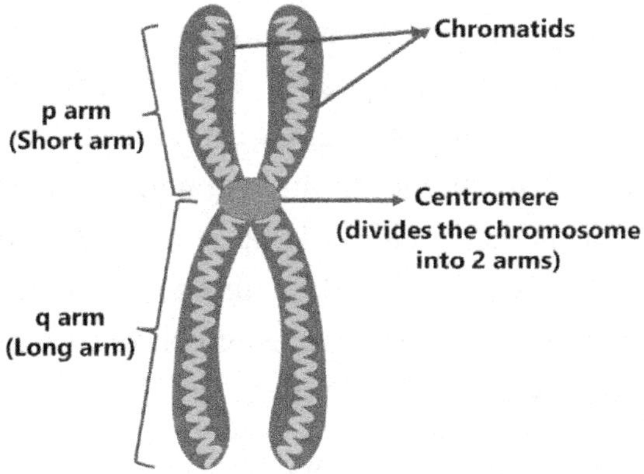

Fig 1-35: Structure of a chromosome

The entire collection of genes that help make you who you are is written across 23 pairs of chromosomes. That's 46 chromosomes altogether. The full set of 46 chromosomes can be found in the nucleus of almost every cell of the body. Each chromosome's DNA ranges from 50-250 million base pairs. A person inherits half of his chromosomes, i.e., 23 chromosomes from his biological mother and the other 23 chromosomes from his biological father, giving the person a total of 46 chromosomes.

Out of these 23 pairs of chromosomes, 22 pairs of chromosomes are called autosomes. And the 23rd pair of chromosomes are referred to as the sex chromosomes, as they determine the gender of an individual. Females possess two X chromosomes, while males possess one X and one Y chromosome. Since human cells have 23 pairs of chromosomes, thus they are called **diploid cells** or 2n, where n stands for the number of chromosomes. Only sperm and eggs are the cells that contain only one homologous chromosome from each pair of the chromosomes and thus are said to be **haploid**, designated as 1n. When a sperm and egg fuse to form a zygote, their genetic material combines to form one complete, diploid set of chromosomes.

A chromosome contains hundreds to thousands of genes. Each chromosome varies in size and number of genes. For instance, chromosome 1 has 3000 genes; chromosome 2 has 2500 genes, and so on. Every gene has a specific and fixed location on the chromosome. This position is called the locus of the gene. For example, the locus of the hemoglobin beta gene designated as *HBB* is 11p15. It tells that the hemoglobin beta gene is located on the p arm or short arm of chromosome 11. The numbers one and five identify the exact site where the gene appears when chromosome 11 is viewed under a microscope with a suitable dye.

Also, every gene is given a unique name. The genes are also allocated symbols, which are short combinations of letters and sometimes numbers as an abbreviated representation of the gene name. For example, a gene associated with cystic fibrosis, located on chromosome 7 is called the cystic fibrosis transmembrane conductance regulator, and its symbol is *CFTR*. Other examples of the naming of genes include the *rpoA* gene, which encodes the α-subunit of RNA polymerase, the *rpoB* gene, which encodes the β-subunit of RNA polymerase, *polA* gene, which encodes DNA polymerase I, etc. Also, remember the gene symbol is always written in *italics*.

1.8. What are alleles?

Now let's understand another important term, allele, and what is the difference between a gene and an allele? We know that an individual has 23 pairs of chromosomes. The two chromosomes of a pair are referred to as homologous since they share similar characteristics, including identical size and shape. The homologous pair of chromosomes in your genome is formed by pairing one homologous chromosome that came from your biological mother and the other homologous chromosome that came from your biological father. The pairing of homologous chromosomes is done in such a way that the gene encoding for the same trait is always carried on a similar place or position on both of the homologous chromosomes.

In other words, homologous chromosomes carry the same type of genetic information: i.e., they have the same genes in the same locations. We have already studied that every gene has a specific and fixed location on the chromosome called the locus of the gene. Therefore in the case of homologous chromosomes, each gene resides at the same specific locus on both the homologous chromosomes (*Fig 1-36*). Though one copy of the gene is inherited from the mother and another from the father, it is not necessary that the individual will have the same versions of genes on the two homologous chromosomes of a pair.

That's because he may have inherited two different gene versions from his mother and his father. If this is the situation when the two copies of a gene differ from each other, they are known as alleles. In other words, you can define alleles as the **different versions or forms of the same gene**.

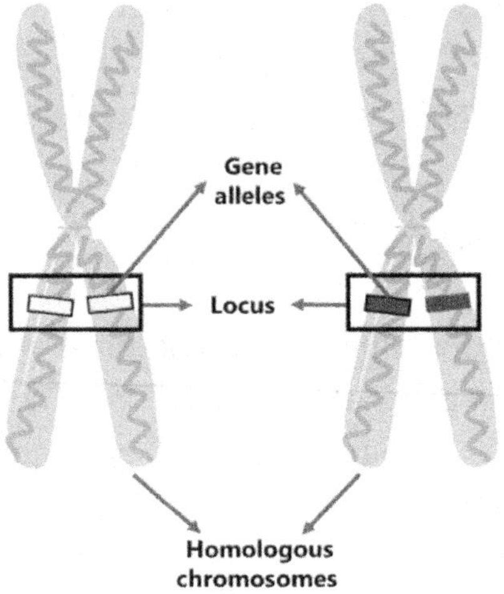

Fig 1-36: Every gene has a specific and fixed location on the chromosome, called the locus of the gene

A gene may have multiple different alleles, though only two alleles can be present at the gene's locus in any individual. For instance, the gene coding for eye color can have the allele for brown color, the allele for blue color, the allele for green color, and for other eye colors. Simply put, a gene specifies which trait is required to be expressed, say eye color, and the alleles of the "eye color" gene give directions for making eyes blue, green, brown, and so on.

Let's understand it through another example. A gene on chromosome 9 determines blood type A, B, AB, or O in a human. It's possible for you to have two identical copies of this gene, one on each of the homologous chromosomes- for example, you may have a double dose of the gene version for the blood type A, which clearly indicates both your parents have genes on chromosome 9 that determine blood type A. On the other hand, you may have two different alleles on your two homologous chromosomes, such as one allele for type A and one allele for type B, giving you an AB blood group. So it will be possible if your one parent's gene on chromosome 9 determines blood type A and your other parent's gene on chromosome 9 determines blood type B.

1.9. Dominant and Recessive Inheritance

Now let's discuss what dominant and recessive inheritance is. But before studying that, let's first discuss the terms homozygous and heterozygous. When the two inherited alleles for a given gene are identical, for instance, if both the alleles on chromosome 9 determine blood type A, then the specific gene is said to be homozygous. Alternatively, when the two alleles for a gene are different, for instance, if the individual has one allele for blood group type A and one allele for blood group type B, then the specific gene encoding for the blood group is said to be heterozygous.

Alleles can be dominant or recessive. A dominant allele will exhibit its effect even in the presence of only one copy of the allele, which can be inherited from either parent. In contrast, for a recessive allele to be expressed in an individual's phenotype, the individual must have two copies of the allele, one inherited from each parent.

Alleles are represented using letters with dominant alleles being represented by uppercase letters and recessive alleles by lowercase letters. For example, in the case of eye color, the allele for the brown eye color is dominant and therefore represented by capital B. The allele for blue eye color is recessive, therefore represented by the lowercase letter b.

For your knowledge, the brown eye color allele makes pigment melanin that turns your eyes brown, but the blue eye color allele does not make a blue pigment. Instead, it makes no pigment, and an eye with the least pigment is blue in color. For example, suppose if an individual has one allele for brown eye color, represented by B and the second allele for blue eye color, represented by b (*Fig 1-37*). Since the allele for brown eyes is dominant, the individual will develop brown color eyes though the allele for blue eyes is also present. Therefore, you can say an individual needs only one copy of the dominant 'brown eye' allele to have brown eyes, although if the individual has two copies of the brown eye allele, he will still have brown eyes.

	B Dominant allele	b Recessive allele
B	**BB** **(Brown eye color)**	**Bb** **(Brown eye color)**
b	**Bb** **(Brown eye color)**	**bb** **(Blue eye color)**

Fig 1-37: Inheritance of dominant and recessive alleles for eye color

On the other hand, the allele for blue eyes is recessive. Therefore to develop blue color eyes, the individual must have two copies of the blue color allele. In other words, he should be homozygous for the blue eye color allele represented by bb in lowercase letters.

Thus, an individual with the genotype BB or Bb will have brown eyes, while an individual with the genotype bb will have blue eyes.

Codominance: If, in case, both alleles are dominant, then it is called codominance. Because of codominance, the resulting characteristic is due to both alleles being expressed equally. An example of this is the blood group AB. If the individual inherits the allele for blood type A from his father and the allele for blood type B from his mother, then, in this case, both the alleles A and B will be expressed equally, giving the individual blood type AB.

1.10. DNA mutations

Any mutation in the gene of one of the 23 nuclear chromosome pairs or in the mitochondrial genome can cause genetic disease. Mutation can be defined as the change in a gene's DNA sequence. Well, you already know that DNA is transcribed to make mRNA, which is then translated to make protein. And that protein performs some critical function in your body. The proper functioning of each cell relies on numerous proteins that perform their tasks in the right places at the right times. Therefore, if changes occur in the DNA sequence of a gene, then there are changes in mRNA, which will eventually produce a defective protein or no protein at all. If the protein affected by the mutation is critical for normal body functions, it can disrupt normal development or lead to a medical condition. Medical conditions resulting from mutations in one or more genes are known as genetic disorders.

Sometimes the mutations are so deleterious that they prevent an embryo from surviving. An individual can inherit a gene mutation from one or both of his parents. In other cases, a mutation can also happen during his lifetime. Such mutations are called **acquired mutations**. These can occur randomly by errors during DNA replication. Or can occur due to some environmental exposure such as cigarette smoke, radiation, or certain chemicals. Such acquired mutations may be a cause of many types of cancers.

The simplest mutation in the human genome is a **point mutation** when the mutation occurs at just one point in DNA. The simple type of point mutation is a substitution, in which one nucleotide gets replaced with another nucleotide, which can disrupt the gene and eventually produce a defective protein. For instance, in the case of sickle cell disease, a genetic disorder of the blood, there aren't enough healthy red blood cells to carry oxygen throughout the body. This genetic disease occurs due to mutations in *HBB* genes. Hemoglobin is composed of four protein subunits, namely two alpha-globin subunits and two beta-globin subunits. The *HBB* gene is responsible for encoding the beta-globin protein. In sickle cell disease, a mutation occurs in the seventeenth nucleotide of the beta-globin gene, where an A nucleotide is replaced with a T nucleotide (*Fig 1-38*).

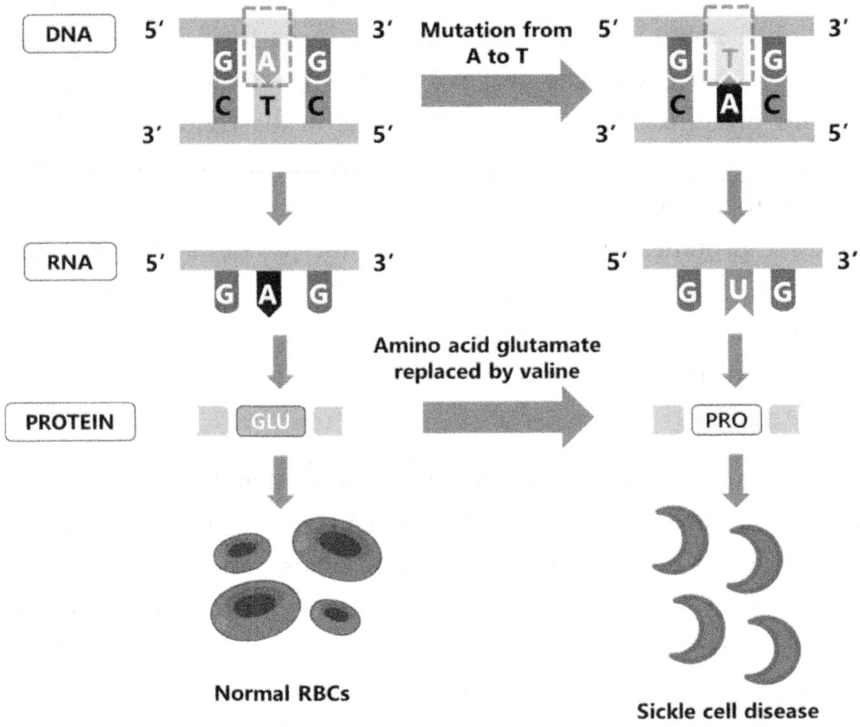

Fig 1-38: A point mutation in the HBB gene, leading to sickle cell disease

When translated into amino acids, this mutation results in the amino acid glutamate being replaced by the amino acid valine in the beta-globin subunit of the hemoglobin protein. A single nucleotide mutation can have dire consequences; the mutated hemoglobin molecules stick together and form abnormal filaments that change the shape of the red blood cells, eventually leading to anemia. Anemia can, in turn, lead to an increased risk of stroke and infection, and severe bone pain.

Genetic diseases can be recessive or dominant. Take the example of sickle cell disease. It is a recessive genetic disease. It means that for the disease to occur, both the copies of the individual's gene *HBB* must carry the mutation at the seventeenth position, where A is replaced by T nucleotide. If the mutation is only on one copy of the gene, then the non-mutated gene can produce enough normal hemoglobin to overcome the negative effects of the mutated hemoglobin. Though individuals with only one mutated copy of the *HBB* gene are still carriers of the sickle cell trait, they can pass the mutated gene to their offspring. When two sickle cell trait carriers have a child together, there is a 25% chance that their child might get a mutated sickle cell allele from both parents, resulting in the child having sickle cell disease (*Fig 1-39*).

The other category of genetic diseases is dominant genetic diseases. In such genetic diseases, a single copy of the mutated gene is enough to cause the disease. One such example is WHIM syndrome, in which a thousandth nucleotide of the *CXCR4* gene is mutated from a C to a T nucleotide. However, only one mutated copy of the gene creates a hyperactive protein that dominates the functioning of healthy protein.

To sum up, sickle cell disease and WHIM syndrome are both examples of genetic diseases caused by simple substitution mutations, i.e., one nucleotide is replaced with another nucleotide. In the case of sickle cell disease, the seventeenth nucleotide of the beta globulin gene is mutated from an A to T. And in the case of WHIM syndrome, the thousandth nucleotide of the *CXCR4* gene is mutated from a C to T nucleotide.

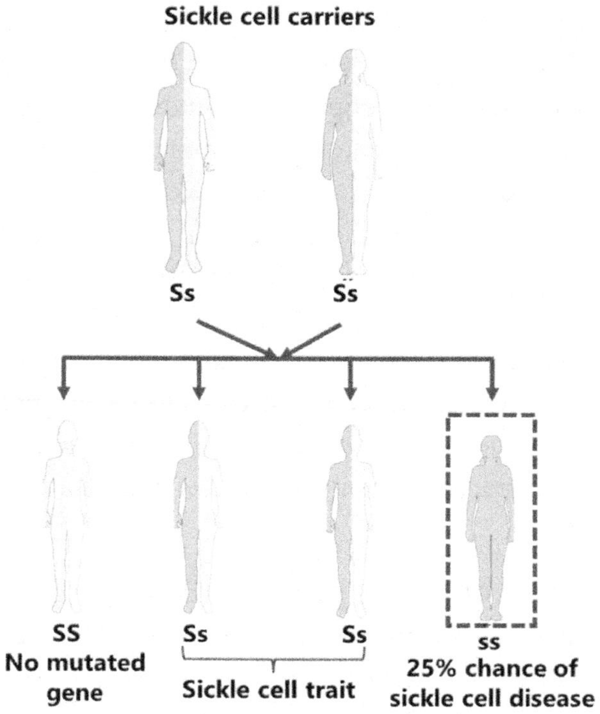

Fig 1-39: Sickle cell disease inheritance

There are many other examples of genetic diseases that can result from insertions or deletions in DNA. An insertion mutation results in a change in the number of DNA bases in a gene by adding a piece of DNA. On the other hand, a deletion removes a piece of DNA. The addition or loss of nucleotides in the gene changes the gene's reading frame (*Fig 1-40*), and a reading frame consists of groups of 3 bases. One group of 3 bases is called a codon, and 1 codon codes for one amino acid.

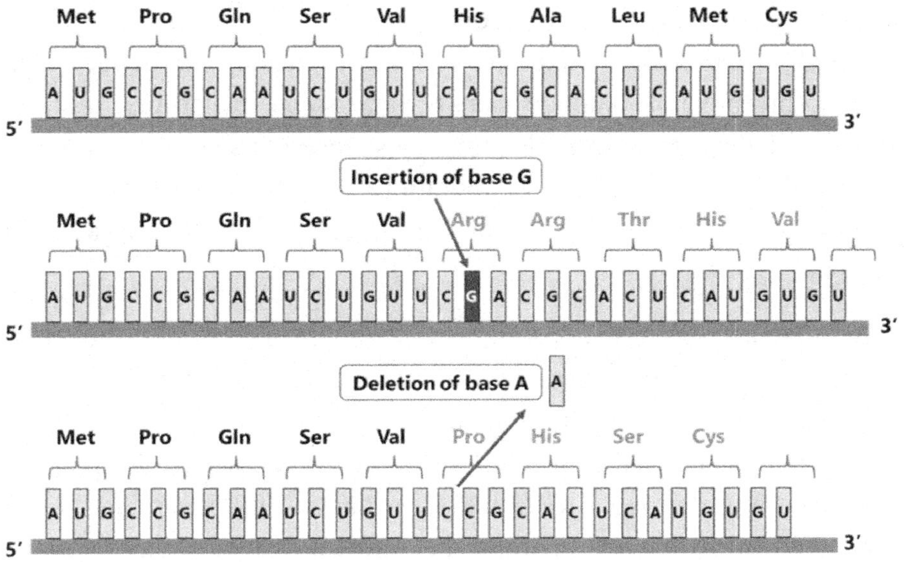

Fig 1-40: Frameshift mutation

Thus the insertions and deletions can lead to a shift in the grouping of these bases, which in turn changes the code for amino acids. As a result, the protein made by the gene may be defective.

An example of insertion mutation is Huntington's disease, a neurodegenerative disorder. In this disease, the mutation occurs in the *HTT* gene in such a manner that the same 3 letters of DNA get repeated too many times. The repetition of nucleotides in the gene causes the brain cells to produce abnormal proteins that gradually damage them.

An example of deletion mutation is cystic fibrosis, a life-threatening disease that primarily affects the lungs. In this disease, the deletion of 3 nucleotides in the *CFTR* gene results in a protein that lacks the amino acid phenylalanine at position 508. The loss of phenylalanine affects the proper folding and function of the CFTR protein, leading to the development of cystic fibrosis.

Mutations can also occur in multiple genes or, in some cases, affect whole chromosomes. When damage to chromosomes occurs, such as when large segments of the chromosome are missing, duplicated, or inverted, it can affect hundreds or even thousands of genes. This makes chromosome damage usually fatal since it disrupts multiple genes and their functions, leading to severe consequences for the organism. In some cases, these mutations can result in the disruption of critical genetic processes that are essential for the organism's survival.

Down Syndrome is a prominent example of a chromosomal abnormality that occurs if an individual has an extra copy of chromosome 21, i.e., the individuals have 3 copies of chromosome 21 rather than two normal copies. Infants born with Down syndrome have decreased muscle tone, a flat face, eyes slanting up, irregular-shaped ears, the ability to extend joints beyond the usual, a large tongue relative to the mouth, etc.

1.11. Genome Sequencing

Thanks to DNA sequencing, because of which we know the genetic causes of many diseases. DNA sequencing is a process that enables scientists to read and record the contents of the human genome nucleotide by nucleotide. We have already discussed that there are 23 pairs of chromosomes within a normal cell. Chromosomes are the structures that house DNA which is made up of four chemical bases: A, G, C, and T, shorthand for adenine, guanine, cytosine, and thymine. Like in the case of computers, where the data is stored in 0 and 1 binary digits, in human DNA, the code of life is stored in the 4 chemical bases or nucleotides A, T, G, and C. Altogether, our DNA contains about 12 billion such chemical bases. Sequencing the genome means reading and recording the contents of the human genome nucleotide by nucleotide and then determining the identity of the 12 billion nucleotides that compose the human genome. The first sequencing methods were developed in the 1970s. After completing the sequence of the human genome in 2001, scientists have identified around 4000 different kinds of DNA mutations that can cause genetic diseases. The sequencing of the genome can be helpful in an enormous number of ways. For instance,

(i) Genome sequencing has the potential to reveal if an individual is at an elevated risk of developing cancer or any other genetic diseases. For example, a mutation in the *APOE* gene increases the chances of Alzheimer's disease by 2 to 3 fold. And people with a mutation in the gene *MYBPC3* may be at risk of suffering from the thickening of the heart muscle. But the good news is that some people with these variants may actually be able to reduce their risk of getting the related disease through lifestyle changes such as diet and exercise and through treatment, including gene editing.

(ii) Genome sequencing can also help design personalized medicines for individuals. For instance, what genes are misregulated in cancer patients allows physicians to select the best chemotherapy and potentially expose the patient to less toxic treatment.

1.12. Polymerase chain reaction

Before understanding how genome sequencing works, let's first understand a very important technique used in molecular biology, which is Polymerase chain reaction or simply referred to as PCR. The polymerase chain reaction is a technique that allows a "target" DNA sequence to be selectively amplified. DNA amplification by PCR consists of 3 steps: denaturation, annealing, and extension.

The first step is **denaturation**, in which the double-stranded (ds) DNA is denatured at 94°C, because of which double-stranded template DNA is separated into two single strands (*Fig 1-41*). The second step is **annealing**, during which the temperature is lowered to 45-60°C. This enables the primers to bind to each single-stranded DNA template.

Typically, two different primers are included in the reaction mixture, with each primer complementing only a short sequence at the 3' end of each strand. These primers are single-stranded sequences and are much shorter than the target region.

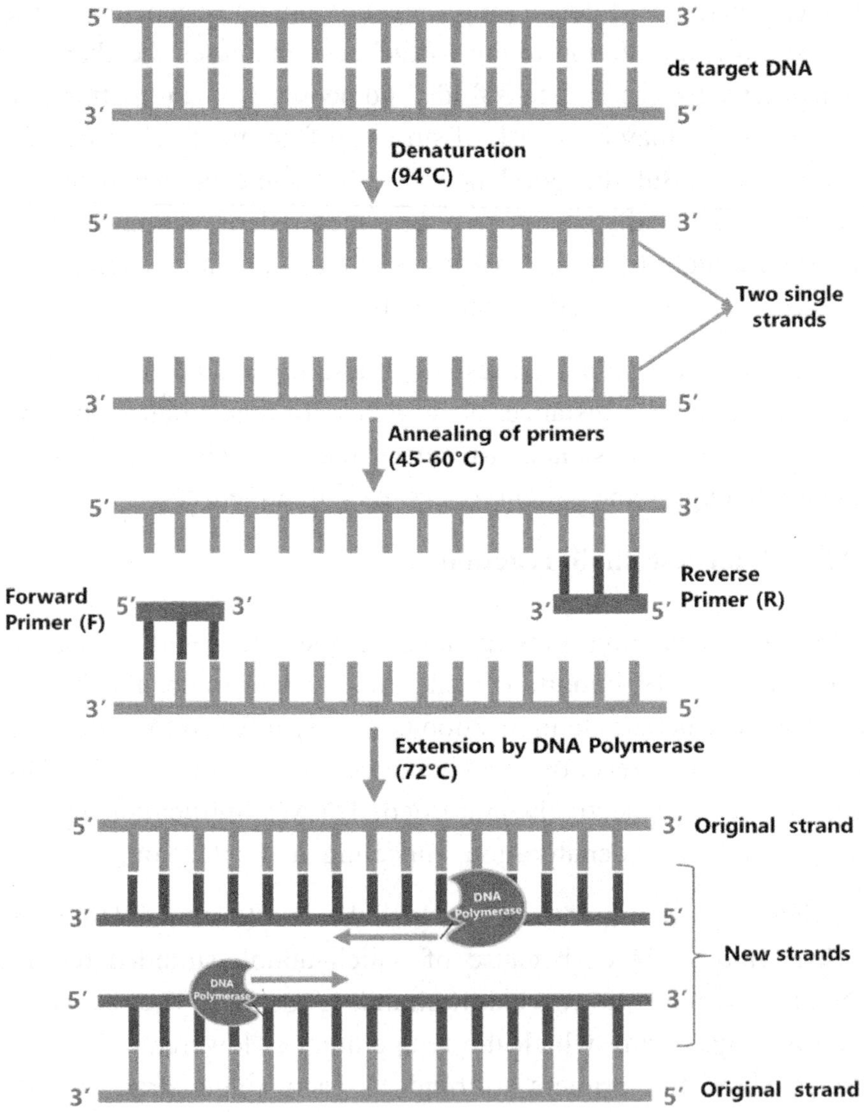

Fig 1-41: Steps involved in Polymerase Chain Reaction (PCR)

The third step is **extension,** during which the temperature is raised to 72°C, allowing the DNA polymerase to synthesize a new DNA strand by adding free nucleotides from the reaction mixture (*Fig 1-41*).

The newly formed strands are synthesized in the 5'-3' direction and are complementary to the template DNA strands. During every extension step, the amount of DNA target sequences is doubled. This results in exponential amplification of the specific DNA target region, as the original template strands plus all newly produced strands serve as template strands for the next round of elongation in every successive cycle.

In the *Fig 1-42,* you can observe that if you start PCR with 1 copy of template DNA, then after 1 cycle of PCR, the number of dsDNA doubles and after 2 cycles of PCR, you will get 4 copies of ds DNA. Each PCR cycle involves denaturation, annealing, and elongation. To generate millions of copies of the DNA target, several cycles of this process are required.

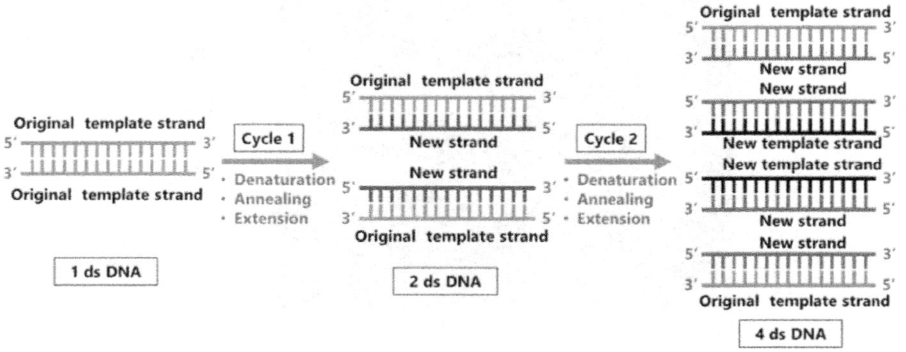

Fig 1-42: Doubling of dsDNA after each cycle of PCR

To calculate the number of DNA copies produced after a specific number of PCR cycles, the formula used is 2^n, where n is the number of cycles. Thus, if you subject your DNA of interest to 5 cycles of PCR, then you will end up with 2^5 or 64 copies of the original double-stranded DNA target region.

Finally, to check whether PCR successfully generated the amplification of the DNA target region, agarose gel electrophoresis is done for size separation of the PCR products (*Fig 1-43*).

The size of amplified PCR products is then determined by comparison with a DNA ladder. DNA ladder is a molecular weight marker that contains DNA fragments of known size and is run on the gel alongside the PCR products.

This is how PCR is performed.

Understanding the PCR technique is so essential because the sequencing of the human genome relies on the PCR technique.

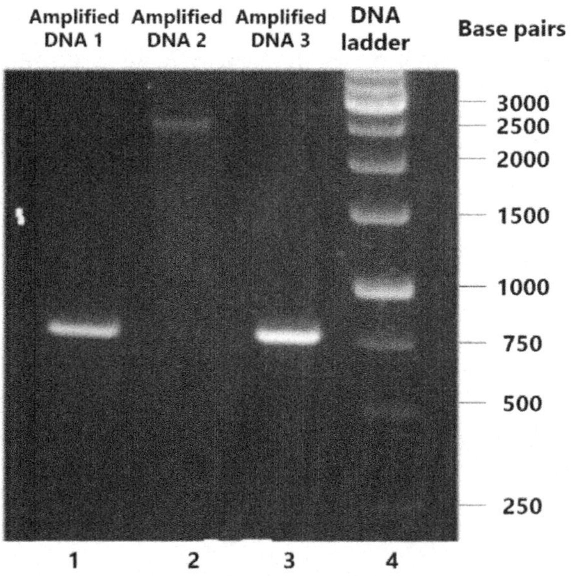

Fig 1-43: Agarose gel electrophoresis

1.13. How is Genome Sequencing done?

The first step is to extract DNA from the individual whose genome is required to be sequenced. The DNA is then subjected to sequencing, allowing scientists to read and record the nucleotide content of the genome.

Sanger sequencing, originally developed in 1970, remains the gold standard for DNA sequencing and is still used today in routine sequencing procedures. The process of Sanger sequencing shares many similarities with polymerase chain reaction (PCR), as it involves copying DNA in vitro. Like PCR, Sanger sequencing involves the three key steps of denaturation, annealing, and extension. To initiate the reaction, the DNA sample is denatured at 94°C, causing it to separate into two single-stranded strands.

The sequencing reaction involves using a high-fidelity DNA polymerase enzyme to generate a complementary copy of a single-stranded DNA template. Each reaction is initiated by a primer complementary to the template and begins DNA synthesis from the 3' end. The reaction employs four different types of nucleotides, including dATP, dTTP, dGTP, and dCTP. It also contains a mixture of four di-deoxynucleotides, ddATP, ddGTP, ddCTP, and ddTTP, which resemble natural deoxynucleotides enough to allow their incorporation into the growing strand.

However, they differ from them in two ways: first, they lack the 3' OH group needed to form the phosphodiester bond between nucleotides during DNA strand elongation (*Fig 1-44*). Thus, the incorporation of di-deoxynucleotides leads to chain termination once they are incorporated into the growing strand. Secondly, each di-deoxynucleotide is attached to a unique fluorescent dye, which enables automatic detection of the DNA sequence.

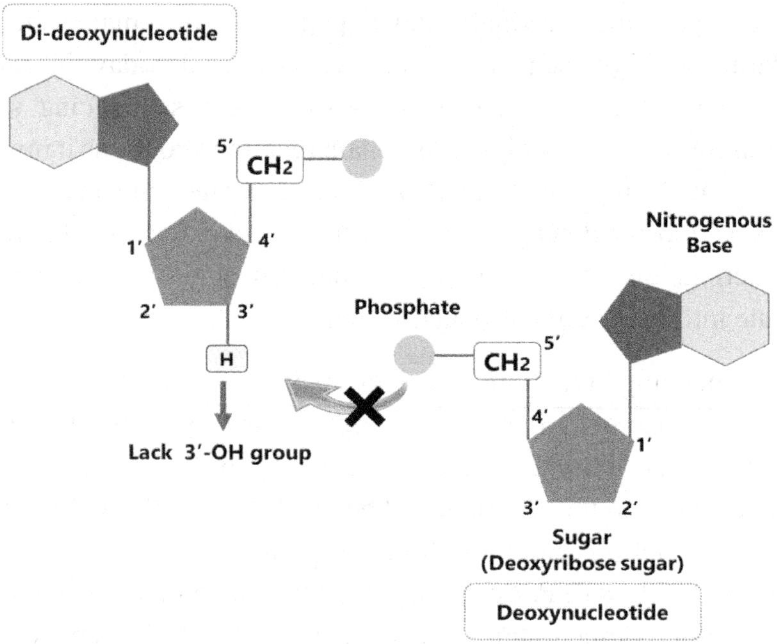

Fig 1-44: Di-deoxynucleotides lack the 3'-OH group required to form phosphodiester bond between nucleotides during DNA strand elongation

After the DNA template is denatured into two single strands, the reaction is cooled to a temperature between 45-60°C so that the primer can bind to the single-stranded template. Upon binding, the temperature is raised back to 72°C, allowing the DNA polymerase to synthesize new DNA starting from the primer.

The DNA polymerase continues to add nucleotides to the chain until it encounters a dideoxy nucleotide instead of a normal nucleotide. At this point, no further nucleotides can be added, and the strand ends with the dideoxy nucleotide. The absence of the 3′ OH group in the dideoxy nucleotides causes the growing DNA to terminate whenever they are incorporated by DNA polymerase. This process is repeated in multiple cycles.

After the cycles are completed, it's nearly certain that a dideoxy nucleotide will have been incorporated at every position of the target

DNA in at least one reaction (*Fig 1-45*). This means that the tube will contain fragments of various lengths, each ending at a different nucleotide position in the original DNA.

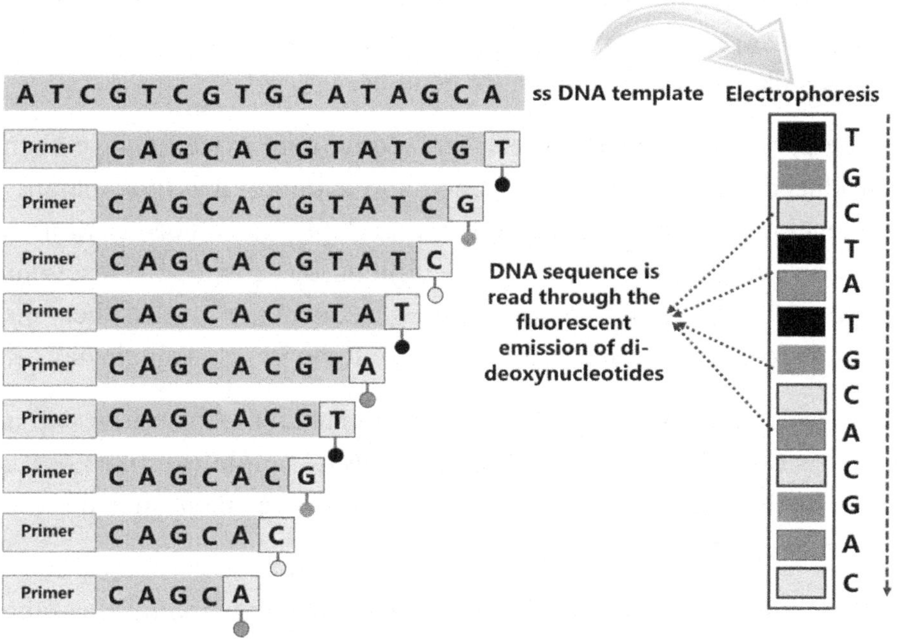

Fig 1-45: Sanger sequencing

The reaction mixtures are then loaded onto the sequencing machine, either by manual placement on slab gels or automatic loading into capillaries. The DNA fragments are then electrophoresed to separate the DNA molecules based on size.

Next, the DNA sequence is read through the detection of fluorescent emissions from the di-deoxynucleotides as they flow through the gel (*Fig 1-45*). As a result, the sequence of the original piece of DNA can be reconstructed by adding one nucleotide after another based on the sequence of colors registered by the detector.

Current Sanger sequencing equipments utilize capillary-based automatic electrophoresis, which typically analyzes 8 to 96 sequencing reactions at the same time. Sanger sequencing provides high-quality sequence data for DNA fragments up to around 900 base pairs. However, it is cost-prohibitive and not suitable for large-scale projects like sequencing an entire genome.

In the past decade, **next-generation sequencing systems** have been introduced for genome sequencing enabling massive parallel sequencing reactions. These systems are highly advanced platforms that can process millions or even billions of sequencing reactions simultaneously. With next-generation sequencing, it has become possible to sequence the entire human genome in just a single day.

To determine the sequence of bases in human DNA, the initial step is to extract the DNA. Once isolated, the DNA sample is broken down into smaller fragments. These fragments are then attached to oligonucleotides called adaptors.

There are two types of adaptors: one has sequencing binding sites, and the other has an additional complementary sequence that can bind to the oligonucleotides fixed on the surface of the flow cell (*Fig 1-46*). The resulting fragments with adaptors attached are then denatured into single-stranded DNAs.

The single-stranded DNA fragments are then anchored onto a solid surface through complementary base pairing with oligonucleotides that have been fixed to the surface.

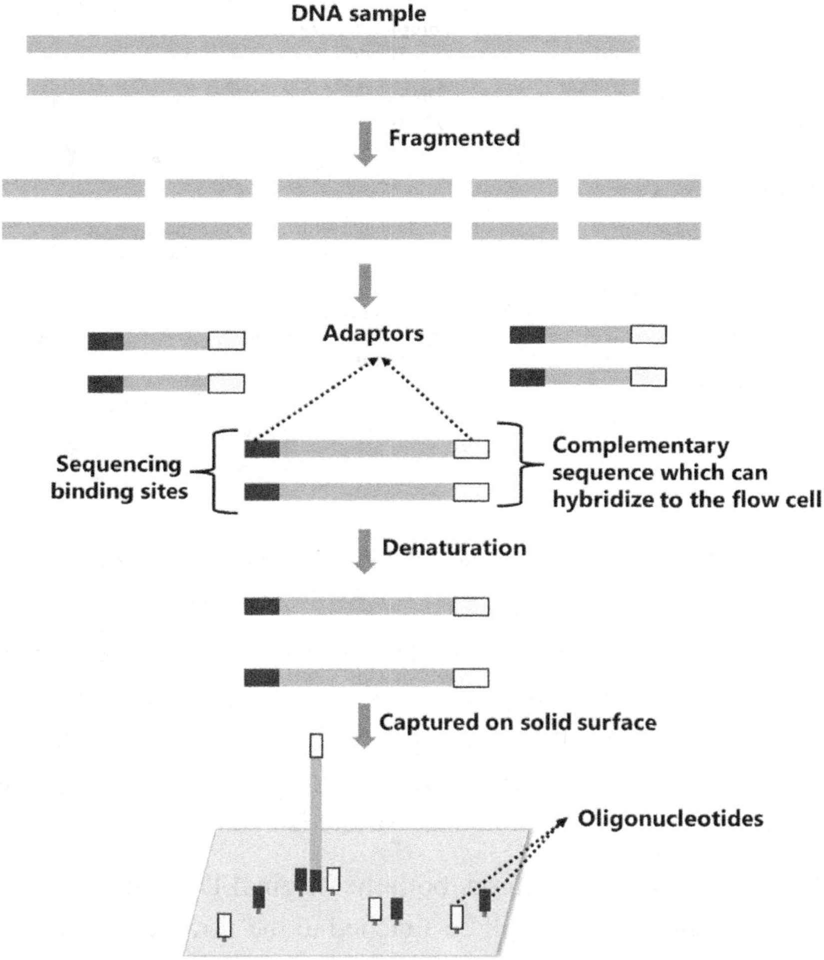

Fig 1-46: Denatured adapters-ligated DNA fragments are captured on a solid surface

In the subsequent step, a polymerase enzyme synthesizes a complementary strand to the single-stranded DNA, using the fixed oligonucleotide on the surface as a primer (*Fig 1-47*). The result is two separate strands of DNA. The strand that is not attached to the surface of the flow cell is then removed through washing, leaving the second strand firmly bound to the fixed oligonucleotide through a covalent bond.

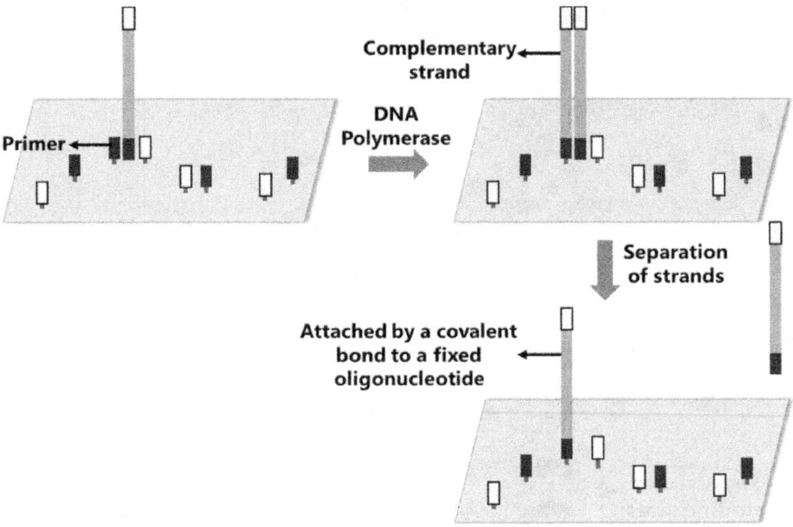

Fig 1-47: The synthesized complementary strand remains attached to the fixed oligonucleotide, and the other strand is washed away

The next step involves constructing bridges. This is achieved through the interaction between the second type of adaptor attached to the DNA fragment and another fixed oligonucleotide (as illustrated in *Fig 1-48*). This interaction triggers another round of polymerase chain reaction, in which a complementary strand is synthesized.

After the strands are denatured, both the original DNA fragment and its complementary strand will be attached to the flow cell surface.

This process is repeated multiple times, with each iteration resulting in the additional formation of bridges with fixed oligonucleotides. Through multiple rounds of polymerase chain reaction, a large number of copies of the original DNA and complementary fragments can be generated.

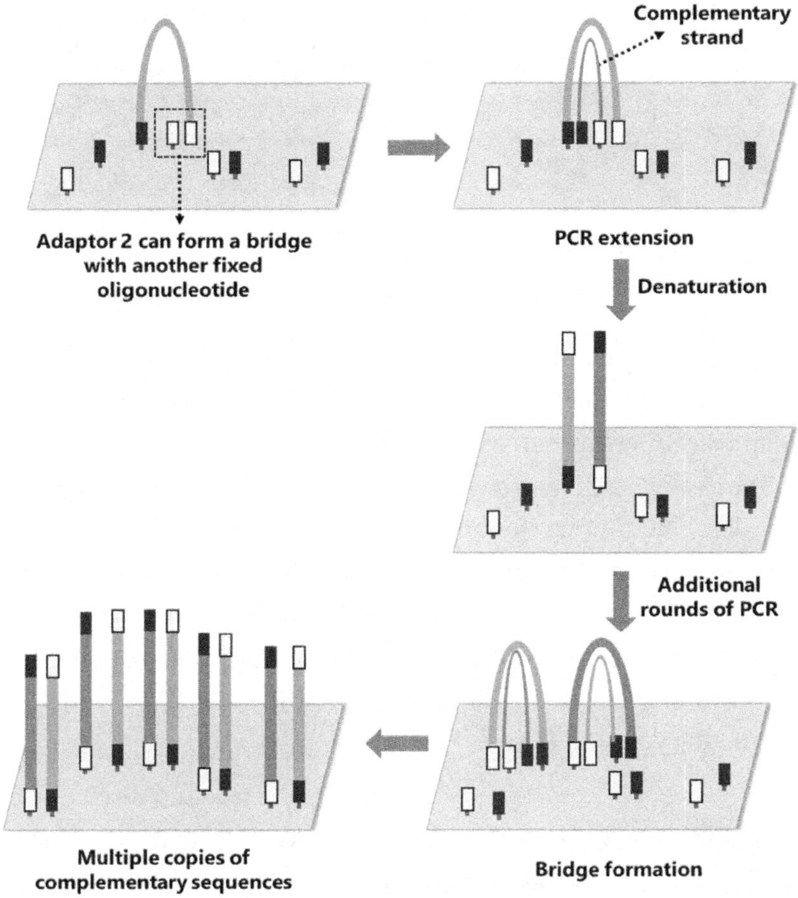

Fig 1-48: After many rounds of PCR, multiple copies of original DNA and complementary fragments are generated

After this step, one of the strands is trimmed at the point of attachment to the oligonucleotide, resulting in individual DNA molecules of a single sequence (*Fig 1-49*). This creates small clusters of DNA on the solid surface, each containing numerous identical copies of a specific DNA sequence.

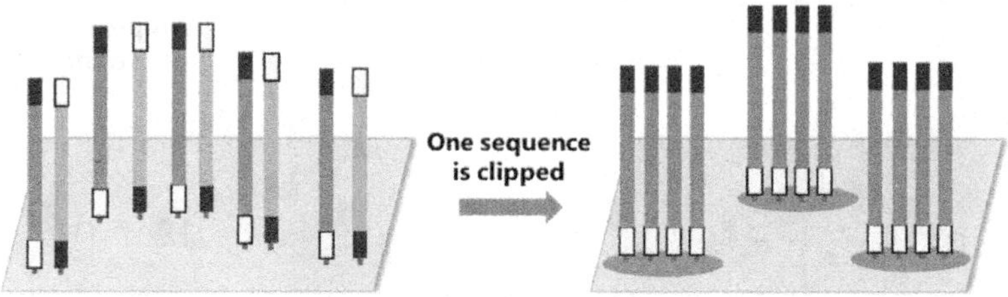

Fig 1-49: Tiny clusters of DNA are formed on the solid surface, with each containing many identical copies of a specific DNA sequence

DNA sequencing then begins by annealing a sequencing primer to the template. The sequencing primer is designed to be complementary to the adapter sequence that was attached to the 3' end of the template DNA. It provides a 3' hydroxyl group for the DNA polymerase to add nucleotides. The four nucleotides (A, G, C, and T) used for sequencing have unique properties that allow for distinction between them. Firstly, each nucleotide is attached to a different fluorescent group. Secondly, they have a blocking group on their 3' end, which prevents any additional nucleotides from being added to the new complementary strand (*Fig 1-50*).

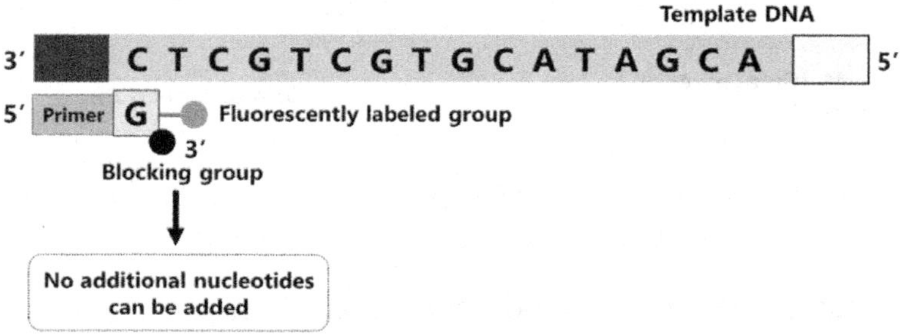

Fig 1-50: Nucleotides used for sequencing have fluorescently labeled groups and blocking groups on their 3' ends

Since each nucleotide added by the DNA polymerase is attached to a different fluorescent group, the laser causes it to emit fluorescence, which is recorded. The color of the fluorescence corresponds to the type of nucleotide incorporated into the growing strand. Once the fluorescence is recorded, the fluorescent label and blocking group are removed from the bound nucleotide, exposing a 3' hydroxyl group. This allows for further nucleotides to be added to the complementary strand in the next step of the process.

And so, the process continues by adding another nucleotide, say adenine, to the growing strand. This nucleotide, due to the presence of a blocking group, prevents the addition of further nucleotides. The fluorescence of this nucleotide is triggered and recorded. This cycle is repeated until the complementary strand is fully synthesized, and the sequences from all the spots are recorded at the same time (*Fig 1-51*).

However, practically, not all of the DNA fragments on the flow cell will incorporate a nucleotide during each cycle, resulting in a set of overlapping fragments that represent the same region of the original DNA template but differ in length and sequence at the ends. Finally, the software combines the overlapping sequenced fragments to create a longer sequence that represents the original DNA template.

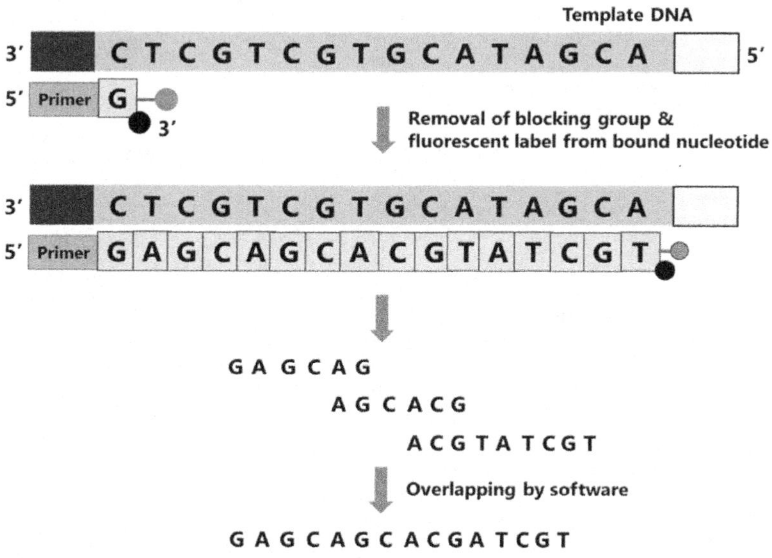

Fig 1-51: The software assembles overlapping sequence fragments into longer pieces

In this manner, the sequencing process is able to determine the overall sequence of a genome. Additionally, there are numerous other next-generation sequencing techniques that identify the **release of pyrophosphate** as nucleotides are added to the new strand of DNA by the polymerase enzyme.

Furthermore, some systems monitor the direct release of protons (H+) during the incorporation of individual bases into the DNA by the polymerase enzyme, which helps to sequence the human genome.

Your genome plays a crucial role in defining who you are and determining how your cells function. By sequencing your genome, you can gain valuable insights into your genetic makeup. The advent of next-generation sequencing has greatly impacted various fields, including the identification of genetic disorders, personalized medicine, and clinical diagnostics.

Practice Test 1

1. Information about life is stored as a code in:
(a) 0 and 1 binary digits
(b) A, G, C, and I bases
(c) A, G, C, and T bases
(d) Sugar and phosphate molecules

2. Which of the following options represents the building blocks of DNA?
(a) Nitrogenous bases, phosphates, and deoxyribose sugar
(b) Nitrogenous bases, phosphates, and ribose sugar
(c) Phosphorus bases, nitrogen, and ribose sugar
(d) Phosphorus bases, nitrogen, and deoxyribose sugar

3. The "rungs" of the DNA ladder are made of:
(a) sugars and phosphates
(b) glucose and sugars
(c) base pairs
(d) phosphates and hydrogen

4. The two strands of DNA are said to be:
(a) Parallel as they both run in the same direction 5'-3'
(b) Antiparallel as they run in opposite directions, one strand from 5'-3' and the other from the 3'-5' direction

5. Base A from one strand always pairs with the base T on the other strand by
(a) Phosphodiester bond
(b) 2 H bonds
(c) Covalent bond
(d) 3 H bonds

6. Base C from one strand always pairs with base G on the other strand by
(a) Phosphodiester bond
(b) 2 H bonds
(c) Covalent bond
(d) 3 H bonds

7. Why is the 3' end of the strand called so?
(a) Because of the exposed hydroxyl group attached to the 3' carbon of the sugar

(b) Because of the exposed phosphate group attached to the 3' carbon of the sugar
(c) Because of the exposed hydroxyl group attached to the 5' carbon of the sugar
(d) Because of the exposed phosphate group attached to the 5' carbon of the sugar

8. Why is the 5' end of the strand called so?
(a) Because of the exposed hydroxyl group attached to the 3' carbon of the sugar
(b) Because of the exposed phosphate group attached to the 3' carbon of the sugar
(c) Because of the exposed hydroxyl group attached to the 5' carbon of the sugar
(d) Because of the exposed phosphate group attached to the 5' carbon of the sugar

9. One strand of DNA reads 5'-AAAGTTCCGTGACT-3'; the complementary strand would read:
(a) 3'-AAAGTTCCGTGACT-5'
(b) 3'-TTTCAAGGCACTGA-5'
(c) 3'-UUUCAAGGCACUGA-5'
(d) 5'- TTTCAAGGCACTGA-3'

10. The enzyme that catalyzes the synthesis of DNA is called
(a) DNA polymerase
(b) DNA gyrase
(c) DNA ligase
(d) Helicase

11. DNA polymerase synthesizes
(a) DNA in the 3'-5' direction
(b) DNA in the 5'-3' direction
(c) mRNA in the 3'-5' direction
(d) mRNA in the 5'-3' direction

12. Which of the following enzymes is not used for DNA replication?
(a) RNA polymerase
(b) DNA gyrase
(c) DNA ligase
(d) Primase

13. Which enzyme synthesizes RNA primer that helps the DNA polymerase add nucleotides to the new strand in the correct order?
(a) single-stranded binding protein
(b) Primase
(c) DNA ligase
(d) Helicase

14. ____________ helps keep the two strands of the DNA double helix separate during replication.
(a) single-stranded binding protein
(b) Primase
(c) DNA ligase
(d) Helicase

15. Which of the following enzymes is responsible for unwinding small sections of the DNA helix just ahead of a replication fork?
(a) DNA polymerase
(b) Helicase
(c) Single-stranded binding proteins
(d) Topoisomerase

16. DNA replication is
(a) conservative
(b) conservative and discontinuous
(c) semi-conservative and semi-discontinuous
(d) semi-conservative and discontinuous

17. The 3' – 5' phosphodiester linkage joins
(a) two DNA strands
(b) two nucleotides
(c) a nitrogenous base with a pentose sugar
(d) two nucleosides

18. Okazaki fragments occur with replicating:
(a) both strands
(b) the lagging strand
(c) the leading strand

19. If a DNA molecule is composed of 40% adenine, then what percentage of guanine would be expected?
(a) 10%
(b) 20%

(c) 40%
(d) 80%

20. A stretch of a chromosome that codes for a trait can be called a:
(a) chromatid
(b) replication fork
(c) gene
(d) base-pair

21. RNA is synthesized on a DNA template strand in a process called ______, which utilizes the enzyme _______
(a) replication, DNA polymerase
(b) transcription, DNA polymerase
(c) transcription, RNA polymerase
(d) translation, RNA polymerase

22. Given the following DNA strand 3’ G G A C T G A T T 5’, which of the following is its complementary mRNA?
(a) 5’ C C T G A C T A A 3’
(b) 5’ C C U G A C U A A 3’
(c) 5’ G G A C T G A T T 3’
(d) 5’ T T A G T C A G G 3’

23. The following RNA strand was produced: 5′ AAAAUGAGUAAG 3′ Which of the following DNA strands could have been used as the template for this RNA?
(a) 3′ AAAATGAGTAAG 5′
(b) 3′ TTTTACTCATTC 5′
(c) 3′ TTTATGTGCTTC 5′
(d) 3′ UUUTACUCAUUC 5′

24. Select all of the correct steps necessary to convert an RNA transcript into mature messenger RNA (mRNA) after it is initially made.
(a) Introns are removed from the sequence.
(b) Exons are joined together to form the mature mRNA.
(c) A tail of C nucleotides is added to the 5’ end of the RNA.
(d) A 3' cap is added to the beginning of the RNA.

25. How are DNA and RNA different?
(a) DNA uses Uracil, while RNA uses Thymine
(b) DNA is double-stranded, while RNA is triple-stranded.

(c) While RNA is capable of traveling within the cytoplasm of a cell, DNA is exclusively located within the nucleus.
(d) DNA contains deoxyribose sugar, while RNA contains ribose sugar.
(e) c and d are correct.

26. Like DNA polymerase, RNA polymerase also requires a primer for mRNA synthesis.
(a) True
(b) False

27. In a cell, DNA transcription is stopped when RNA polymerase reaches:
(a) the end of DNA
(b) terminator
(c) stop codon
(d) restriction site

28. A researcher isolated a specific gene and its corresponding mRNA. Upon comparison, the mRNA was found to contain 1,000 fewer bases than the DNA sequence. Did the researcher isolate the wrong DNA?
(a) Yes, because the mRNA should contain more bases than the DNA sequence as the bases flanking the gene are also transcribed
(b) Yes, because mRNA is made from a DNA template and should be of the same length as the gene sequence
(c) No, because the mRNA got partially degraded after it was transcribed
(d) No, because the final mRNA contains only exons, the introns get removed

29. The first mRNA codon to specify an amino acid is always
(a) TAC
(b) AUG
(c) UAG
(d) UAA

30. Which DNA template strand sequence corresponds to the first amino acid insertion in a protein?
(a) TAC
(b) AUG
(c) UAG
(d) UAA

31. Which of the following statements is false?
(a) The genetic code is overlapping.
(b) The genetic code is universal.
(c) Degenerate codons specify the same amino acids.
(d) The genetic code is a triplet.

32. The amino acid sequence of a polypeptide chain determines the ____ structure of the protein.
(a) Primary
(b) Secondary
(c) Tertiary
(d) Quaternary

33. Out of the ____ different possible codons, ____ specify amino acids and ____ specify stop signal.
(a) 180, 20, 60
(b) 64, 61, 3
(c) 61, 60, 1
(d) 20, 17, 3

34. An antibiotic interferes with the ability of the ribosome to move. What affect would its exposure have on a bacterial cell?
(a) Protein synthesis will be affected
(b) RNA synthesis will be affected
(c) DNA replication will be affected
(d) Intron splicing will be affected

35. Which of the following statements explains the genetic code as degenerative?
(a) a given amino acid may have more than one codon
(b) each codon specifies more than one amino acid
(c) the first two bases specify the amino acid
(d) All of the above

36. What is the maximum number of amino acids that a protein could contain if it was produced from the transcription and translation of a gene composed of 30-nucleotides?
(a) 10
(b) 15
(c) 60
(d) 90

37. The length of the DNA molecule exceeds the dimensions of the nucleus of a eukaryotic cell. Then how is DNA accommodated?
(a) DNAse digestion
(b) Deletion of non-essential genes
(c) Supercoiling in nucleosomes
(d) Elimination of repetitive DNA

38. How many types of histone proteins are present in the chromosomes?
(a) 3
(b) 4
(c) 5
(d) 6

39. The histone proteins are present in molar ratios of:
(a) 2H1:2H2A:2H2B:2H3:2H4
(b) 2H1:2H2A:2H2B:1H3:1H4
(c) 1H1:2H2A:2H2B:2H3:2H4
(d) 1H1:2H2A:2H2B:1H3:1H4

40. How many copies of the H2B histone would be present in chromatin containing 50 nucleosomes?
(a) 100
(b) 50
(c) 25
(d) 200

41. A diploid cell contains 6 billion (6×10^9) base pairs of DNA. Considering each nucleosome encompasses 200 bp. How many nucleosomes will be present in a cell?
(a) 2×10^6
(b) 3×10^7
(c) 3×10^8
(d) 2×10^9

42. An allele is defined as:
(a) A synonym for a gene
(b) Gene that is not made up of DNA
(c) a homozygous genotype
(d) one of several possible forms of a gene

43. A given gene may have multiple different alleles.
(a) True
(b) False

44. Phenotype refers to the ________ of an individual.
(a) actual physical appearance
(b) genetic makeup
(c) recessive alleles
(d) dominant alleles

45. An example of codominant allele expression is the human ABO blood group system.
(a) True
(b) False

46. If two individuals who are carriers of a recessive genetic disease choose to have children, what is the likelihood that their offspring will also be carriers of the disease?
(a) 1 out of 4
(b) 2 out of 4
(c) 3 out of 4
(d) 4 out of 4

47. If a woman who is homozygous and healthy marries a man who is heterozygous for a genetically inherited recessive disease and they decide to become parents, then what is the probability that they will have a healthy child?
(a) 1 out of 4
(b) 2 out of 4
(c) 3 out of 4
(d) 4 out of 4

48. If two parents are homozygous for a genetically inherited recessive trait, then what is the probability that they will have an offspring who does not have this trait in their phenotype?
(a) 0%
(b) 25%
(c) 50%
(d) 100%

49. In sickle-cell disease, a glutamate → valine substitution results in the formation of HbS molecules, which:

(a) Stick together and cannot adequately carry O_2
(b) Have very high-affinity binding for O_2
(c) Stabilize the RBCs against oxidative damage
(d) Cause high levels of repulsion between neighboring HbS molecules

50. ATTCGAGCC- Original
ATTTGAGCC - Mutated
It is an example of a
(a) Insertion- Frameshift mutation
(b) Deletion- Frameshift mutation
(c) Point mutation
(d) None of the above

51. For a DNA sequence CCT GTG ACC AAA ACT, what type of mutation would occur if a Thymine gets inserted between Adenine and Cytosine in the triplet ACC?
(a) Insertion- Frameshift mutation
(b) Deletion- Frameshift mutation
(c) Point mutation
(d) None of the above

52. The purpose of PCR is to_________.
(a) make more copies of DNA primers to increase DNA transcription
(b) make more RNA to increase protein synthesis
(c) recycle DNA using thermocyclers
(d) amplify and make many copies of a DNA sequence

53. How many double-stranded DNA molecules are obtained from one copy of ds DNA after 4 cycles of PCR?
(a) 4
(b) 8
(c) 16
(d) 32

54. Primers used for the process of PCR are _________.
(a) Single-stranded RNA sequences
(b) Single-stranded DNA sequences
(c) Double-stranded RNA sequences
(d) Double-stranded DNA sequences

55. When DNA is heated, primers anneal to DNA strands.
(a) True

(b) False

56. For DNA amplification to occur, which of the following are needed?
(a) RNA primers
(b) thermostable DNA polymerase
(c) loose ribonucleotides
(d) All of the above

57. Which of the following is called the dideoxynucleotide chain termination method?
(a) Maxam-Gilbert method
(b) Sanger's method
(c) Edman's method
(d) Automated sequencing method

58. How many ddNTPs are used in the Sanger method?
(a) 1
(b) 2
(c) 3
(d) 4

59. What distinguishes each dideoxynucleotide used in Sanger sequencing?
(a) Its attached fluorescently labeled groups
(b) Its blocking group on the 3' end
(c) Its ability to release H+
(d) Its ability to release phosphate

60. What is the function of the primer in sequencing reaction?
(a) To isolate the DNA
(b) To provide a 3' hydroxyl group for the DNA polymerase to begin adding nucleotides
(c) To fluoresce
(d) To block additional nucleotides from being added to the new complementary strand

61. What is the purpose of next-generation sequencing?
(a) To determine the sequence of a genome
(b) Personalized medicine
(c) To identify genetic diseases
(d) All of the above

Answers:

1. (c) 2. (a) 3. (c) 4. (b) 5. (b) 6. (d) 7. (a) 8. (d) 9. (b) 10. (a) 11. (b) 12. (a) 13. (b) 14. (a) 15. (b) 16. (c) 17. (b) 18. (b) 19. (a) 20. (c) 21. (c) 22. (b) 23. (b) 24. (a) and (b) 25. (e) 26. (b) 27. (b) 28. (d) 29. (b) 30. (a) 31. (a) 32. (a) 33. (b) 34. (a) 35. (a) 36. (a) 37. (c) 38. (c) 39. (c) 40. (a) 41. (b) 42. (d) 43. (a) 44. (a) 45. (a) 46. (b) 47. (d) 48. (a) 49. (a) 50. (c) 51. (a) 52. (d) 53. (c) 54. (b) 55. (b) 56. (b) 57. (b) 58. (d) 59. (a) 60. (b) 61. (d)

Chapter 2: Gene Editing: ZFNs and TALENs

2.1. Introduction to gene editing

Though genome sequencing represents a huge development in the study of genetic disease, it is ultimately a diagnostic tool, not a form of treatment. Undoubtedly genome sequencing allows the researchers to see how genetic diseases are written in the language of DNA, but to cure genetic diseases, the gene itself has to be repaired to permanently reverse the disease-causing mutation rather than just giving drugs to patients to alleviate the adverse effects of the disease temporarily. For instance, in the case of sickle cell disease caused by a mutation in the *HBB* gene at the 17th position, where A is replaced by T nucleotide, current treatments include frequent blood transfusions, drug hydroxyurea, and bone marrow transplants. Although these therapies can help manage the symptoms, they do not address the underlying genetic defect. To truly cure genetic diseases, researchers are exploring new technologies that enable them to fix the defective gene by rewriting the mutated genetic code.

The technique that aims to treat the genetic disease by rewriting the mutated genetic code is called **gene editing** or **genome editing**. For example, gene editing of mutated *HBB* gene replaces the incorrect T nucleotide at position 17th with the correct nucleotide A. The repaired gene translated into amino acids results in the functional beta-globin subunit of the hemoglobin protein. And you know hemoglobin protein is the major oxygen transport component of red blood cells. Thus, correcting the sickle cell disease permanently.

Simply put, gene editing or genome editing works in a way similar to Microsoft Word -it's like using the cursor to correct a typo by removing an incorrect letter and inserting the correct one.

Genome editing inserts, deletes, or removes DNA using artificially engineered nucleases or molecular scissors. There are generally 3 families of artificially engineered nucleases: Zinc Finger nucleases, TALENs, and CRISPR/Cas system.

2.2. Zinc-finger nucleases (ZFNs)

Zinc-finger nucleases are basically types of artificial restriction endonucleases enzymes. A restriction endonuclease is an enzyme that recognizes a specific sequence of nucleotides and then produces a double-stranded cut in the DNA. For instance, natural EcoRI restriction endonuclease recognizes a specific GAATTC sequence within DNA and then cuts at this specific sequence. But the trouble with using natural restriction enzymes for genome editing is that their recognition sequences are really short, between 4-8 base pairs. Therefore, any restriction enzyme can cut in hundreds or thousands of places in the genome. But for genome editing purposes, a restriction enzyme is required that can target a longer desired DNA sequence, ideally long enough for its unique occurrence in the genome. And the Zinc Finger nucleases are one of those enzymes that can cleave the DNA sequence you want to edit.

As the name suggests, Zinc Finger nuclease has zinc finger repeats and a nuclease.

Zinc Finger repeats constitute the DNA-binding domain, and the nuclease constitutes the DNA cleavage domain of the Zinc Finger nuclease (Porteus & Carroll, 2005; Durai et al., 2015; Urnov et al., 2010). Now let's first talk about the **Zinc finger repeats** that constitute the DNA binding domain of Zinc Finger nuclease. Each Zinc Finger repeat is a structural motif of proteins. The hydrogen bonding between the amino acids within a peptide chain determines the secondary structure of all proteins, which is composed of basic units such as α-helix or β-sheet. But in a protein, the supersecondary structures formed by the combination of the two or three basic secondary structures, α-helix or β-sheets, are known as "motifs."

Like in the case of one zinc finger motif, approximately 30 amino acids fold to form a supersecondary structure containing two antiparallel β sheets and an α helix (*Fig 2-1*). In the motif, two Cys and two His residues are coordinated by a zinc atom.

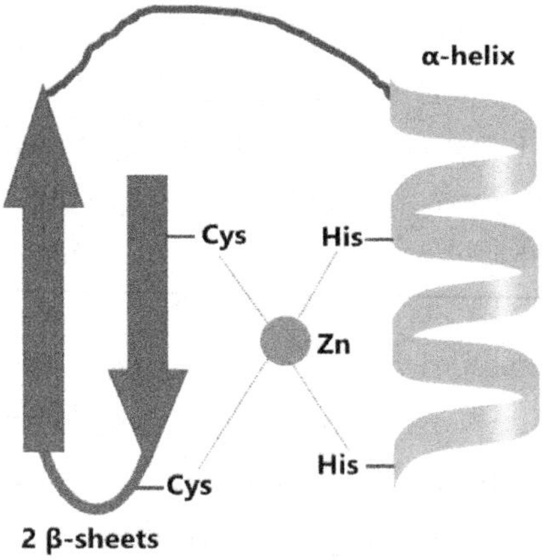

Fig 2-1: Zinc finger motif

The amino acid residues within the α helix of the motif are responsible to make sequence specific DNA contacts in the major groove of the DNA. Thus by changing one or more of the six critical residues located within or adjacent to the α-helix can alter the DNA-binding specificity of a single zinc finger motif. A single zinc finger motif can recognize and bind to only 3 nucleotides, so to recognize 9 nucleotides within a DNA target, 3 zinc finger motifs are required. In many species, including humans, proteins containing zinc finger motifs act as transcription factors, i.e., regulate gene expression by binding selectively to gene promoters. An example of a human gene that codes for a zinc finger transcription factor is the *EGR1* gene.

For designing ZFN that can target the desired DNA, a polypeptide containing at least 3 Zinc finger motifs is designed such that the motifs are able to recognize and bind to the 9 nucleotides within desired target DNA sequences. But Zinc fingers, by themselves, can just bind to the target DNA but can't cleave it.

So, if you want to design a protein that can selectively bind a promoter to activate a gene, then only zinc finger motifs are required. But if you want to use zinc fingers to create targeted breaks in DNA, you need a pair of scissors to go with them. Therefore a DNA-cleaving domain is required to be fused to the zinc fingers.

The DNA cleavage domain of ZFN is derived from FokI restriction endonuclease enzyme obtained from the bacteria *Flavobacterium okeanokoites*. FokI enzyme itself consists of an N-terminal DNA-binding domain and a non-specific DNA cleavage domain at the C-terminal. The recognition site of the enzyme is 5'-GGATG-3'. Once the enzyme is bound to ds DNA via its DNA-binding domain at the recognition site, the DNA cleavage domain is activated. Activated cleavage domain then cleaves DNA, without further sequence specificity. It cleaves the first strand 9 nucleotides downstream and the second strand 13 nucleotides upstream of the nearest nucleotide of the recognition site.

This implies the presence of two separate protein domains within FokI: one for sequence-specific recognition of DNA and the other for the non-specific endonuclease activity (*Fig 2-2*).

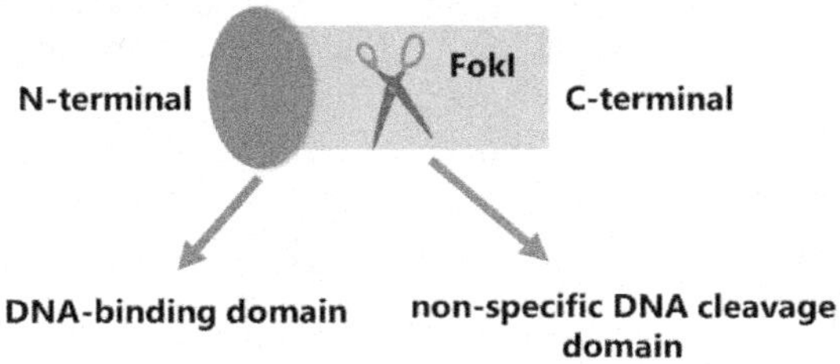

Fig 2-2: Domains of FokI enzyme

Thus the engineered zinc finger nucleases are formed by fusing the FokI cleavage domain to the Zn finger repeats, which are DNA-binding proteins (Miller et al., 2007; Ramalingam et al., 2011). Studies have also shown that the nuclease domain of the FokI enzyme requires dimerization for endonuclease activity on the target DNA.

Therefore, a FokI monomer is fused to each of the two Zinc finger proteins. This facilitates the zinc finger nuclease to effectively bind to the 18 bp recognition site within the desired DNA sequence of the target genome. Then dimerized FokI nucleases can cleave the target DNA and produce a double-stranded break (*Fig 2-3*).

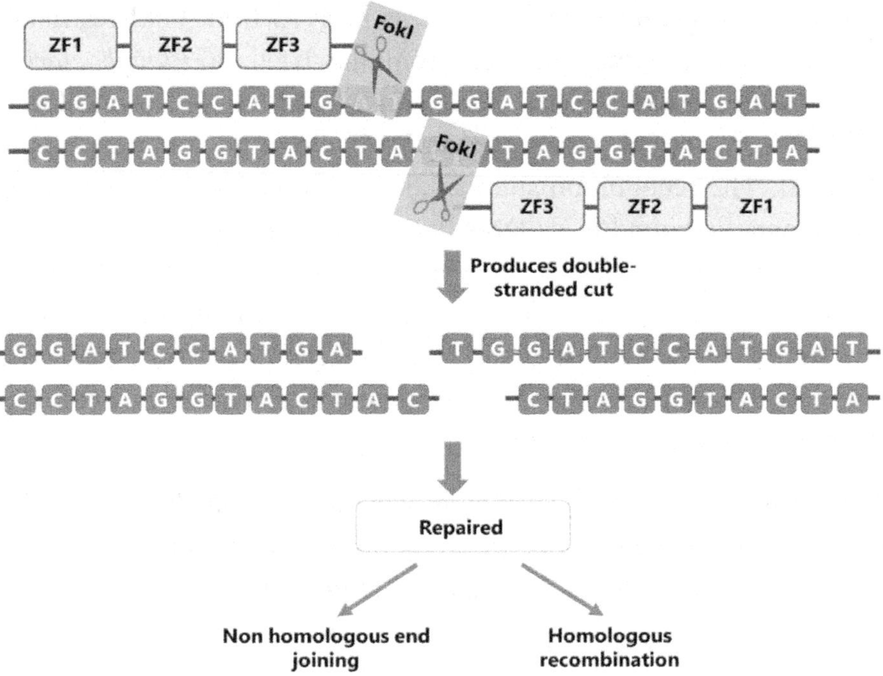

Fig 2-3: Zinc finger nuclease creates a ds cut in the target DNA

The double-stranded break can be repaired either by non-homologous end joining or by homologous recombination (Urnov et al., 2005; Carroll, 2011; Chen et al., 2011; Whyte et al., 2011). And both of these repair processes can be exploited to achieve targeted genome editing, such as gene knockout and gene addition.

Non-Homologous End Joining: Non-Homologous End Joining is a rapid and efficient double-stranded repair mechanism that involves a simple ligation of the two DNA ends that result from a double-stranded break. The FokI-mediated DNA cleavage leaves overhanging ends, which can either be filled in completely or incompletely by the host polymerase enzyme or chewed back by limited exonuclease activity (*Fig 2-4*). Then the ligase enzyme ligates the two DNA ends to complete the repair process. As a result of this repair process, small insertions or deletions of nucleotides can be generated in the region, flanking the ZFN-mediated cleavage site.

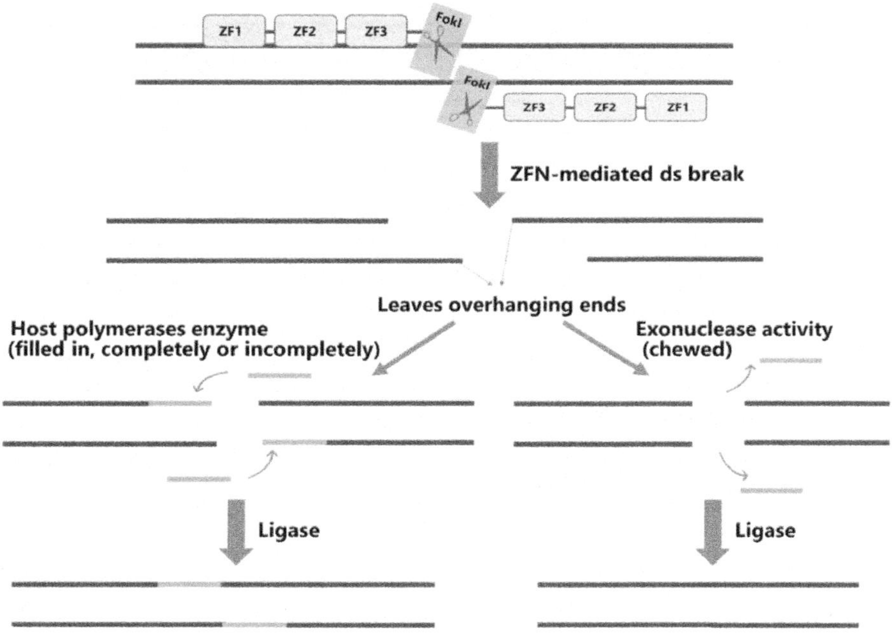

Fig 2-4: Repair of ZFN-mediated ds break by non-homologous end joining

When the zinc finger nucleases are targeted to the protein-coding regions, the outcome is often a shift in the reading frame, usually leading to a null allele or non-functional allele of the targeted gene. The zinc finger nucleases combined with repairing by non-homologous end joining can be used for introducing mutations or for knocking out a gene. Knocking out a gene means making the gene inactivate or inoperative. Let's understand it with an example. In a study, researchers designed a pair of zinc finger nucleases targeted to the first coding exon of *CCR5* (*Fig 2-5*). CCR5 is the major co-receptor for HIV entry and is present on the human T cell. When these ZFNs were introduced into human T cell lines, the zinc finger nucleases disrupted the *CCR5* gene. This ultimately resulted in decreased expression of CCR5, as well as protection from HIV infection because it blocked the entry of HIV into T cells.

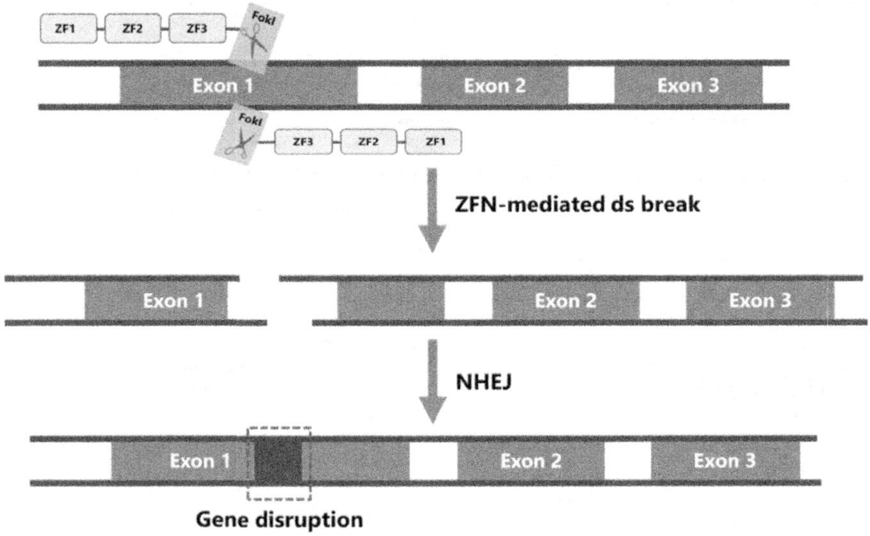

Fig 2-5: The first coding exon of the CCR5 gene gets disrupted by ZFN

Homology-directed repair (HDR) pathway: Alternatively, the homology-directed repair pathway, abbreviated as HDR, utilizes the presence of a template DNA for repairing double-stranded breaks (*Fig 2-6*). The template DNA carries the desired sequence of the gene of interest and is flanked by approximately 1 kbp homology on either side of the mutation. The template DNA then replaces the target double-stranded locus to restore genetic information in an error-free manner.

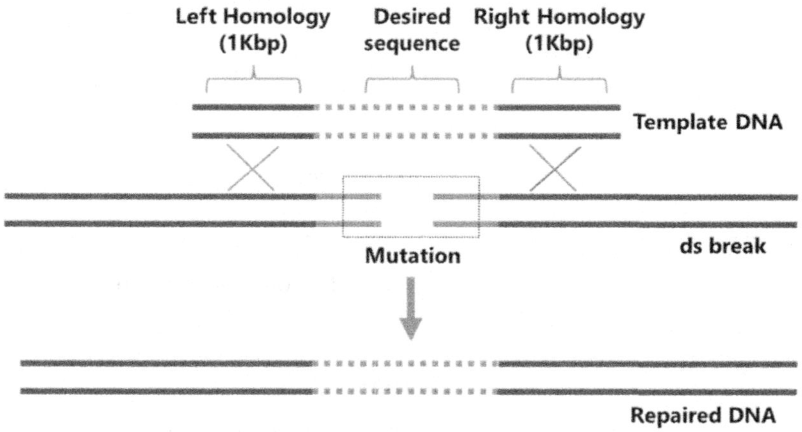

Fig 2-6: HDR utilizes the presence of a template DNA for repairing ds breaks

When ZFN binds to the targeted DNA, and the FokI enzyme introduces double-stranded breaks in it (*Fig 2-7a*), the introduction of double-stranded breaks activates the DNA damage repair by homology-directed repair. Therefore, after ZFN cleavage, the ends of the target DNA are digested by the 5′-3′ exonuclease enzyme, leaving 3' overhangs (*Fig 2-7b*).

One of the resulting single-stranded 3′ ends invades homologous sequences in the template DNA (*Fig 2-7c*). The invading 3′ end is then extended by DNA polymerase represented by a dashed line (*Fig 2-7d*).

After some synthesis, the extended end withdraws and anneals to the other end at the original break (*Fig 2-7e*). The gaps are filled in, and the continuity of the strands is restored by ligation (*Fig 2-7f*).

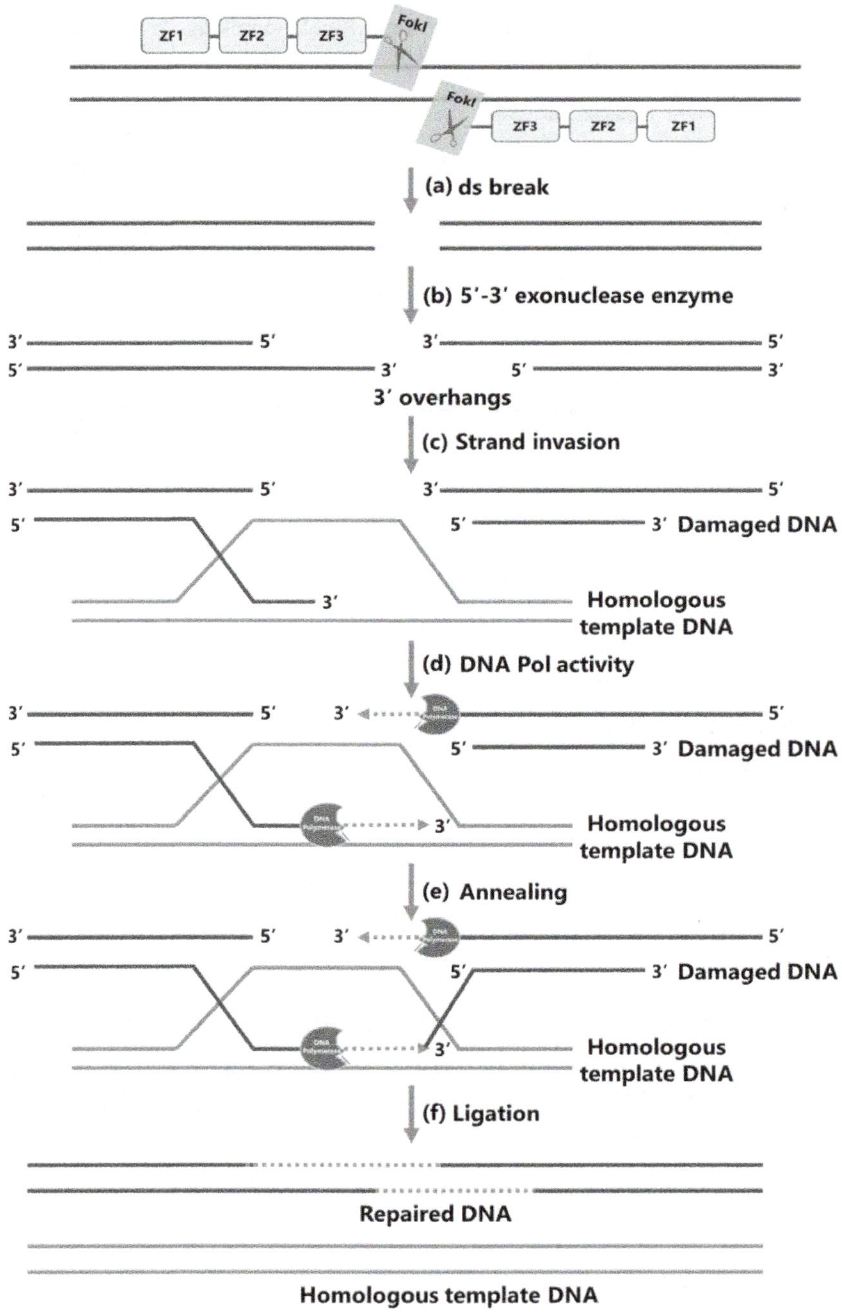

Fig 2-7: Repair of ZFN mediated ds break by HDR pathway

In a study, the researchers designed ZFNs targeting the *IL2Rγ* gene, which is mutated in X-linked severe combined immune deficiency syndrome or SCID. Specifically, the ZFNs were designed to target exon 5 of the *IL2Rγ* gene. To introduce the desired changes, a plasmid containing the unmutated or wild-type exon with approximately 1 kbp homology on either side of the mutation was introduced into the cells.

Co-transfection of ZFNs and the plasmid carrying desired DNA into cells replaced the mutated *IL2Rγ* exon 5 with wild-type exon 5. Thus, fixing the gene responsible for causing X-linked severe combined immune deficiency syndrome.

2.3. Transcription Activator Like Effector Nucleases (TALENs)

The other gene editing technique is Transcription Activator Like Effector Nucleases abbreviated as TALENs. This technique is similar to Zinc Finger Nucleases in some aspects like ZFNs, TALENs too have 2 domains: one is the DNA binding domain and the second is the DNA cleavage domain. DNA cleavage domain in both cases is the non-specific nuclease FokI enzyme to cleave the genome at a specific targeted site (Wood et al., 2011). But the major difference in the two techniques is the difference in their DNA binding domains. As already discussed, the DNA binding domain in the case of the zinc finger nuclease technique is derived from the zinc finger motifs. On the contrary, the DNA-binding domain of TALEN is derived from transcription activator-like effector proteins or TALEs. These proteins are secreted by plant pathogen Xanthomonas bacteria, and these proteins act as transcription factors to alter the transcription of genes in host plant cells, thus promoting their successful infection.

A TALE protein consists of a tandem array of repeated segments flanked by N and C terminal domains. Each repeated segment consists of about 34 amino acids (*Fig 2-8*). The repeated segments adopt a unique two-helix-shaped structure, with the hypervariable amino acid residues at positions 12 and 13 of each repeated segment making base-specific contacts with the DNA in the major groove (Joung et al., 2013; Sun et al., 2013).

Therefore, the amino acid sequence of all the repeats is mostly the same except for the amino acids at positions 12 and 13.

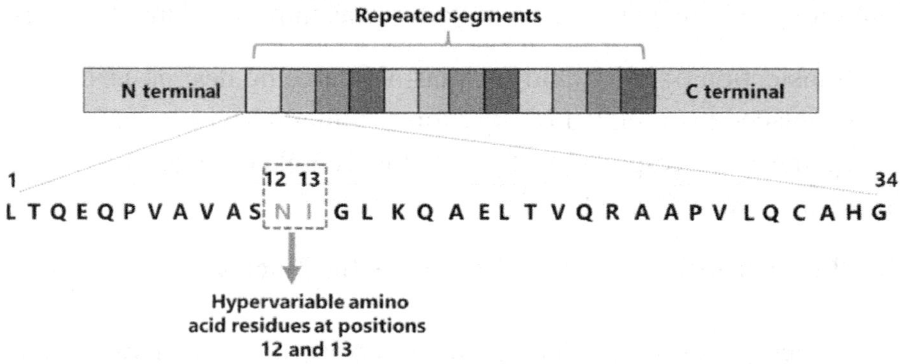

Fig 2-8: A TALE protein

The variable pair of amino acids at positions 12 and 13 is most likely responsible for the binding of one segment of TALE protein to a specific nucleotide in DNA. For instance

- A pair of histidine (H) at position 12 and aspartate (D) at position 13 can bind to cytosine (C) in DNA.
- A pair of asparagine (N) and glycine (G) can bind to thymine (T) in DNA.
- A pair of asparagine (N) and isoleucine (I) can bind to Adenine (A) in DNA.
- And a pair of two asparagine (N) residues can bind to Guanine in DNA.

In short, each TALE repeat in an array binds to a single base of DNA. On the contrary, in the case of zinc finger nucleases, each zinc finger repeat recognizes a DNA triplet. So, a large number of tandem repeated segments are required for the TALE protein to efficiently bind to a particular nucleotide sequence in DNA.

These tandem repeats of TALE protein can be engineered to recognize and bind to the desired target DNA. If you know the nucleotide sequence of the genomic region you wish to alter, then a TALE protein can be synthesized by assembling the sequence of repeats that are able to recognize the desired target DNA (*Fig 2-9*). The sequence of amino acids of variable pairs at positions 12 and 13 of all the repeats must be designed so that they can recognize and bind to the nucleotides of the desired DNA because amino acids at positions 12 and 13 are responsible for binding to DNA. The rest sequence of repeats remains the same.

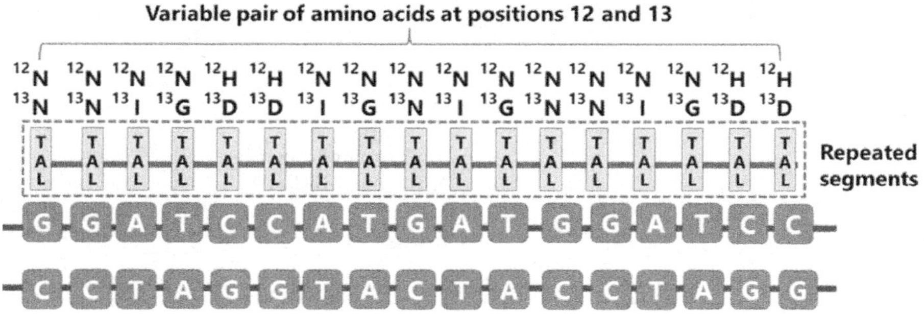

Fig 2-9: Multiple repeats of TALE protein can target a target DNA sequence

Thus with 17 repeats of TALE protein, you can uniquely target a 17 base pair DNA sequence.

Similar to Zinc Fingers, TALE protein with repeated arrays, by themselves, can just bind to the target DNA but can't cleave it. So for using it for genome editing purposes, a DNA-cleaving domain is required to be fused to the TALE to form TALEN. The DNA cleavage domain of the TALENs technique is also derived from the endonuclease enzyme FokI, like in the case of zinc finger nucleases.

While understanding Zinc Finger nucleases, we have discussed that the nuclease domain of the FokI enzyme requires dimerization for endonuclease activity on the target DNA. Thus, for fulfilling this condition, 2 TALE protein constructs are required, and then a FokI monomer is fused to each of the two TALE protein constructs (*Fig 2-10*). This also facilitates the TALE protein to bind to DNA effectively. Once the TALE protein has bound to the target DNA, the dimerized FokI nucleases can cleave the target DNA by introducing a double-stranded break.

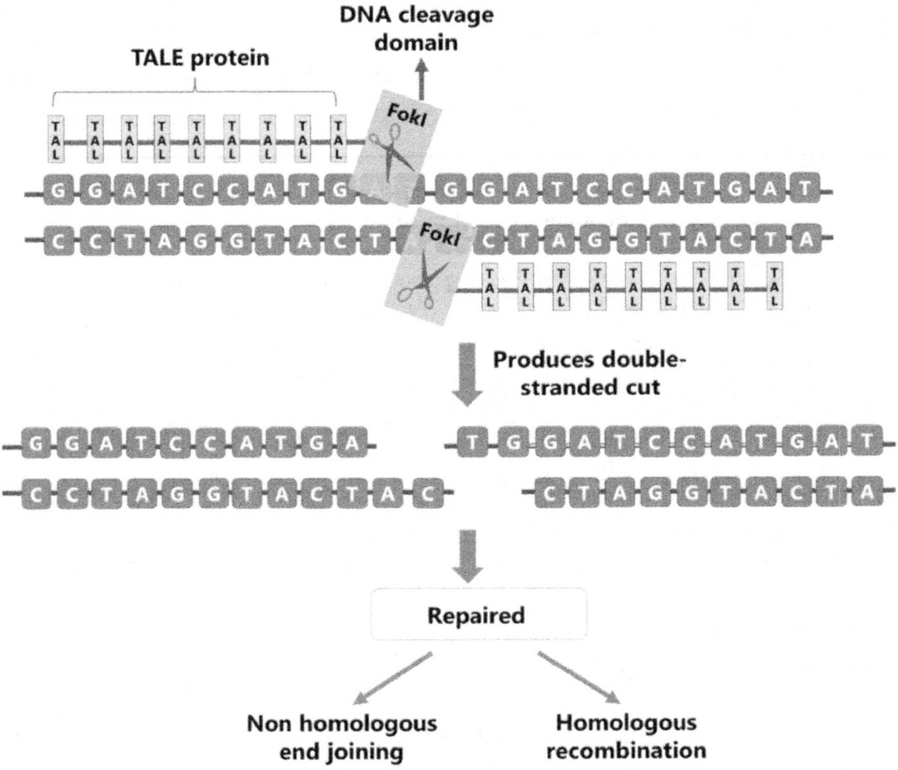

Fig 2-10: TALEN creates a ds cut in the target DNA

The double-stranded break can be repaired either by Non-homologous end joining or by homologous recombination repair mechanisms. These repair mechanisms can then be exploited to achieve targeted genome engineering, such as gene knockout, gene addition to insert DNA of interest, or altering the code of a gene (Tesson et al., 2011; Panda et al., 2013; Sung et al., 2013; Zu et al., 2013). TALENs may also be useful for gene editing in humans for treating genetic disorders like sickle cell anemia or cystic fibrosis.

2.4. Limitations of ZFNs and TALENs

TALENs are more specific than ZFNs and produce fewer off-target mutations than ZFNs (Gaj et al., 2013; Pattanayak et al., 2014). The off-target activity may lead to unwanted double-strand breaks and consequently yield chromosomal rearrangements and/or cell death. This makes ZFNs more toxic to cells than TALENs (Mussolino et al., 2014).

Despite their advantages, TALENs have some downsides compared to ZFNs. One such drawback is their larger size, as one TALE repeat is required for each nucleotide. The typical size of a cDNA encoding a TALEN is around 3 kbp, which is larger than the approximately 1 kbp size of a cDNA encoding a ZFN. This larger size can make it more difficult to deliver and express a pair of TALENs in cells compared to ZFNs. Furthermore, the larger size of TALENs can make them less attractive for therapeutic applications, especially those requiring delivery via viral vectors with limited cargo size. For instance, adeno-associated virus, which is a commonly used viral vector, can only accommodate cargo sizes of less than 5 kbp. Therefore, the larger size of TALENs may limit their application in some therapeutic settings.

A more recent gene-editing technology CRISPR-Cas operates differently from TALENs and ZFNs and also offers certain advantages over TALENs and ZFNs. In the following chapters, we will delve into this technology and explore its advantages over TALENs and ZFNs.

Chapter 3: CRISPR/Cas- Adaptive Immune system of Bacteria & Archaea

To understand why CRISPR technology is of great interest for gene editing purposes, we need first to understand the background of CRISPR, where it comes from, and how it works.

3.1. CRISPR/Cas- Defense system in bacteria & archaea

Like us humans, bacteria, and archaea are also susceptible to viruses. The viruses that infect and replicate within bacteria and archaea are known as bacteriophages or phages. To enter a bacterial cell, bacteriophages attach to the specific receptors like polysaccharides, proteins, etc., present on the surface of bacterial cells and then inject their genetic material inside the bacterial cell. Once the viral genome is inside the bacterial cell, it uses the bacterial cell's machinery inorder to replicate and synthesize viral proteins. The viral proteins and genome assemble to form new virions within the cell. Then the newly formed virions are released by causing the lysis of bacterial cells, thus ultimately leading to the death of the bacteria. In order to protect themselves from viral or phage attacks, bacteria have developed various defense systems. For instance,

(i) Restriction-modification system: It is one of the efficient and common bacterial strategies to prevent phage attack. The restriction-modification system is composed of a restriction endonuclease and a cognate methyltransferase (*Fig 3-1*). The restriction endonuclease enzyme cleaves the double-stranded DNA of viral DNA by recognizing specific short DNA sequences called recognition sites. But these restriction enzymes can also cleave the bacterial genome. Thus, to prevent the destruction of its own genome by restriction enzymes, the methyltransferase (MTase) enzyme of the restriction-modification system causes methylation of bacterial DNA. Therefore, the restriction enzymes cannot cleave the methylated DNA, so only the phage genome is targeted (Pleška et al., 2016; Dupuis et al., 2013). But some phages can usually overcome this defense by modifying their own DNA by adding methyl groups to it, thus blocking the restriction enzymes.

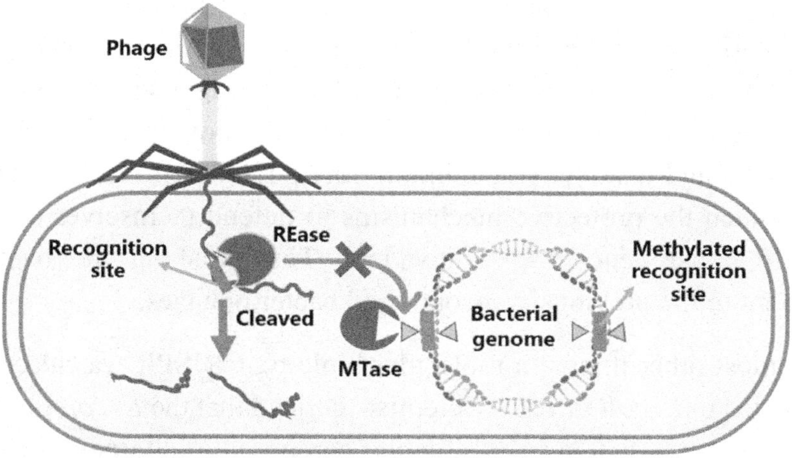

Fig 3-1: Restriction-modification system

(ii) Abortive infection: Another defense system used by bacteria to prevent phage attack is "abortive infection." This is an altruistic mechanism in which an infected bacterium kills itself, thereby eliminating phage propagation. Undoubtedly, this mechanism is lethal to the single infected bacterium, but the death of the infected cell is a sacrifice to protect the surrounding clonal population from attack by phages (Lopatina et al., 2020).

(iii) CRISPR/Cas system: The CRISPR/Cas system is a relatively new type of defense system identified in bacteria (Horvath and Barrangou, 2010; Bhaya et al., 2011; Marraffini, 2015). Nearly all known archaea and about half of bacteria possess CRISPR.

The CRISPR system is a kind of adaptive immune system of humans. A vital feature of the adaptive immune system is the generation of a memory of past infections that allows the immune system to rapidly mount a more efficient and robust response upon recurrent infections.

Both the adaptive branch of the mammalian immune system and CRISPR-Cas immunity have these key features. For example, memory B cells in humans generate an accelerated and more robust antibody-mediated immune response in the case of reinfection with an antigen, eventually leading to its neutralization. The analogous system in bacteria CRISPR/Cas system allows the bacteria and archaea to create memory of the invader phage by capturing a small snippet of the phage's genetic material. This would allow the bacteria to recognize and distinguish incoming 'foreign' DNA, from 'self' DNA and launch the protective mechanisms to defend themselves against the previously encountered invaders. Thus, making the bacteria resistant to the previously encountered bacteriophages.

Like most other things in molecular biology, CRISPR was also first identified in *E. coli* in 1987. Scientists noticed that the *E. coli* genome contains an unusual genetic structure composed of alternating repeat and non-repeat DNA sequences, whose biological significance was unclear at that time. The CRISPR system's function was brought to light 20 years later, in 2007, when scientists identified and sequenced CRISPR loci in *Streptococcus thermophilus* strains. When the non-repeat DNA sequences, also called spacers of the CRISPR loci, were sequenced, it was found that the spacers were homologous to bacteriophage and plasmid sequences. Thus, leading to the hypothesis that CRISPR may be a defense mechanism of bacteria against foreign elements. To test this possibility, an incredible experiment was designed by scientists. A phage or virus-susceptible wild-type *Streptococcus thermophilus* strain was infected with two different virulent bacteriophages. It resulted in the generation of nine different phage-resistant *Streptococcus thermophilus* strains. The sequencing of CRISPR loci in these strains showed that new spacers had been inserted in the CRISPR locus of the wild-type strain *(Fig 3-2)*. The new spacer sequences were found to have similarities with sequences present in the genome of the phages used in the experiment. This finding provided further evidence to support the hypothesis that, through the use of CRISPR, bacteria can incorporate new spacers from viral genome sequences in response to viral infection. This integration process enables the bacteria to create memory banks that store genomic information from past encounters with invading viruses. Because of which bacteria can become resistant to those viruses.

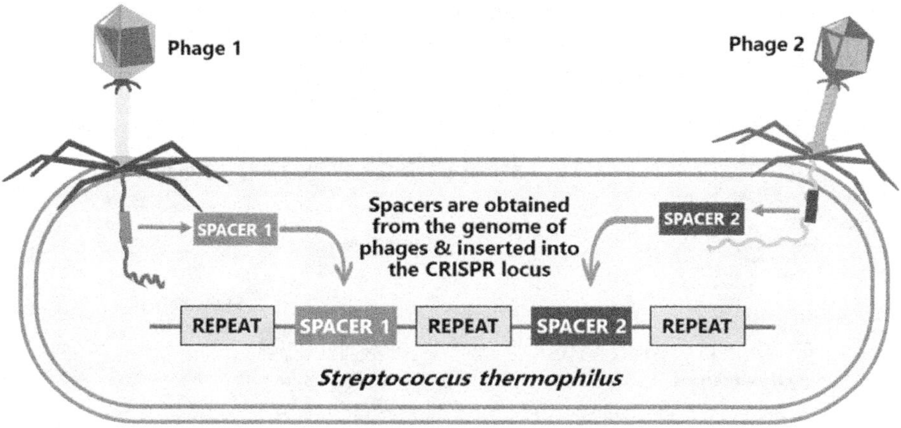

Fig 3-2: Integration of spacers from viral genome into the CRISPR locus of Streptococcus thermophilus

3.2. Components of the CRISPR-Cas system

CRISPR stands for Clustered Regularly Interspaced Short Palindromic Repeats. The full form of CRISPR tells about two main parts found in CRISPR. First, there are repetitive sequences, which are short segments of DNA, 20-40 bp in length. Also, these repeats are palindromic. A palindrome is a sequence of letters that read the same either from the left or right. Like in the example of the phrase "NEVER ODD OR EVEN." In palindromic repeat sequences, each repeat is arranged in a palindromic fashion, meaning that the repeat's sequence on one strand is identical to the opposite strand's sequence when both are read in their respective 5' to 3' direction. For instance, in the repeat sequence shown in *Fig 3-3*, when either strand is read from 5' to 3' direction, the nucleotide sequence remains the same, i.e.,

GGATCC. Because of this, the two halves of the sequence are complementary, allowing them to base pair. Thus, when these palindromic sequences are transcribed to form RNA, they can create hairpin shaped secondary structures.

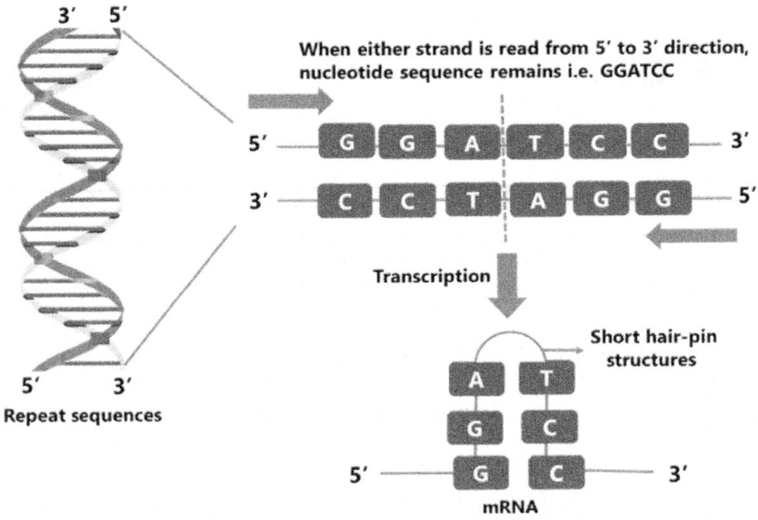

Fig 3-3: Repeat sequences are arranged in a palindromic fashion in CRISPR loci

Additionally, the repeat sequences are highly conserved within a CRISPR locus meaning that these repeat sequences within a CRISPR locus are all identical one after another after another and so on.

The full form of CRISPR "Clustered **Regularly-Interspaced** Short Palindromic Repeats" suggests that the short palindromic repeat sequences are regularly interspaced. It means that the spacer DNA, or referred to as spacers, are present between the repeat sequences. And these spacers are not identical. Each segment of the spacer is unique. Interestingly these unique sequences match with the DNA found in the viruses. These spacer sequences are some kind of acquired immune system in bacteria that protect the bacteria from viruses. The bacteria use these spacers as recognition elements to find matching virus genomes and destroy them. In other words, these spacers function as an immunological memory bank, storing sequences from previous encounters with invading organisms. The number of spacers within a CRISPR array can range from as few as one to several hundred, depending on the species.

Another feature associated with CRISPR loci is a region rich in adenine (A) and thymine (T), known as the leader sequence, located upstream to CRISPR loci. These leader sequences have a significant role that we will discuss while studying the CRISPR-Cas system's mechanism.

The CRISPR array also consists of CRISPR-associated genes, generally called *cas* genes. These *cas* genes make Cas proteins *(Fig 3-4)*.

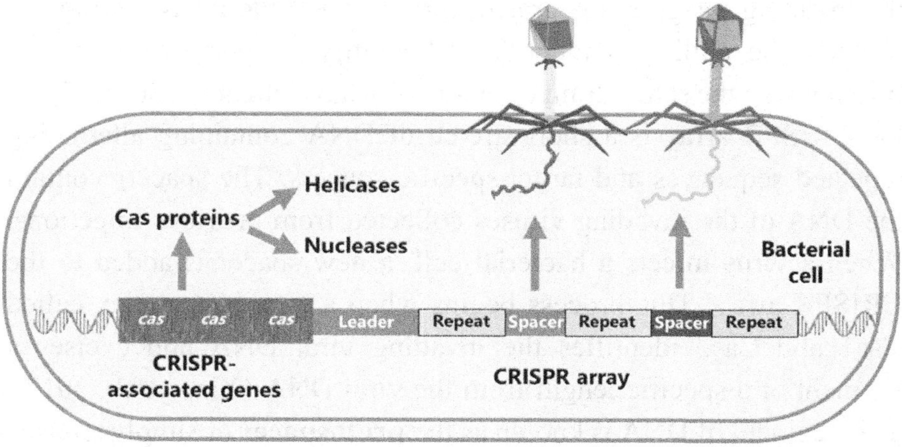

Fig 3-4: Components of the CRISPR-Cas system

The Cas proteins can be helicases, the proteins that unwind DNA, or the nucleases that cut the target DNA. The CRISPR system works through the cooperation of many diverse Cas proteins. These Cas proteins constitute the backbone of the CRISPR system. Thus, the system is generally called the CRISPR-Cas system.

3.3. Mechanism of the CRISPR-Cas system in bacteria

Let's take a scenario where the bacteriophage infects a bacterial cell by attaching itself to the cell's surface and injecting its viral genome into the cell. The viral DNA then gets integrated into the bacterial genome, and using the host machinery, the phage replicates inside the bacterial cell. To encounter the infection, the bacterial CRISPR-Cas system plays a vital role. The CRISPR-Cas mediated defense process occurs in three steps (Rath et al., 2015; Amitai and Sorek, 2016; Hille et al., 2018).

(i) The first step for CRISPR-Cas mediated defense is known as adaptation or acquisition. This step is required to provide genetic memory to the bacterial cell. During this stage, the genetic material of the invading phage is integrated into the CRISPR locus, giving the bacteria the ability to identify and modify its response to future invasions by the same strain of phage. We have already discussed that the CRISPR array is a short stretch of DNA containing alternating repeated sequences and target-specific spacers. The spacers contain the DNA of the invading viruses collected from previous infections. When a virus infects a bacterial cell, a new spacer is added to the CRISPR array. This process begins when a protein complex called Cas1 and Cas2 identifies the invading viral DNA and excises a segment of a specific length from the viral DNA (Xiao et al., 2017). This segment of DNA is known as the **protospacer** or simply **spacer**. Then the protospacer is added to the front of the CRISPR array between the two repeat sequences *(Fig 3-5)*. By this mechanism, bacteria generate the embedded memory of the invading virus.

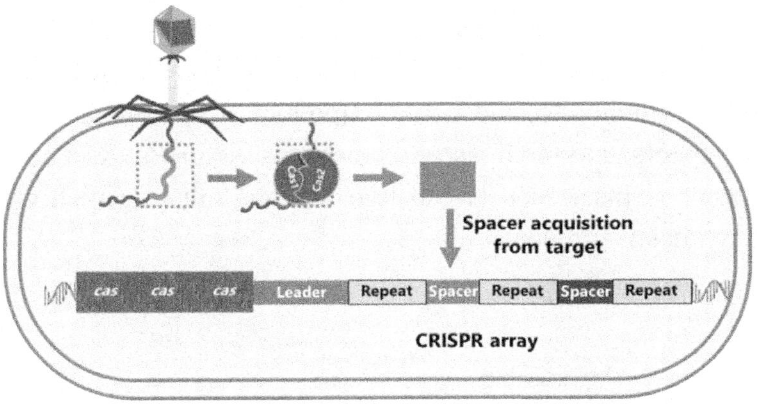

Fig 3-5: Protospacer (or called spacer) acquisition from the target genome between two repeat sequences

Mechanism of spacers' integration in the CRISPR array: To integrate protospacers (spacers) into the CRISPR array, leader sequence, and leader proximal repeat are required. The spacer acquisition is usually directional, with new spacers added to the leader-proximal end of the array. Further, there is also a requirement for generating one new repeat because the protospacer is added into the front of the CRISPR array between two repeats. For this, the leader-proximal repeat serves as a template for the synthesis of the new repeat. Simply put, the leader's presence is crucial for adding spacers to the front of the CRISPR array *(Fig 3-6)*.

The palindromic repeats form a cruciform structure that recruits Cas1 and Cas2 proteins to mediate protospacer DNA integration. Cas1 and Cas2 proteins have both nuclease and integrase activity. The nuclease activity mediates the excision of protospacer DNA from the viral DNA. On the other hand, integrase activity of the Cas1-Cas2 complex mediates the integration of protospacer DNA into the CRISPR array. The 3'-OH groups present in the protospacer facilitate a nucleophilic attack on the 5' ends of the first repeat of the CRISPR array, leading to the creation of a new spacer within the array. As a result, the

CRISPR array is expanded and contains an incomplete single-stranded DNA repeat on either side of the new spacer. The polymerase enzyme then generates the new complementary strands to generate dsDNA repeats, and the ligase enzyme seals the nicks. As a result, the protospacer is inserted between two repeats, out of which one is a newly synthesized repeat.

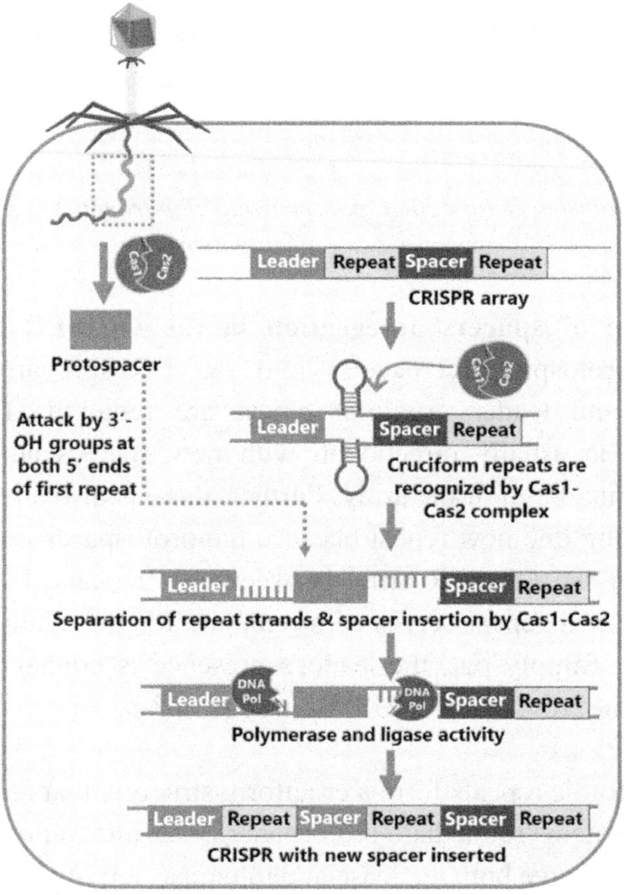

Fig 3-6: Mechanism of integration of spacers in the CRISPR array

(ii) The second step of the CRISPR-Cas mediated defense process is expression. Once the protospacer from the invading DNA is integrated into the CRISPR locus, the CRISPR locus is transcribed into a precursor-CRISPR RNA, written as pre-crRNA *(Fig 3-7)*. CRISPR transcription initiates in the leader region. The leader contains promoter elements and binding sites for RNA Polymerase and other regulatory proteins required for transcription. A long primary transcript, the precursor-CRISPR RNA or pre-crRNA, is generated, which contains a series of secondary structures or hairpins because of the palindromic repeats. The pre-crRNA is subsequently processed into smaller mature crRNAs by Cas proteins. Each piece of crRNA corresponds to a single spacer flanked by a partial repeat sequence. Because the cleavage of the pre-crRNA occurs in the repeat sequences, mature crRNA consists of a spacer flanked by partial repeats.

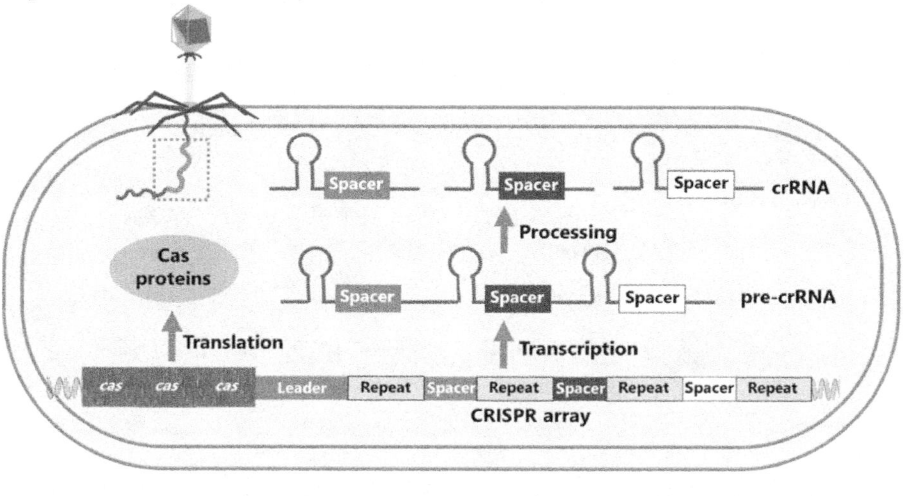

Fig 3-7: CRISPR RNA (crRNA) expression

(iii) The third step of the CRISPR-Cas mediated defense process is interference. In this step, the mature crRNA forms a complex with one or more Cas proteins to form an effector search complex. Then this search complex patrols the inside of the cell, looking for invading phages or other foreign genetic material that matches the crRNA sequence, similar to the way a detective uses a fingerprint to identify a criminal. Once the match is found, i.e., the sequence of crRNA base pairs with the complementary protospacer sequence in the invading virus's nucleic acids, the specific Cas proteins chop up the invading genetic material of the phage and thus destroying the phage *(Fig 3-8)*. This is how the CRISPR Cas system protects bacteria from viral infection.

The crRNA is also called **guide RNA** or simply written as **gRNA** because it guides the interference machinery to its protospacer targets. Thus, the CRISPR-Cas system is also called RNA-guided targeting of the viral genome.

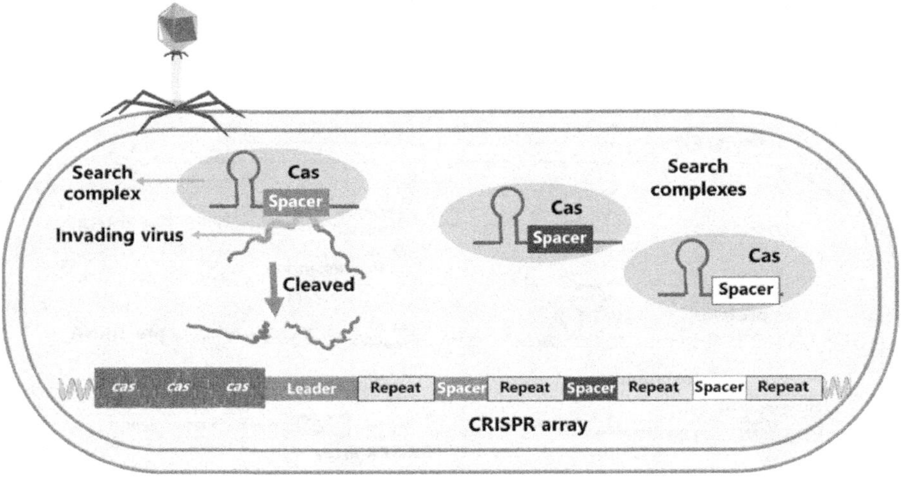

Fig 3-8: crRNA-Cas search complex chops up the invading viral genome

But imagine if the phage invading the bacteria is not new, and the protospacer matching to the viral genome is already present in the bacterial genome. In that scenario, after infection, the adaptation step doesn't occur. Instead, the second step expression initiates directly, in which the CRISPR locus is transcribed into a precursor-CRISPR RNA, and *cas* genes are translated to form Cas proteins. The pre-crRNAs are processed to form mature crRNAs, which then form complexes with Cas proteins to create a complete search complex.

Once the sequence of crRNA base pairs with the complementary protospacer sequence in the invading virus's nucleic acids, the specific Cas proteins chop up the invading genetic material of the phage and thus destroy the phage. Therefore protecting the bacteria from the invading viral infection.

3.4. Types of CRISPR-Cas systems

Though the CRISPR-Cas system's general mechanism remains the same, the CRISPR-Cas system can be classified into six major types-Type 1, 2, 3, 4, 5, and 6. This classification is based on differences in the processing of pre-crRNA, the interference step, and the requirement of different Cas proteins (Makarova et al., 2011; Makarova and Koonin, 2015;).

Type 1 CRISPR-Cas system: In the Type 1 CRISPR-Cas system, pre-crRNA cleavage is carried out by Cas endoribonuclease Cas6 to form mature crRNA *(Fig 3-9)*. The mature crRNA then interacts with the Cas protein complex consisting of 5 Cas proteins Cse1, Cse2, Cas7, Cas5e, and Cas6e subunits. This Cas proteins-crRNA complex is called Cascade, which locates foreign genetic material. If Cascade binds to foreign DNA, its conformation changes and causes the recruitment of a Cas nuclease Cas3 during the interference step, which then degrades the target DNA.

Type II CRISPR-Cas9 system: In type I CRISPR-Cas system, Cas endoribonuclease Cas6 carries out the pre-crRNA cleavage. But in the case of the type II CRISPR-Cas system, this process involves the expression of a transactivation RNA or tracrRNA (Chylinski et al., 2014; Le Rhun et al., 2019; Jiang and Doudna, 2017).

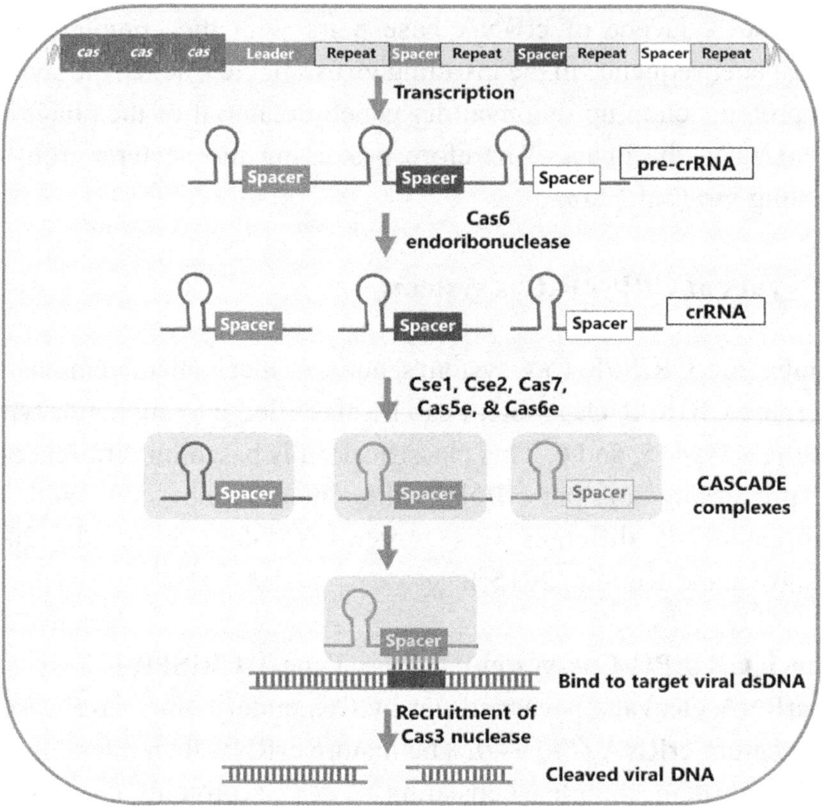

Fig 3-9: Type I CRISPR-Cas system

TracrRNA includes an "anti-repeat" region that hybridizes with repeat sequences in the pre-crRNA transcript (Chylinski et al., 2013). The resulting duplex is then cleaved in the repeat sequences by RNase III in a Cas9-dependent reaction (Deltcheva et al., 2011).

Then crRNA and tracrRNA form a complex with Cas9 protein to form a complete search complex *(Fig 3-10)*. Cas9 is literally a **programmable protein** because it has a program defined by the crRNA guide sequence, which is a 20-letter nucleotide sequence derived from the CRISPR spacer sequence. It means, the 20-letter nucleotide sequence directs the Cas9 protein to recognize a piece of viral DNA that matches the 20-letter sequences of crRNA. If the sequence of crRNA matches the invading virus's DNA sequence, then Cas9 cuts the viral DNA by introducing a double-stranded break.

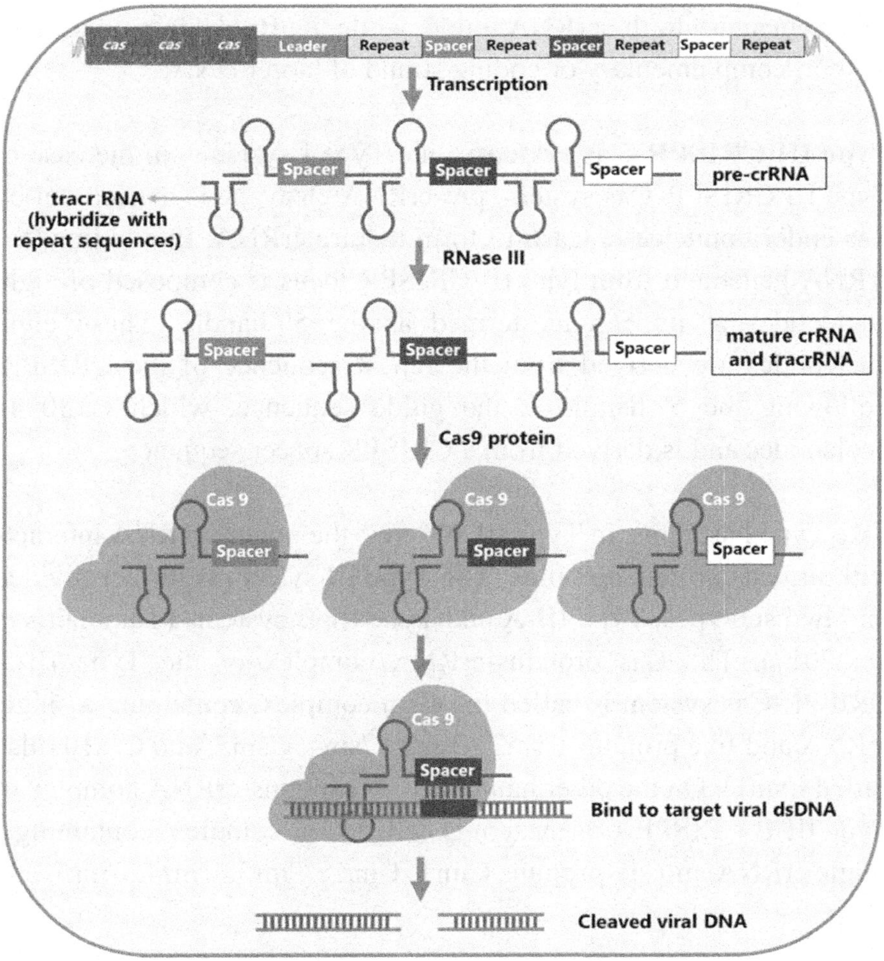

Fig 3-10: Type II CRISPR-Cas system

Like type 1, in the type II system, the target nucleic acid is dsDNA. The Cas9 protein has two lobes; one lobe is for target recognition. And the other lobe contains nuclease activity. The recognition lobe is essential for binding crRNA and target DNA. On the other hand, the nuclease lobe cleaves the target DNA by generating double-stranded breaks (DSBs). Additionally, the nuclease lobe contains HNH and RuvC nuclease domains, and these domains cause double-stranded breaks in the target DNA. The HNH domain cleaves the DNA strand complementary to the crRNA guide, while the RuvC domain cleaves the non-complementary or coding strand of target DNA.

Type III CRISPR-Cas system: Like Type I systems, in the case of Type III CRISPR-Cas system, pre-crRNA cleavage is carried out by Cas endoribonuclease, Cas6 to form mature crRNA. In addition, the crRNA generated from type III CRISPR locus is composed of eight nucleotides at its 5′ end, termed as the 5′ handle. These eight nucleotides are derived from the repeat sequence of the CRISPR. Following the 5' handle is the guide sequence, which is 30–45 nucleotides and is derived from a CRISPR spacer sequence.

Like Type I systems, in Type III systems, the mature crRNA interacts with the Cas protein complex. The Type III system is further divided into two subtypes, Type III-A and Type III-B systems (Tamulaitis et al., 2017). The Cas proteins-crRNA complex of the Type III-A CRISPR-Cas system is called the Csm complex, containing a single crRNA and five proteins Csm2, Csm3, Csm4, Csm5, and Cas10 (also called Csm1). On the other hand, the Cas proteins-crRNA complex of Type III-B CRISPR-Cas system is called Cmr complex, containing a single crRNA and six proteins Cmr1, Cmr2, Cmr3, Cmr4, Cmr5, and Cmr6.

The mechanism of targeting invading nucleic acids by the type III system is different from type- I and II systems. In Type III-A, invading DNA targeting requires directional transcription across the target sequence or protospacer sequences to produce an RNA transcript that is complementary to the crRNA (Samai et al., 2015; Pyenson et al., 2017). While transcription, denaturation of the target double-stranded

DNA occurs in the transcription bubble. Then during the interference step, the crRNA binds to the targeted RNA transcript; this binding leads to the activation of Cas10 and Csm3 nucleases of the Csm complex required to cleave denatured target nucleic acid and the RNA transcript *(Fig 3-11)*. Cas10 cleaves the non-template or coding DNA strand, and Csm3 protein chops RNA within the protospacer region. The Csm complex also recruits Csm6 protein that degrades nonspecific transcripts in the vicinity.

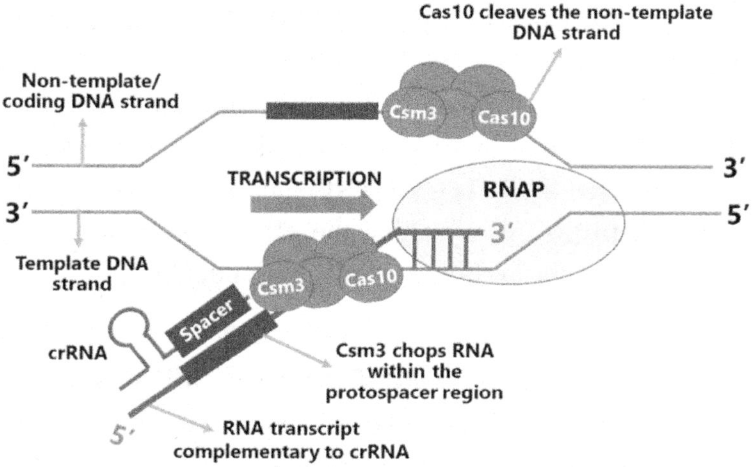

Fig 3-11: Type III-A CRISPR-Cas system

Cas proteins-crRNA complex of **Type III-B CRISPR-Cas system**, also called Cmr, can cleave an ssRNA target that is complementary to the CRISPR RNA. The Cmr effector complex consists of six proteins Cmr1-6 and a crRNA. Cmr complex can cleave the target RNA at multiple sites within the region of base-pairing.

Thus, target nucleic acid is usually dsDNA in type I, type II, and type III-A systems. In contrast, the type III-B system targets complementary single-stranded RNA or ss DNA of the invading virus.

Type 4 CRISPR-Cas system: Type IV CRISPR-Cas system is a newly discovered system and is not well characterized. Type 4 systems occur within plasmids and lack the genes like Cas1 and Cas2 that encode the CRISPR-Cas mediated defense system's first step adaptation (Pinilla-Redondo et al., 2020). The type IV CRISPR–Cas system has Cas proteins like Cas7 (also called Csf2), Cas5 (also called Csf3), and a smaller version of Cas8 referred to as Csf1. Moreover, Type IV CRISPR–Cas also encodes a DinG family helicase Csf4 and a type IV-specific Cas6-like protein Csf5. A recent structural and biochemical analysis of a type-4 CRISPR–Cas system demonstrated the role of the Cas6-like protein in both the maturation of crRNAs and in the subsequent formation of a Cascade-like crRNA-guided effector complex, composed of Csf1, Csf3, Csf5, and multiple copies of Csf2. As in other CRISPR–Cas systems, the effector complex survey the cellular environment searching for matching nucleic acid targets.

Type 5 CRISPR-Cas12 system: Type 5 CRISPR-Cas system uses the Cas12 enzyme. Cas12 has several critical differences from Cas9: Cas12 causes a 'staggered' cut in double-stranded DNA, producing ends with a single-stranded overhang instead of the 'blunt' cut made by the Cas9 enzyme. Also, Cas12 requires only a CRISPR RNA (crRNA) for successful targeting. By contrast, Cas9 requires both crRNA and a transactivating rRNA to target the invading virus's DNA successfully.

Type 6 CRISPR-Cas13 system: The type 6 CRISPR-Cas system utilizes the Cas13 enzyme. Cas13 is an RNA-guided RNA endonuclease, meaning it does not cleave DNA but only single-stranded RNA. Cas13 is guided by its crRNA to the target ssRNA, and once bound, it cleaves the target RNA.

An additional interesting feature of Cas12 and Cas13 enzymes is that they show **trans or collateral-cutting activity**. It means that on finding the target, the cleavage activity of Cas12 and Cas13 enzymes is not just restricted to the target DNA or RNA, but they can also cut any single-stranded non-targeted nucleic acid molecules in the vicinity. For instance, after binding and cutting the target DNA, Cas12 becomes further active and chops any single-stranded DNA molecule present in its vicinity. On the other hand, after cleaving the target RNA, the activated Cas13 enzyme chops any single-stranded RNA molecule present in its vicinity.

Although this collateral DNAse and RNase activity might appear to be a disadvantage in terms of specific gene editing, but it has made these enzymes a powerful tool for the development of CRISPR-based diagnostics. We will study about this in detail later in the book.

3.5. Protospacer Adjacent Motif or PAMs

CRISPR systems in bacteria were evolved to defend bacteria against viruses. After the integration of protospacers into the CRISPR locus, the CRISPR locus is transcribed to form precursor-crRNA. After maturation, crRNA forms a complex with Cas proteins to create a search complex. And if the sequence of crRNA matches with the sequence of the DNA of the invading virus, then the Cas protein cuts the viral DNA. Thus, cleavage of viral DNA destroys the invading viral genome and therefore protects the bacteria from the viral infection. But the **viral DNA targeted by the search complex is the same sequence as the protospacer DNA** in the CRISPR array. Because protospacers are the excised segments of the invading viral DNA.

So how exactly is Cas protein able to distinguish between itself and the enemy? This is where PAM comes in. The PAM stands for Protospacer Adjacent Motif. PAM is a specific sequence of nucleotides, around 2-6 bp, that follows the protospacer sequence in a viral genome. It is imperative to understand that PAM is a component of the invading virus or plasmid but is not a component of the bacterial CRISPR locus. In the interference step of CRISPR system-mediated defense, specific Cas proteins recognize and bind the PAM sequence. The PAM recognition is necessary to facilitate the double-stranded DNA target's unwinding, thereby triggering the base pairing between the crRNA and the DNA target followed by cleavage with the Cas proteins (Gleditzsch et al., 2019).

On the contrary, Cas nucleases cannot successfully bind to or cleave the target DNA sequence if the PAM sequence does not follow it. For instance, in Type II CRISPR-Cas systems, Cas9 recognizes the PAM sequence **-NGG-** when read in the 5’-3' direction' where N is any nucleotide followed by 2 Guanine residues (Palermo et al., 2017; Cencic et al., 2014). This PAM sequence must be present for the Cas9 protein to know that it's ok to latch onto and cut the viral DNA *(Fig 3-12)*. On the other hand, the spacer sequences within the CRISPR array are not followed by "GG." It means that the Cas9 cannot bind to the CRISPR array and thereby avoids cutting the bacterium's own genome.

The PAM sequence "NGG " is associated with the Cas9 nuclease of *Streptococcus pyogenes*, whereas different PAM sequences are associated with the Cas9 proteins of different bacteria. For Instance, *Staphylococcus aureus* Cas9 recognizes "NNGRRT" PAM sequence where R is any purine may be A or G nucleotide. And *Campylobacter jejuni* Cas9 recognizes NNNVRYAC PAM sequence where V is A, G, or C nucleotide; and Y is T or C nucleotide.

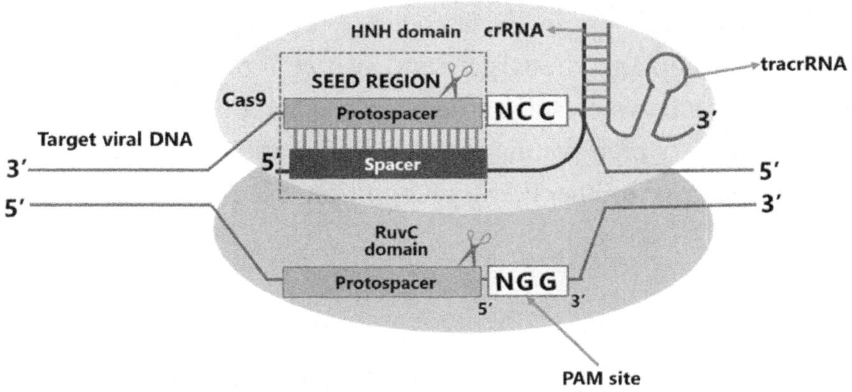

Fig 3-12: Protospacer Adjacent Motif (PAM) for Cas9 nuclease of Streptococcus pyogenes

We have already discussed that the type-1 CRISPR-Cas system utilizes the Cascade complex in the interference step and recruits Cas 3 protein to degrade viral DNA. In *E. coli,* the Cascade complex consists of a mature crRNA and five Cas proteins, Cse1, Cse2, Cas7, Cas5e, and Cas6e subunits. During the Interference step, the Cse1 subunit of the Cascade complex recognizes the PAM region of the viral DNA. It then positions the Cas3 protein adjacent to the PAM to ensure the cleavage of viral DNA.

Both Type I and II CRISPR-Cas systems search for foreign genomes during the interference step, requiring the presence of a PAM sequence and perfect protospacer-crRNA complementarity in the so-called seed region, which is located adjacent to the PAM. Thus the seed region can be defined as the region of target-crRNA complementarity at positions 1–5 and 7–8 of the 5′ end of the crRNA spacer sequence. Therefore, a conserved protospacer-adjacent motif or PAM, which is present in the target, but not the CRISPR locus, allows for distinction between foreign and host DNA.

On the other hand, no PAMs have been detected for Type III CRISPR-Cas systems. The crRNA formed in the type III system contains a 5'-

handle. This handle originates from the CRISPR repeat sequence, and the 3' spacer contains the target sequence *(Fig 3-13)*. Therefore, the discrimination between self and non-self is achieved by the presence of a 5' handle, which results in "self-inactivation," a fundamentally different process to the PAM recognition used by Type I and II systems.

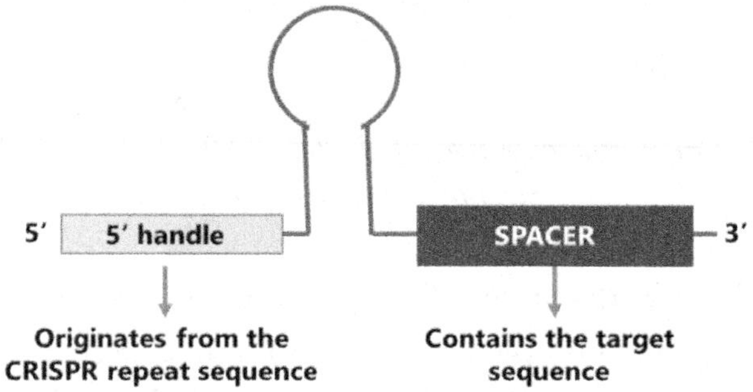

Fig 3-13: cRNA in type III CRISPR-Cas system

PAMs also serve an additional role in the protospacer acquisition step. While studying the CRISPR-Cas system's mechanism, we have already discussed that Cas1 and Cas2 complex is required in capturing new protospacers from the invading viral DNA. During this acquisition process, Cas9 works with Cas1 and Cas2 to find a PAM sequence, and then Cas1 and Cas2 remove the protospacer next to it. Protospacer picking with PAMs guarantees that when the same virus infects again, and Cas9 is armed with a matching crRNA, nothing will stop it from destroying the enemy DNA.

Chapter 4: CRISPR-Cas system as a tool for genome editing

4.1. Why is the CRISPR-Cas9 system suitable for gene editing?

Over the past decade, the CRISPR-Cas9 system has become a widely used genome editing technique due to its speed, affordability, precision, and ease of use. As previously mentioned, the CRISPR-Cas systems are categorized into six types (I, II, III, IV, V, and VI) based on differences in pre-crRNA processing, interference steps, and the required Cas proteins. However, the type II CRISPR-Cas system, also known as the CRISPR-Cas9 system, has gained significant attention in the scientific community for its usefulness in genome editing. This preference stems from the fact that:

(i) Type I, III, and IV CRISPR-Cas systems require several Cas proteins for cleaving the target viral genome. For instance, the Cascade complex consists of 5 Cas proteins in type I systems, while the Csm and Cmr effector complexes of type III systems also consist of around 5 Cas proteins for targeting the viral genome. Whereas in the type II CRISPR/Cas systems, only one Cas9 protein is required to target and cleave the viral genome.

(ii) Additionally, the Cas9 protein, the part of the search effector complex, participates in the processing of pre-crRNAs to form mature crRNAs. On the other hand, in type I and type III systems, Cas6 protein, which is not a part of the search effector complex, is required for the processing of pre-crRNAs to form mature crRNAs.

Thus due to the simplicity and requirement of only the Cas9 enzyme for processing crRNA and forming search effector complex, the Type II CRISPR-Cas9 system has been widely adopted as a gene editing tool.

4.2. Components of the CRISPR-Cas9 system

Now let's discuss the components of the CRISPR-Cas9 system that makes it a wonderful gene editing tool. We have already discussed that in the case of type 2 CRISPR systems, the generation of mature crRNA requires 2 RNAs: one is pre-crRNA itself, and the second is transactivation RNA or tracrRNA. TracrRNA includes an "anti-repeat" region that hybridizes with repeat sequences in the pre-crRNA transcript. The resulting duplex is then cleaved in the repeat sequences by RNase III in a Cas9-dependent reaction. Then crRNA and tracrRNA form a complex with Cas9 proteins to form a complete search complex. The duplex of crRNA and tracrRNA guide the Cas9 protein to target and cleave the invading DNA based on sequence complementarity. It means that if the sequence of crRNA matches with the sequence of the DNA of the invading virus, then Cas9 cuts the viral DNA.

Thus the simplicity of type II CRISPR-Cas system, with only three required components: Cas9 along with the crRNA and tracrRNA makes this system easy to be used for genome editing. This potential was realized in the year 2011 by the Doudna and Charpentier labs. And the researchers discovered that this system could be simplified compared to what nature has done, where there are 2 separate molecules of RNA: crRNA and tracrRNA, that provide the program for Cas9 protein (Doudna, and Charpentier, 2014). Acting like a set of "molecular scissors," Cas9 can precisely cut the two strands of DNA at a specific site in the genome, allowing for the removal or addition of bits of DNA.

The simplified engineered CRISPR-Cas9 system was developed by linking the two RNAs crRNA and tracrRNA to form a single guide RNA abbreviated as sgRNA or simply called guide RNA, written as gRNA *(Fig 4-1)*.

Thus the gRNA consists of 2 regions: one is a program or 20-letter nucleotide sequence, and the other region is a handle-shaped RNA scaffold.

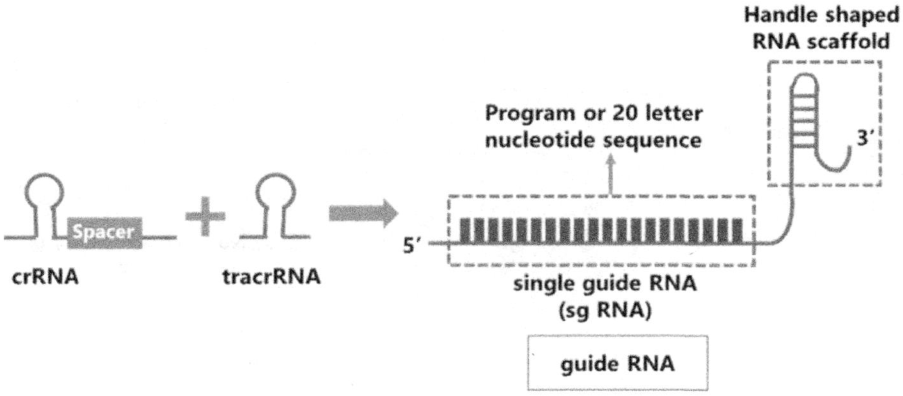

Fig 4-1: guide RNA (gRNA)

The program or 20-letter nucleotide sequence in the gRNA is complementary to a small sequence of target DNA of the host that is required to be edited. In other words, if the target DNA has a sequence ATGCGC when read in the 5' to 3' direction, then the gRNA sequence is complementary and reads TUCGCG in the 3'-5' direction. Furthermore, since the guide sequence is made up of RNA, it has a U base that stands for uracil instead of a T that stands for thymine.

The other region in the gRNA is the handle-shaped RNA scaffold which interacts with the Cas9 protein. The gRNA-Cas9 complex then binds to the target DNA sequence in the host genome. After the gRNA binds at the desired position in the genome, the nuclease domains of the Cas9 enzyme introduce double-stranded breaks in the desired DNA *(Fig 4-2)*. The HNH nuclease domain cleaves the DNA strand that is complementary and paired with the 20-letter nucleotide sequence in sgRNA. In contrast, the RuvC nuclease domain cleaves the non-complementary or coding strand. Thus, cleaving the target

DNA by producing double-stranded breaks in it. Therefore, the 20-letter guide sequence in the **gRNA ensures that the Cas9 enzyme cuts at the right point in the genome.**

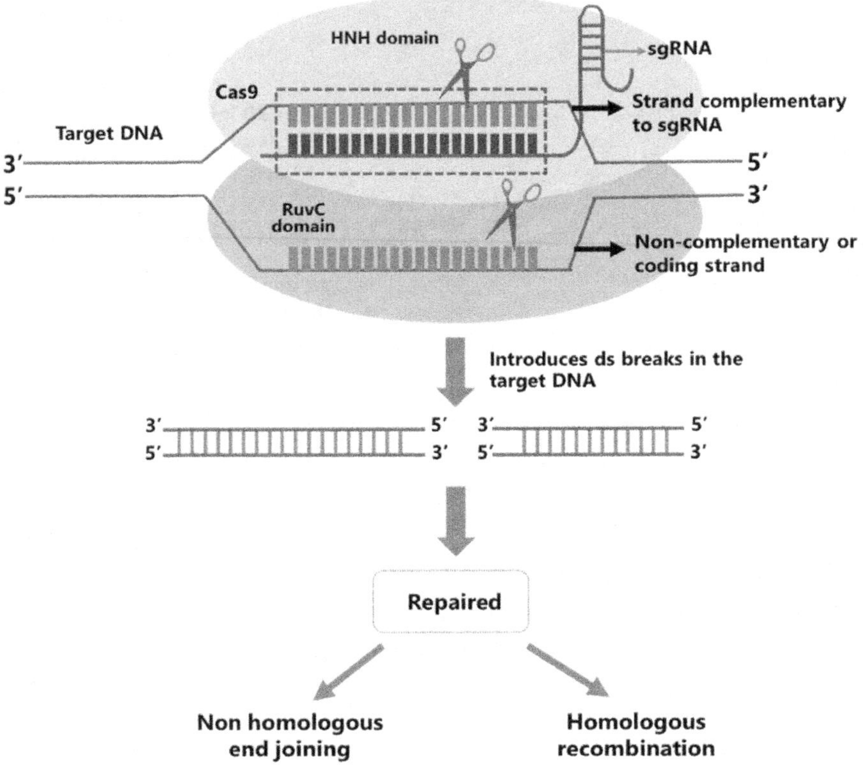

Fig 4-2: CRISPR-Cas9 system for genome editing

Once the DNA is cut, researchers could use the cell's own DNA repair machinery to add or delete pieces of genetic material to make changes to the target DNA.

The beauty of the CRISPR-Cas9 gene editing tool is the easy design of its gRNA. The handle for binding of Cas9 protein remains the same in every gRNA, but one can change its program or 20-letter guide sequence to direct Cas9 to a different target DNA for its modification.

Thus, Cas 9 is also referred to as a "**programmable endonuclease**" because one can program it to cut a specific DNA by providing a target sequence or predesigned 20-letter sequence in the gRNA (Jinek et al., 2012).

4.3. Repair of double-stranded breaks by Non-homologous end joining (NHEJ) repair pathway

When Cas9 creates double-stranded breaks, the cell's inherent repair mechanisms can fix them via two pathways: non-homologous end-joining repair or homology-directed repair (Hefferin and Tomkinson, 2005; Davis and Chen, 2013). Knowing how these pathways work is important, as this could be relevant when planning your experiment.

Let's first discuss the Non-homologous end-joining repair pathway, abbreviated as NHEJ. This pathway does not require template DNA for repairing double-stranded breaks. Several steps are involved in this repair pathway:

(i) The initial step in the NHEJ repair pathway is the recognition and binding of the Ku heterodimer Ku70/Ku80 to the double-stranded break or DSB *(Fig 4-3)*.

(ii) Once the Ku70/Ku80 heterodimer is bound to the DSB ends, it serves as a scaffold for the recruitment of protein kinases called DNA-PKcs, and some accessory factors like PAXX and XLF to form a tight complex. This protein-complex holds ends of double-stranded breaks together.

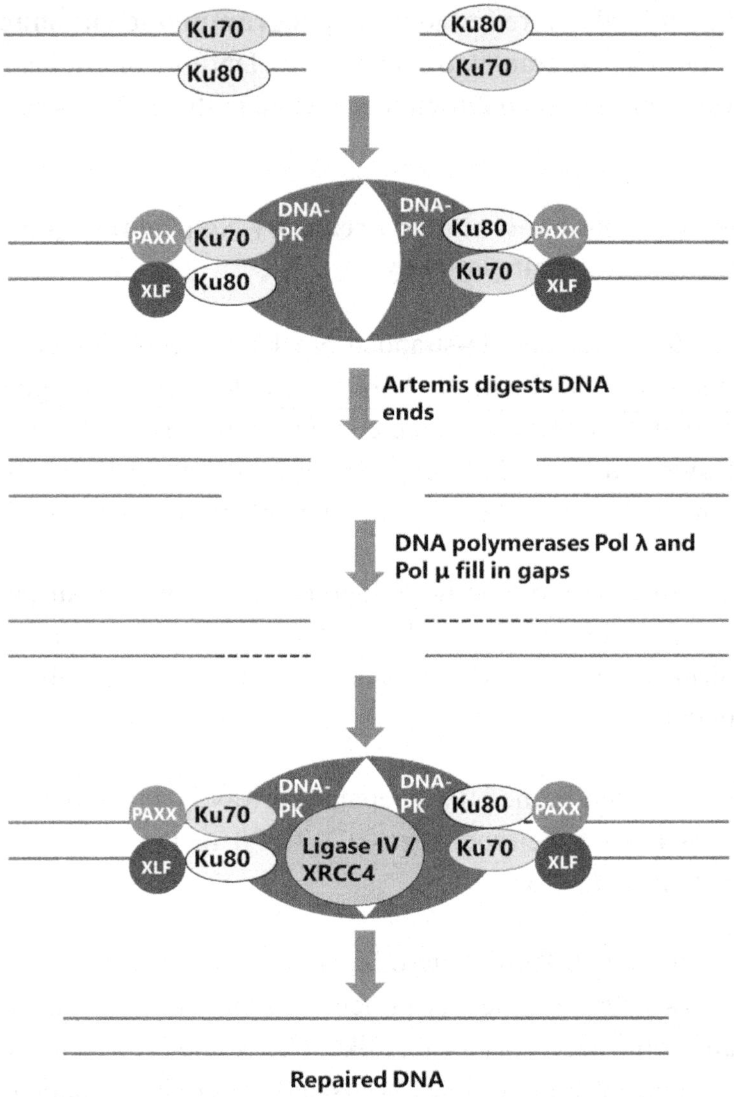

Fig 4-3: Mechanism of NHEJ pathway

(iii) The next step is processing the broken DNA ends, which involves removing and adding nucleotides to create ligatable ends. Different DNA end processing enzymes are required for this, like the Artemis enzyme that can digest DNA ends, DNA polymerases Pol λ and Pol μ that can fill in gaps, and various nucleases that can remove blocking end groups and make the ends ligatable. These DNA end processing enzymes are recruited to the DSBs by a Ku70/Ku80 scaffold.

(iv) After the processing, ligase IV enzyme, along with its cofactor XRCC4 is recruited, which grabs the broken DNA ends to seal and ligate the broken DNA together, thus repairing the double-stranded break.

It's worth noting that the process of repairing the DNA breaks in this pathway is prone to errors, often resulting in the introduction of insertion or deletion mutations in the targeted gene. The term "INDEL" is used in molecular biology to describe the insertion or deletion of nucleotides in an organism's genome.

The addition of nucleotides to a gene results in a change in the number of DNA bases by adding a segment of DNA. Conversely, a deletion removes a segment of DNA. When nucleotides are either added or removed from a gene, it alters the gene's reading frame. As we've previously covered in our discussion of translation, a reading frame comprises sets of three bases known as codons, with each codon coding for one amino acid. Therefore, insertions and deletions can cause a shift in the grouping of these bases, ultimately altering the code for amino acids *(Fig 4-4)*. As a result, the protein made by the gene may be defective.

Frameshift mutations can even lead to the formation of premature stop codons within the target gene's open reading frame (ORF), preventing protein production.

In other words, NHEJ-mediated repair of Cas9-generated breaks makes the gene inactive or inoperative.

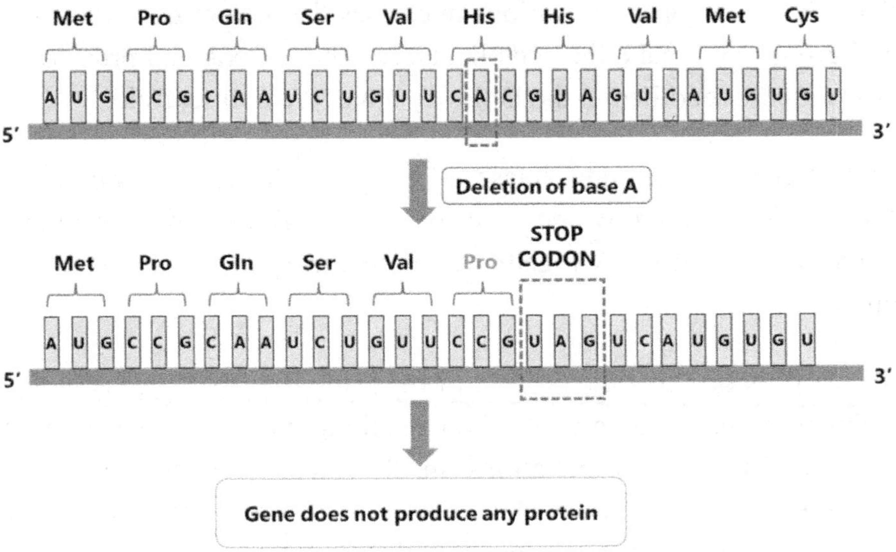

Fig 4-4: Deletion of base leads to premature stop codon; making the gene inactive

Making the gene inactive or inoperative is called **knocking out of a gene**. Therefore, if the intention is to generate a knockout of the desired gene by the CRISPR-Cas9 system, then a non-homologous end-joining repair pathway should be induced in the host cells.

4.4. Repair of double-stranded breaks by Homology directed repair (HDR) pathway

The second repair pathway, Homology Directed repair, abbreviated as HDR, is the more accurate mechanism for double-stranded break repair. Unlike NHEJ, HDR requires a template DNA that has the homology to the flanking regions of the double-stranded break. This method of repair is highly accurate and can be used to introduce specific nucleotide changes into the target gene (Morrical, 2015; Song and Stieger, 2017; Yang et al., 2020).

Let's now discuss the steps involved in the homology-directed repair:

(i) First protein that comes and encounters the damaged DNA is the MRN complex. The MRN complex consists of 3 different proteins MRE11, RAD50, and Nbs1. These 3 proteins bind to the broken DNA ends. The function of the MRN complex is DNA end resection. It means that the MRN complex, after binding, dissects 1000 bases from the 5' end, leaving the 3' overhangs *(Fig 4-5)*. The single-stranded 3' overhangs then serve both as a substrate for proteins required for strand invasion and a primer for DNA repair synthesis.

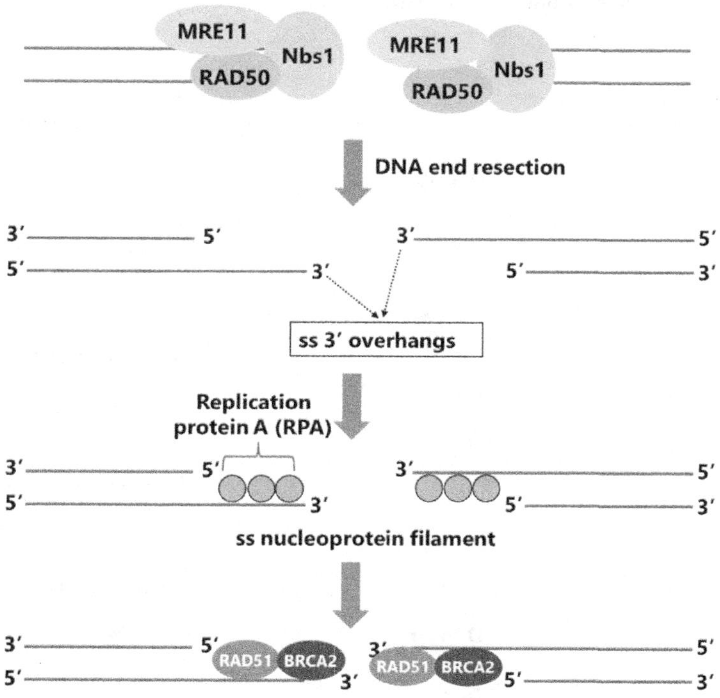

Fig 4-5: Initial steps of HDR pathway's mechanism

(ii) The single-stranded 3' overhangs are then stabilized by small proteins called replication protein A or RPA. The function of RPA protein is to protect the single-strand DNA from nucleases and also to prevent them from coiling back again. This complex of DNA with replication protein A is called single-strand nucleoprotein filament.

(iii) The next protein that will encounter this damaged DNA is RAD51. RAD51, along with BRCA2 protein, replaces the RPA protein. RAD51 protein helps search for homologous template DNA and also aids in invading one of the resulting single-stranded 3′ ends in the homologous repair template *(Fig 4-6)*.

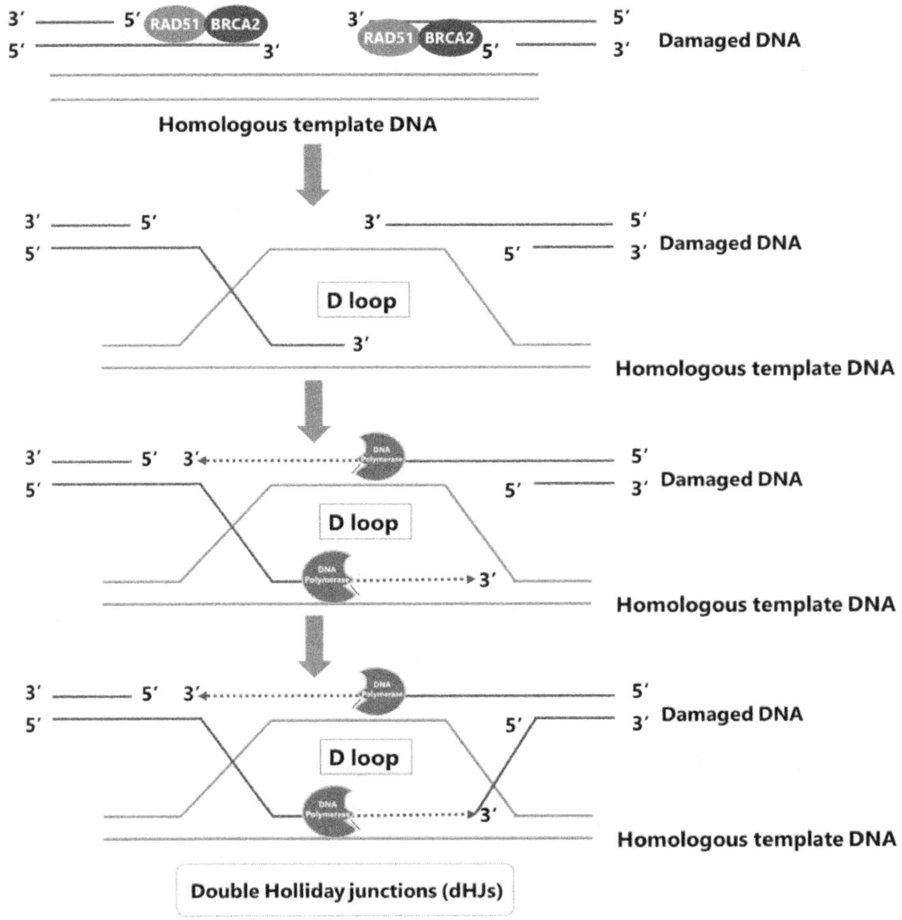

Fig 4-6: Formation of double Holliday junctions

This invasion forms a displacement loop or D-loop, in which the overhang binds with one strand of the template and displaces the other. The 3' ends of the broken DNA are then extended by DNA polymerase using the homologous DNA as a template *(Fig 4-6)*.

This ultimately leads to the formation of double Holliday junctions (dHJs). dHJs are four-stranded branched structures that are formed when elongation of the invasive strand and DNA synthesis from the second end of the double-stranded break occurs.

(iv) Once the process of DNA synthesis is complete, the Holiday junctions are resolved via cleavage so that the repaired DNA and homologous template DNA go into their original double-stranded DNA state. For this, there are 2 different options: Non-cross over Homology directed repair and cross over Homology directed repair.

If the holiday junctions are resolved via Non-cross over Homology directed repair, then both the repaired DNA and the homologous template DNA will be like they were before. For this, both the HJs are resolved in the same manner. In *Fig 4-7a*, you can see the resolution of each junction occurs horizontally. Black arrows represent the sites of cleavage. This cleavage pattern leads to non-crossover homology-directed repair, which restores the original DNA sequence without any genetic exchange between the two homologous strands.

On the other hand, if the holiday junctions are resolved via crossover Homology directed repair, then the repaired DNA and homologous template DNA exchange their parts with each other. For this, both the Holiday junctions are resolved dissimilarly. For example, one junction is resolved on the crossing strand, and the other junction is resolved on the non-crossing strand *(Fig 4-7b)*. Cleavage in this manner leads to crossover homology-directed repair, which causes the repaired DNA and homologous template DNA to exchange their genetic material.

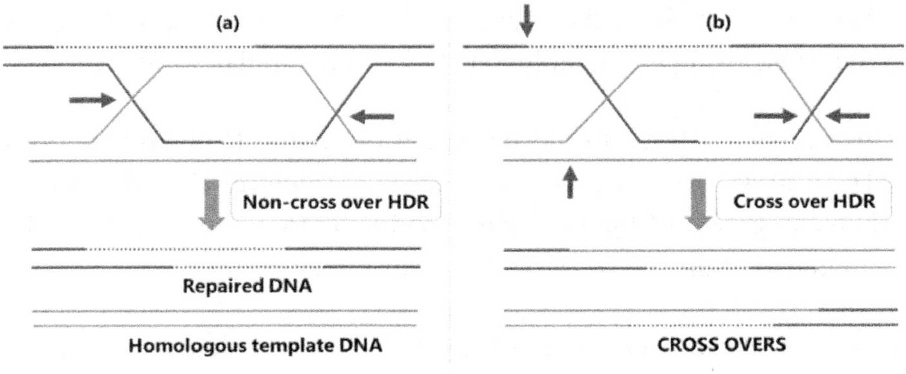

Fig 4-7: Resolution of holiday junctions via (a) non-cross over Homology directed repair and (b) crossover Homology directed repair

Homology Directed repair pathway is highly accurate and can be used to introduce specific nucleotide changes into the target gene by inserting or replacing a DNA sequence near the break site using the donor repair template DNA. Thus homology-directed repair pathway can be used for gene knockout, gene tagging, introducing specific mutations in the desired gene, knock-in, or for promoter studies.

Homology-directed repair (HDR) pathways in cells are typically less efficient than the non-homologous end joining (NHEJ) repair pathway. To increase the efficiency of HDR for Cas9-generated breaks, it is necessary to suppress one of the key molecules in NHEJ, such as KU70, KU80, or DNA ligase IV. This suppression can improve the efficiency of HDR by 5-10 fold.

Chapter 5: CRISPR-Cas system as a tool for genome editing

5.1. Off-target effects of SpCas9

Among the different bacteria harboring CRISPR-Cas systems, the human pathogen *Streptococcus pyogenes* was one of the first bacteria in which CRISPR type II loci was characterized. Furthermore, it was *Streptococcus pyogenes* Cas9 also abbreviated as SpCas9, that was first used to cleave and edit a specific region of the genome of a human cell.

For SpCas9 to cleave the target DNA, the Cas9-gRNA complex binds to the specific sequence called the protospacer of target DNA. The Cas9 protein itself binds the PAM sequence NGG and partially denatures the target DNA. The 20 nucleotide guide sequence of gRNA, then base pairs with the 20-ntd protospacer sequence of target DNA. Following this step, the Cas9 enzyme introduces double-stranded breaks in the target DNA, and eventually, insertions and deletions are induced into the target locus via non-homologous end joining, or more precise changes are introduced via a homology-directed repair pathway.

Although SpCas9 is revolutionizing the fields of genome engineering and genetic therapy, it does have certain limitations.

(i) The first limitation is its off-target activity which leads to modifications at unintended sequences in the genome (Zhang et al.,2015; Tyoko et al., 2016). This can occur when the guide sequence of gRNA, programmed to direct Cas9 to the target DNA, base pairs with another closely matched sequence site in the genome, which differs by only a few nucleotides *(Fig 5-1)*. As a result, the Cas9 enzyme can cleave unwanted DNA sequences that show imperfect complementarities with the associated gRNA, resulting in off-target genome modification.

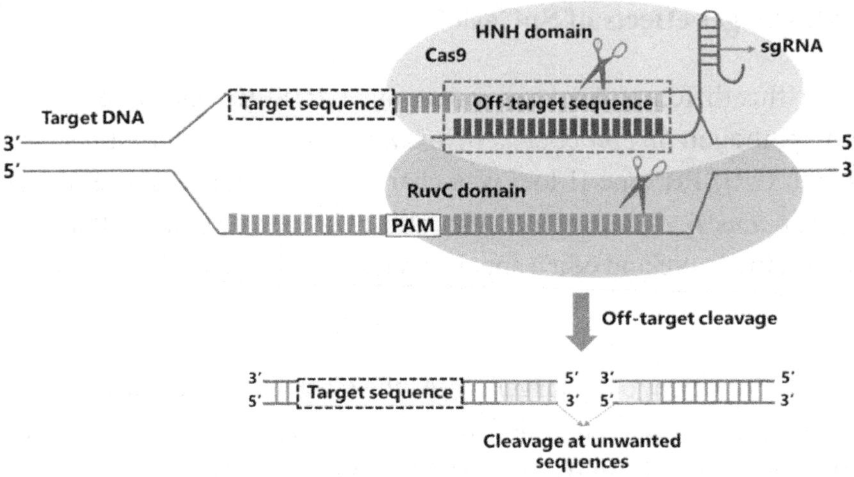

Fig 5-1: Off-target cleavage of target DNA by SpCas9

The base pairing within the guide RNA: protospacer duplex can be divided into 2 regions: One is the PAM-proximal region, and the other is the PAM-distal region *(Fig 5-2)*.

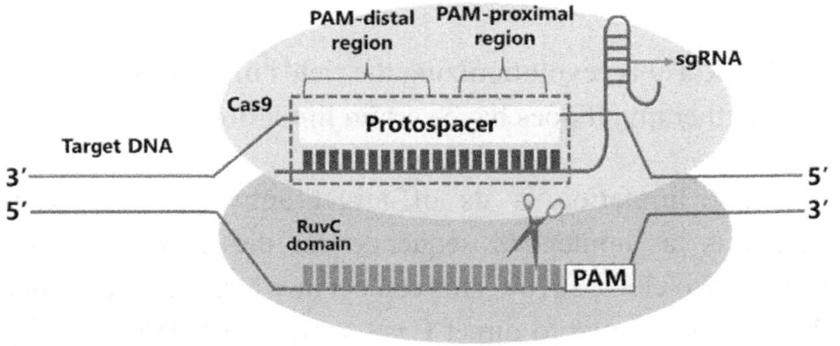

Fig 5-2: PAM-proximal and PAM-distal region regions

SpCas9 enzyme is able to tolerate certain mismatches in the PAM-distal region. On the other hand, the enzyme cannot tolerate mismatches in the PAM-proximal region because mismatches in this region reduce or eliminate the cleavage activity of Cas9. Toleration of mismatches leads to the off-target activity of the Cas9 enzyme when used for genome editing applications, as SpCas9 can create DSBs at sites with limited homology to the intended target DNA. These off-target effects cause detrimental effects in the cell as they could lead to lethal genetic mutations that can cause loss of gene function or generation of cancer cells in animals or humans and generation of undesirable phenotypes in plants.

(ii) The second limitation of the SpCas9 enzyme is the stringent requirement for the PAM sequence "NGG" which limits the target DNA that can be manipulated using this method.

With the objective to overcome these limitations, over the past few years, several variants of the SpCas9 enzyme have been developed (Cebrian-Serrano and Davies, 2017).

5.2. Cas9 nickase

The first variant of SpCas9 that has been widely used for gene editing purposes is Cas9 nickase (Shen et al., 2014; Trevino and Zhang, 2014; Gao et al., 2017). As already discussed, the Cas9 enzyme has two nuclease domains: HNH and RuvC. The HNH domain cleaves the target DNA strand complementary to the guide RNA, while the RuvC domain cleaves the non-complementary or coding strand of the target DNA. Both of the nuclease domains of the Cas9 enzyme function together to generate blunt-ended double-stranded breaks (DSBs) in the target DNA *(Fig 5-2a)*.

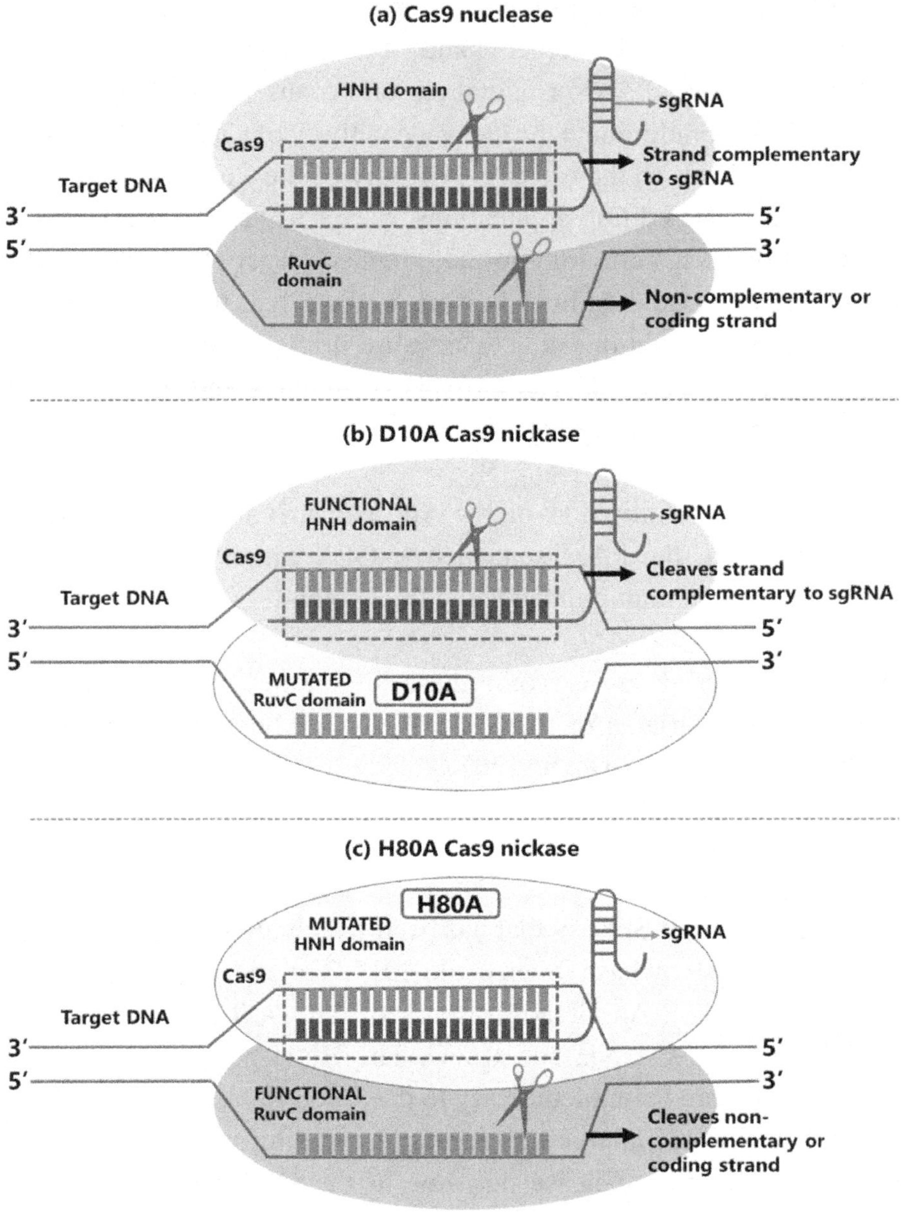

Fig 5-2: (a) SpCas9 with functional HNH and RuvC domains, (b) D10A Cas9 nickase with mutated RuvC domain, and (c) H80A nickase with mutated HNH domain

The double-stranded breaks are then repaired either by a non-homologous end-joining repair pathway or by a homology-directed repair pathway.

Inactivating one of the nuclease domains of the Cas9 enzyme results in the formation of **Cas9 nickase** mutants that cleave or in other words, introduce a single nick in one strand of the target dsDNA.

For instance, when asparagine residue at the 10th position is replaced with alanine in the RuvC domain, it inactivates the RuvC domain, because of which the inactive RuvC domain is unable to cleave the coding strand of the target DNA. As a result, the Cas9 nickase containing the mutated RuvC nuclease domain and a functional HNH domain can generate a nick in the target DNA strand that is complementary to the gRNA *(Fig 5-2b)*.

Conversely, replacing the histidine residue at the 80th position with alanine in the HNH domain results in the inactivation of the HNH domain, which prevents it from cleaving the target DNA strand that is complementary to the guide RNA. As a result, the Cas9 nickase containing the mutated HNH domain and a functional RuVC domain can generate a nick only in the non-complementary or coding strand of the target DNA *(Fig 5-2c)*.

As already discussed, the Cas9 enzyme along with 1 guide RNA generates a blunt-ended double-stranded break in the target DNA. But the paired Cas9 nickases generate two nicks in the target double-stranded DNA, instead of blunt double-stranded breaks in the target DNA *(Fig 5-3)*.

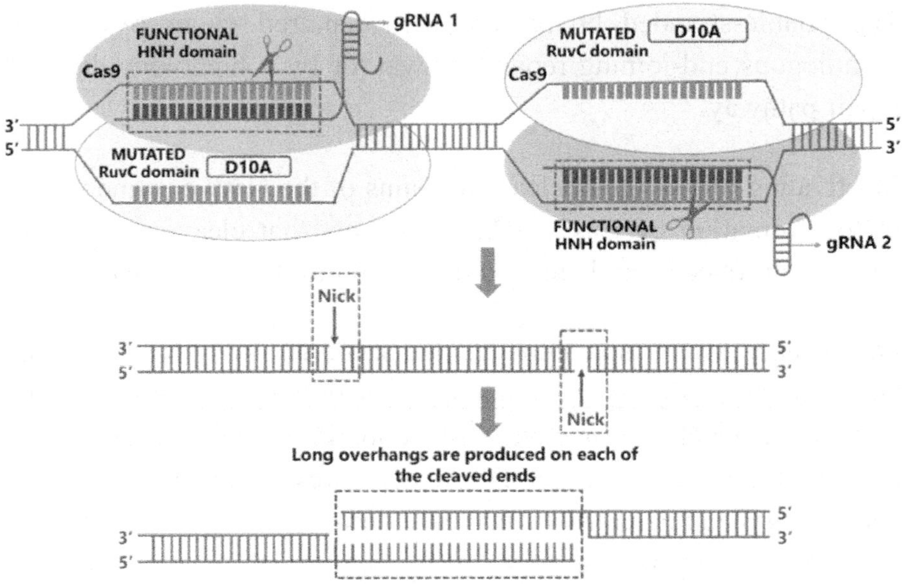

Fig 5-3: Two nicks are produced in the target DNA with the help of paired D10A Cas9 nickases and two gRNAs

For generating 2 nicks in the target DNA, 2 guide RNAs are required, one for each Cas9 nickase mutant. These 2 guide RNAs target the opposite strands of the target DNA, thus guiding the two Cas9 nickase mutants to generate nicks in both strands. These double nicks create a double-strand break (DSB), in which long overhangs are produced on each of the cleaved ends instead of blunt ends.

The nicks in the genome are repaired with high fidelity as compared to blunt-ended double-stranded breaks. The two nicks created by paired Cas9 nickases can be repaired by the homology-directed repair pathway using the single-stranded oligodeoxynucleotides as template DNA. The template single-stranded oligodeoxynucleotides have the homology to the 30-40 bases of the DNA strand that is required to be edited *(Fig 5-4)*.

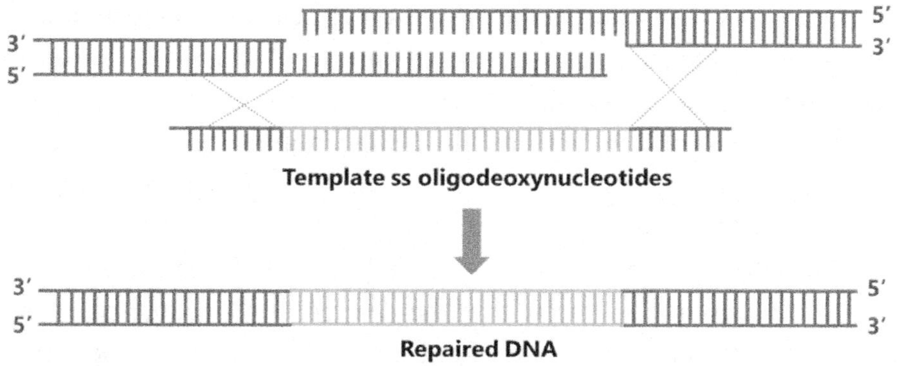

Fig 5-4: Repair of two ss nicks created by paired Cas9 nickases with HDR pathway using the template DNA strand

The advantage of using the Cas9 nickase over the wild-type SpCas9 enzyme for gene editing purposes is its targeting efficiency. Since mismatches at the 5' end of the gRNA are tolerated, therefore the use of the SpCas9 along with 1 guide RNA usually leads to unintended off-target effects.

But with the use of two guide RNAs that are adjacent on the opposite strands of the target site with paired Cas9 nickases, the off-target effects of the Cas9 enzyme are minimized. The use of 2 guide RNAs doubles the sequence length required for target DNA recognition; thus, the chance of cleavage of off-targets gets significantly reduced. Though one Cas9 nickase with 1 guide RNA might still bind an off-target site, it is unable to introduce a double-stranded break in it. Thus the double-nicking strategy of Cas9 nickase can significantly reduce the cleavage and editing of unwanted off-targets in the genome.

5.3. FokI fused catalytically inactive dead Cas9

The second variant of SpCas9 is FokI fused catalytically inactive Cas9 (Guilinger et al., 2014; Saifaldeen et al., 2020). We have already discussed previously that in the case of other gene-editing techniques, Zinc Finger Nucleases and TALENS, the DNA binding domain is fused to the nuclease domain of the FokI enzyme. Also, the nuclease domain of the FokI enzyme requires dimerization for endonuclease activity on the target DNA *(Fig 5-5a, b)*. Thus a FokI monomer is fused to each of the two Zinc finger and TALE proteins. This facilitates the zinc finger nuclease and TALENS to effectively bind to the longer recognition site within the desired DNA sequence of the target genome. Then dimerized FokI nucleases can cleave the target DNA and produce a double-stranded break.

On the other hand, Cas9 is a monomeric nuclease that acts when coupled to its guide RNA to produce a double-stranded break at its target site *(Fig 5-5c)*.

The dimeric and monomeric nature of nucleases are responsible for the off-target activity. Studies have shown that the likelihood of off-target binding events is considerably smaller when the dimeric nuclease is used. Like in the case of ZFNs and TALENs, the FokI nuclease only becomes active after dimerization, which means that both monomers need to bind to the target site simultaneously in order to create a double-stranded break. This dimerization is less likely to occur at off-target sites, as the chances of both monomers binding at an unintended location are lower.

On the other hand, monomeric nucleases like Cas9 can bind to off-target sites more easily than dimeric nucleases because they only require a single unit to bind and cut DNA. This increases the risk of off-target effects. Therefore, the dimeric nature of nucleases like FokI can result in enhanced specificity in genome editing applications.

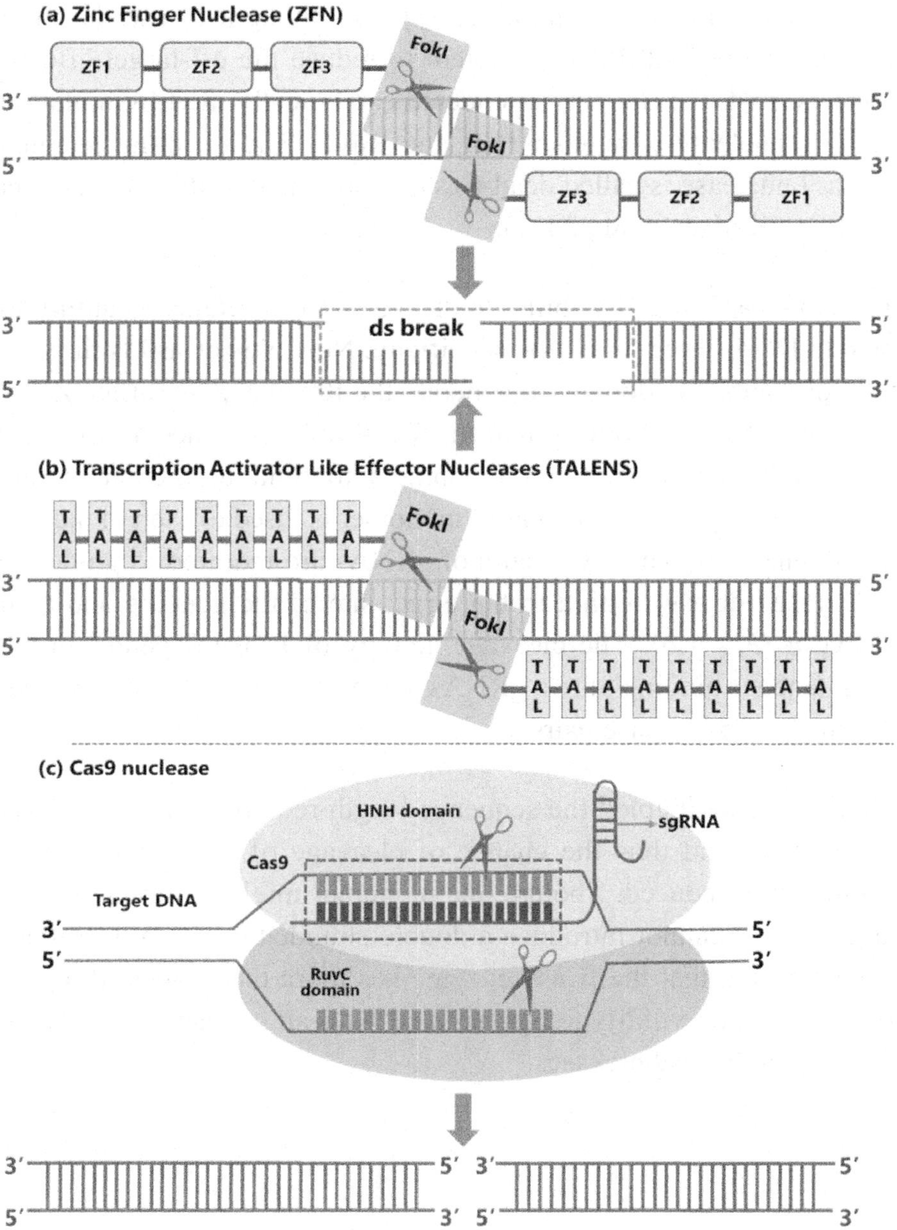

Fig 5-5: Cleavage by enzymes (a) ZFN, (b) TALENS, and (c) Cas

With these principles in mind, researchers thought to use the FokI enzyme in the CRISPR-Cas system to reduce the off-target effects. For this, first, the SpCas9 is rendered catalytically inert by mutating both its nuclease domains, HNH and RuvC. The resulting mutated SpCas9 nuclease is called dead-Cas9, abbreviated as dCas9 (Tadić et al., 2019; Karlson et al., 2021).

Though dCas9 loses its ability to cleave DNA, it retains its ability to bind to DNA based on gRNA specificity. Now, this dCas9 is fused to the Fok1 nuclease domain to generate the fCas9 enzyme. Like ZFNs and TALENs, the FokI domain in fCas9 only becomes active after dimerization, requiring two monomers to bind to the target site simultaneously. Therefore a FokI monomer is fused to each of the two dCas9 enzymes. Thus two pairs of gRNAs are required to guide the FokI fused dCas9 to bind to the target site of the genome and then cleave it *(Fig 5-6)*. The cleavage activity of fCas9 depends on the simultaneous binding of two gRNAs to their target DNA with a spacer distance of 15–25 base pairs.

This approach doubled the sequence length required for target DNA recognition, and thus the chance of cleavage of off-targets can be significantly reduced. Though one monomer might still bind an off-target site, it cannot introduce a double-stranded break in it. Studies have reported that the fCas9 enzyme has more than 140-fold higher specificity than wild-type SpCas9 and with an efficiency similar to that of paired Cas9 nickases.

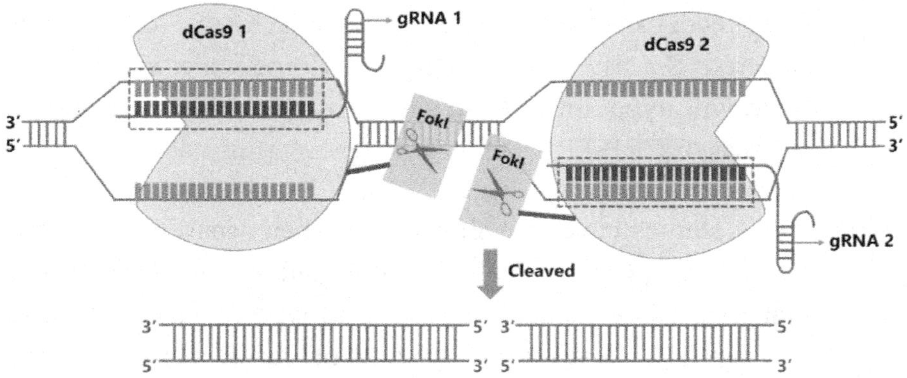

Fig 5-6: Cleavage by two fCas9 enzymes with two gRNAs

5.4. Cas9 nucleases with novel PAM specificities

The Cas9 nucleases with new PAM specificities represent the third type of SpCas9 variant. The PAM sequence, which stands for Protospacer Adjacent Motif, plays a critical role in determining the CRISPR-Cas specificity. SpCas9 typically recognizes the PAM sequence 5'-NGG-3', where N is any nucleotide followed by two Guanine residues. The presence of this PAM sequence is necessary for the Cas9 protein to bind and cleave the target DNA. It has also been reported that the SpCas9-sgRNA complex can recognize alternative PAM sequences such as NAG and NGA to a certain extent. Nonetheless, using such alternative PAM sequences increases the likelihood of off-target mutagenesis.

Therefore, the requirement for an NGG PAM is, however, considered quite stringent, but this also limits the DNA sequences that can be targeted, particularly in A/T-rich regions of the genome. One potential solution to widen the capability of Cas9 to target a wide range of DNA sequences is to engineer Cas9 variants with novel PAM specificities (Kleinstiver et al., 2015; Siksnys and Gasiunas, 2016). This can be done by introducing mutations into the PAM-interacting domains of wild-type SpCas9. By introducing mutations, engineered SpCas9 variants can recognize the PAM sequences NGCG or NGAG. Thus the engineered SpCas9 variants show genome-wide specificities compared to wild-type SpCas9 and show great promise in expanding the target areas of CRISPR technology.

Additionally, studies have also shown that when aspartic acid at the 1135th residue of SpCas9 is replaced by glutamic acid (D1135E), it has a much greater DNA specificity than wild-type SpCas9, and it also shows substantially lower genome-wide off-target activity at the target sites containing alternative PAM sequence NGG. Furthermore, the mutated SpCas9 shows less off-target activity when targeting DNA sequences with non-canonical NAG and NGA PAM sequences. Thus the engineered SpCas9 variants have the potential to expand the DNA targeting scope of CRISPR systems.

5.5. High Fidelity SpCas9

The fourth type of SpCas9 variant is High Fidelity SpCas9. For the generation of high fidelity SpCas9 variants, certain mutations are introduced into the SpCas9 enzyme to reduce off-target interactions (Kleinstiveret al., 2016; Kulcsár et al., 2017; Guo et al., 2019). Thus, making the requirement for complete complementary with the guide RNA more stringent; in other words, no mismatches between the guide RNA and target DNA can be tolerated by Cas9.

Studies of the crystal structure of the SpCas9–sgRNA–target DNA complex have shown that the SpCas9–sgRNA complex might possess more energy than is needed for the optimal recognition and binding to its intended target DNA site. Possessing more energy than required might be responsible for off-target binding activity and, thus, off-target cleavage.

When the guide RNA reaches its target DNA sequence, the double-stranded DNA strands separate as the guide RNA binds competitively to its target DNA strand. The recognition domain (REC) of SpCas9 has a groove lined with positively charged amino acid residues that interact and form hydrogen bonds with the negatively charged phosphate DNA backbone. Specifically, the mutations are introduced in the positively charged residues of SpCas9. These mutations destabilize any mismatches between the guide RNA and the substrate DNA, thus weakening the interactions between the SpCas9 variant enzymes and guide RNAs with non-exact complementarity with the target DNA sequence. Hence, the engineered SpCas9 variant enzymes display lower off-target activity while still retaining the on-target activity.

Few high-fidelity SpCas9 variants generated by introducing mutations are SpCas9-HF1, Enhanced SpCas9, Hyper-accurate Cas9, and HifiCas9.

SpCas9-HF1 enzyme is generated by introducing alanine substitutions in amino acids at 497, 661, 695, and 926 positions (N497A/R661A/Q695A/Q926A) of the WT SpCas9 enzyme. Introducing mutations reduces the genome-wide off-target effects.

Enhanced SpCas9, abbreviated as eSpCas9, is generated by introducing alanine substitutions in amino acids at 848, 1003, and 1060 positions (K848A/K1003A/R1060A) of the WT SpCas9 enzyme.

Hyper-accurate Cas9, abbreviated as HypaCas9 enzyme, is generated by introducing alanine substitutions in amino acids present in the REC3 domain of the WT SpCas9 enzyme. The amino acids at positions 692, 694, 695, and 698 (N692A, M694A, Q695A, and H698A) are substituted with the alanine amino acid (Liu et al., 2020; Rabinowitz and Offen, 2021). The REC3 domain is a part of the recognition lobe of the Cas9 enzyme, which senses mismatches and regulates the nuclease activity of the HNH nuclease domain.

Mutations in the amino acids at positions 692, 694, 695, and 698 of the REC3 domain make it more stringent and prevent the cutting of mismatched DNA sequences by the HNH nuclease domain. Thus reducing the off-target cleavage activity of the SpCas9 enzyme.

Though these enzymes SpCas9-HF1, Enhanced SpCas9, and Hyper-accurate Cas9 reduce the off-target cleavage activity of the SpCas9 enzyme, they all have 1 limitation. The combination of multiple mutations into a single Cas9 variant results in over-engineering of the protein that often lowers the on-target catalytic activity of the enzyme by 20-30%.

To overcome this limitation, **High Fidelity Cas9, abbreviated as HiFiCas9** enzyme, was generated that has the potential to reduce global off-target effects while retaining high on-target activity. The enzyme is generated by replacing the arginine amino acid at position 691 with alanine (R691A). The enzyme with a single mutation exhibits reduced off-target activity while maintaining the robust on-target editing activity of the enzyme.

5.6. Cas9 orthologs

While the commonly used wildtype (WT) *Streptococcus pyogenes* SpCas9 can tolerate multiple mismatches in the target sequence, other naturally occurring Cas9 orthologs from *Staphylococcus aureus* (SaCas9), *Neisseria meningitidis*, and *Campylobacter jejuni* are reported to have higher specificity in genome editing compared to SpCas9. Additionally, these orthologous enzymes show different substrate specificities and recognize different PAM sequences (Gasiunas et al., 2020).

***Staphylococcus aureus* Cas9 (SaCas9):** For instance, Cas9 of *Staphylococcus aureus,* abbreviated as SaCas9, has similar levels of gene targeting compared to SpCas9. Unlike SpCas9, which recognizes

PAM sequence NGG, SaCas9 recognizes PAM sequence NNGRRT sequence where R is any purine, may be A or G nucleotide (Tan et al., 2019; Hu et al., 2020).

The longer PAM sequence NNGRRT of SaCas9 also presents some disadvantages. The range of putative target sites would be reduced compared to the SpCas9, which recognizes a more simple and commonly occurring sequence NGG. To address this limitation, mutagenesis of the SaCas9 has been done to generate SaCas9 variants with different PAM specificities. For instance, mutating the amino acids at positions 782, 968, and 1015 generates the KKHSaCas9 variant, which can recognize the NNNRRT PAM sequence. Additionally, the variant shows similar efficiency and specificity as the wild-type SaCas9.

***Neisseria meningitidis* Cas9 (NmCas9):** Cas9 of *Neisseria meningitidis*, abbreviated as NmCas9, recognizes PAM sequence NNNNGATT. In addition, the guide RNA associated with NmCas9 recognizes a 24 ntd protospacer, which, together with its more complex PAM, implies a high level of specificity for gene editing purposes (Hou et al., 2013; Lee et al., 2016).

***Campylobacter jejuni* Cas9 (CjCas9):** Cas9 of *Campylobacter jejuni*, abbreviated as CjCas9, recognizes NNNVRYAC PAM sequence where V is either an A, a G, or a C nucleotide; and Y is either a T or a C nucleotide (Kim et al., 2017; Nakagawa et al., 2022).

Thus, the use of Cas9 orthologs with different PAM recognition sequences widens the range of potential sequences that can be targeted for gene editing purposes.

Chapter 6: Viral vectors for delivery of CRISPR components

6.1. Introduction to viral vectors

Well, the Cas9-gRNA construct can not be directly inserted into the cell. Instead, a carrier known as a vector is used to deliver the CRISPR constructs into the target cell. The vectors can be viral or non-viral. Let's first talk about viral vectors. Typically viruses are used as vectors because of their uncanny ability to infect the cells. The viruses can't reproduce without getting inside the living host cells and hijacking their machinery. Outside the host cell, the virus is metabolically inert. But inside the host cell, the virus becomes metabolically active. As already discussed, viruses can not synthesize proteins on their own, because they lack ribosomes, therefore, they use ribosomes of the host cells to synthesize viral proteins. Viruses also use enzymes and building blocks of the host cell, such as amino acids, and nucleotides, for replication, protein synthesis, etc. Viruses are particles made up of an outer protein coat and an inner core of genetic material in the form of DNA or RNA.

(i) To initiate infection, the virus binds to the membrane of the host cell and then injects its genetic material into the cell. This genetic material contains basic 'instructions' for producing more viral copies, hacking the body's normal production machinery to serve the needs of the virus *(Fig 6-1)*.

(ii) Once the DNA or RNA of the virus enters the host cell, it takes over the cellular machinery of the cell and turns it into a virus factory, thus producing more and more viruses.

(iii) The virus progeny then infects other surrounding cells, and with this, a new cycle of infection starts.

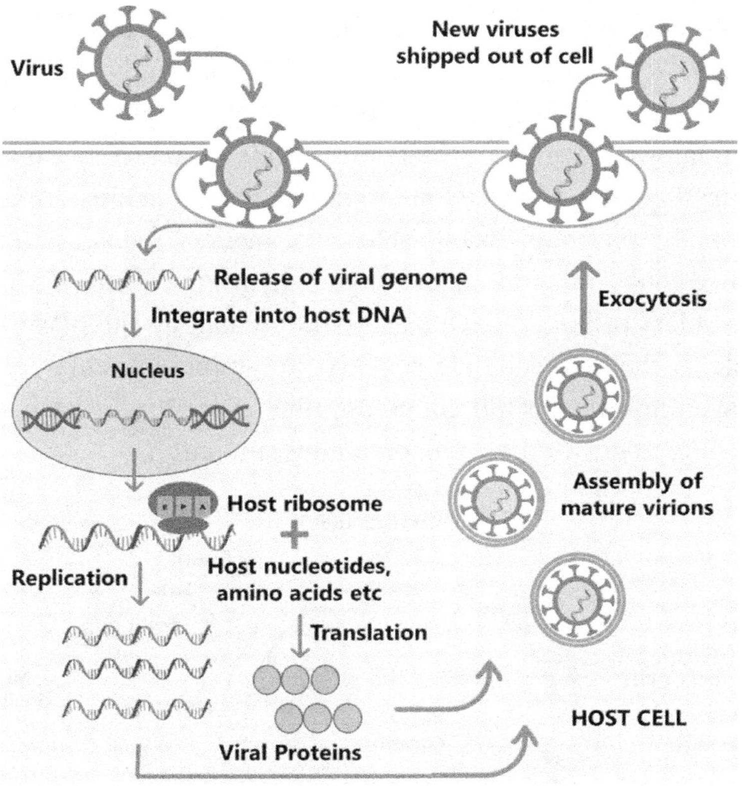

Fig 6-1: Mechanism of viral infection

The unique ability of viruses to infect the cells and integrate their genetic material into the host cell can be utilized for transferring functional genes into human cells. However, before they can be used, these viruses are modified so that they do not cause disease when used in humans. These modified viruses are called viral vectors. They are engineered to bind to the host cell and inject the genome without causing any harm to the patient.

Designing a viral vector involves knowing the genome sequences of the virus required for assembly, packaging, and replication into the targeted cells. In addition, plasmid constructs are necessary for the construction of viral vectors.

Plasmids are double-stranded, small circular pieces of DNA sequences capable of replicating independently from host chromosomal DNA. Any plasmid construct contains an origin of replication or ori, which instructs the plasmid to undergo replication, allowing for the amplification of the genes of interest *(Fig 6-2)*. Additionally, plasmids encode selectable markers, such as antibiotic resistance genes (ampicillin, kanamycin, hygromycin, etc.), which are used to select cells that have successfully taken up the plasmid. For example, if the host cells are sensitive to ampicillin, only cells that have taken up the plasmid containing the ampicillin resistance gene will be able to grow on media containing ampicillin.

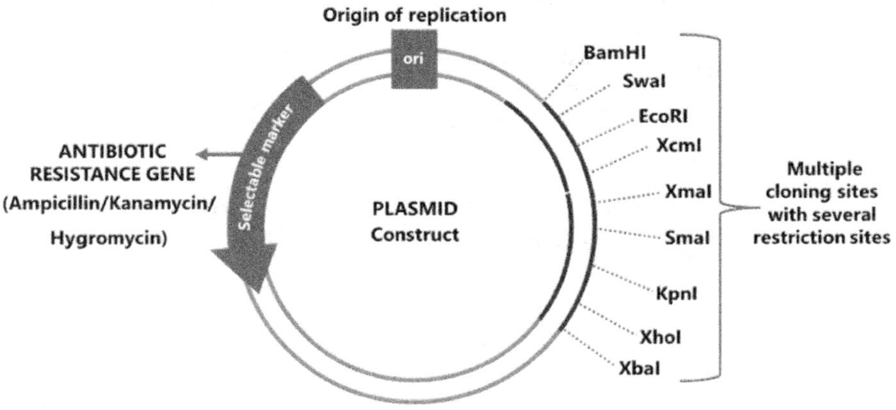

Fig 6-2: Plasmid construct

Plasmids must also contain multiple cloning sites with several restriction sites to insert the desired gene into the plasmid. Restriction sites are recognized and cleaved by the restriction endonuclease enzyme, because of which the circular plasmid gets opened up. If the desired gene is also cut with the same restriction enzyme, say EcoRI, and mixed with the EcoRI digested vector, then through the sticky ends, the gene and vector join together to form recombinants *(Fig 6-3)*. Thus the desired gene gets inserted into the plasmid.

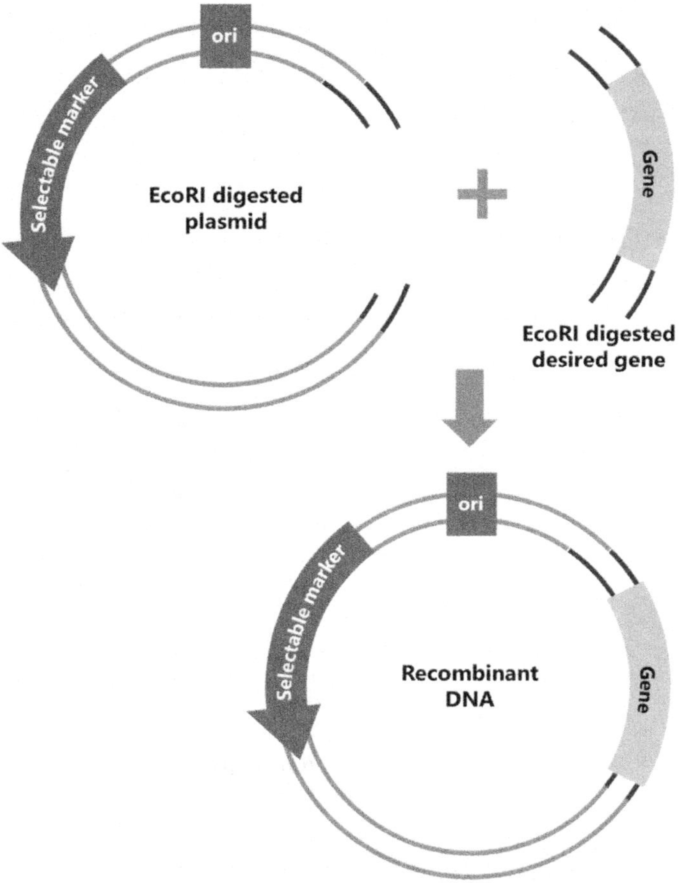

Fig 6-3: Plasmid cloning by restriction enzyme digest

For the construction of any viral vector, at least two plasmid constructs are required. One plasmid construct contains the **transgene**. Transgene is the desired gene that has to be inserted into the target cells. In the case of CRISPR-Cas system, the transgene can be the gene encoding gRNA or Cas9 enzyme. And the other plasmid construct contains the sequences essential for the packaging of virus particles.

For the assembly of viral particles containing transgene as their genome, packaging cell lines are required. The two plasmid constructs, one containing transgene sequence and the other containing packaging sequences, are inserted into the packaging cell lines *(Fig 6-4)*. In the packaging cell line, the packaging sequences are expressed, resulting in the generation of viral particles, which contain the transgene as their genome.

The expression of the transgene and packaging sequence in 2 constructs reduces the pathogenicity and immunogenicity of the virus particles. And these viral vectors formed are replication-deficient.

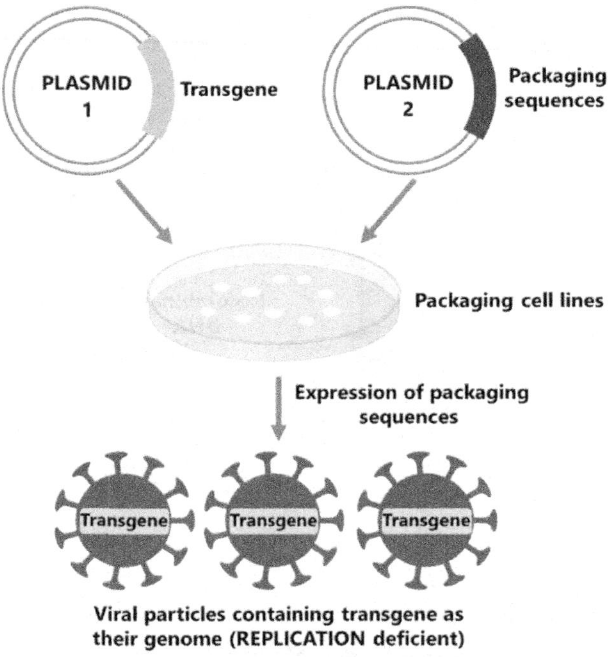

Fig 6-4: Expression of the transgene and packaging sequences in 2 separate plasmid constructs

Suppose the genes required for packaging are expressed in the same plasmid containing transgene, and the viral vector containing this plasmid infects the host cell. Then in the host cell, along with the expression of the transgene, the expression of packaging proteins will also occur, because of which new viral particles will be formed inside the host cell *(Fig 6-5)*.

The new viral particles formed can lyse the target cells and then infect other cells.

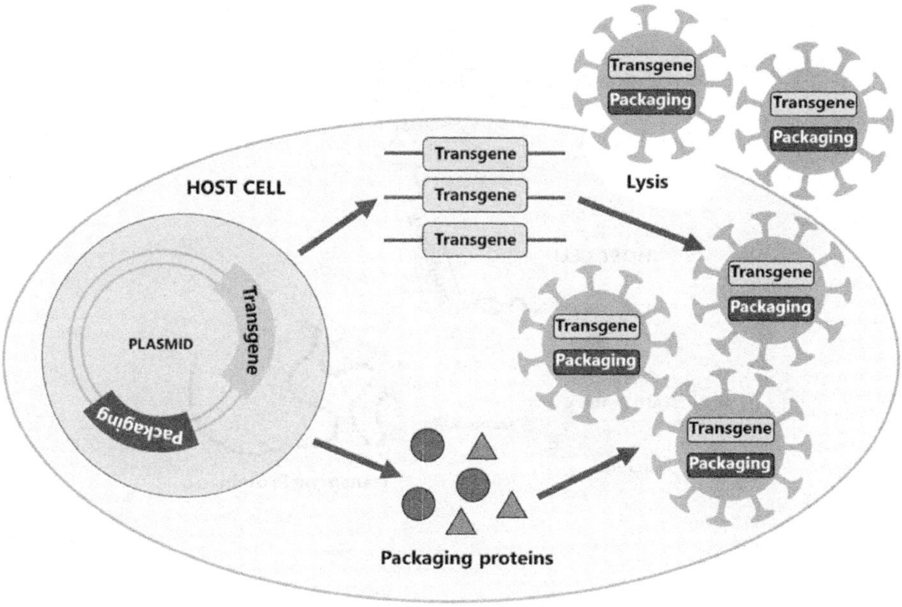

Fig 6-5: Expression of the transgene and packaging sequences in one plasmid construct

To eliminate the possibility of viral replication and viral infection, the genes necessary for the packaging of viruses are expressed in different plasmids so that the viral particles formed in the packaging cells would contain only transgene as their genome. These viral vectors would not contain any packaging sequences, because of which new viral particles will not be formed in the host cell.

These replication-deficient viruses are then allowed to infect the target cells, and after binding to the target cells, the viral vector introduces the gene of interest into the target cells *(Fig 6-6)*.

Then the transgene expression produces the desired protein required for the normal functioning or editing of the cells. Additionally, since the transgene does not contain any sequence for the synthesis of viral proteins, the viral particles will not be formed. Hence the viral vector is unable to infect the host cells.

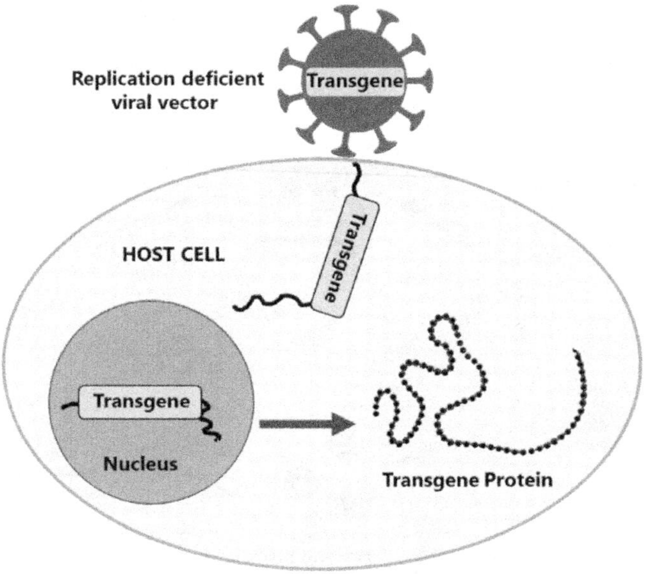

Fig 6-6: Replication deficient viral vector is unable to infect the host cell

There are many viruses that have the potential for gene delivery, but the efficient and most widely used ones are adenovirus, adeno-associated virus, and lentivirus. These viral vectors can broadly be divided into two types: **integrating vectors and non-integrating viral vectors.** Integrating vectors integrate the transgene into the genome of the host cell, which can result in the insertion of the transgene at a random location in the host genome. In case the transgene is inserted within one of the functional genes of the host cell, it can disrupt the gene and cause it to lose its function *(Fig 6-7a)*.

Moreover, if the transgene is inserted near the gene that regulates cell division, it may lead to uncontrolled cell division, which can cause cancer *(Fig 6-7b)*.

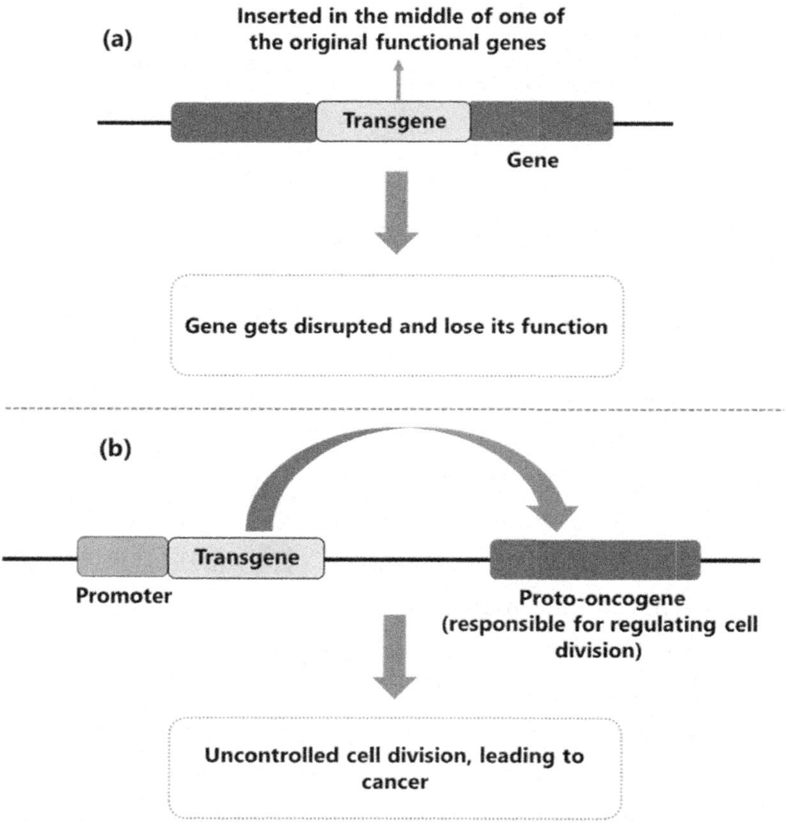

Fig 6-7: Insertion of a transgene (a) into the middle of the functional gene of the host cell and (b) near the gene responsible for regulating cell division

Whereas non-integrating viral vectors do not integrate the transgene into the genome of the host cell *(Fig 6-8)*. In fact, the transgene remains in an episomal state in the nucleus of the host cell, and the instructions in this transgene are transcribed just like any other gene of the host.

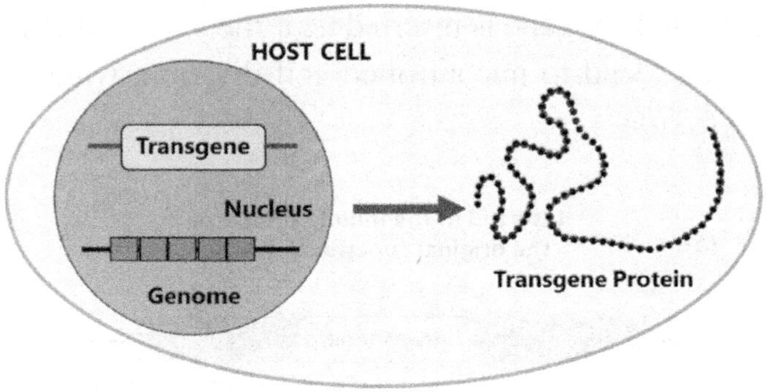

Fig 6-8: Non-integrating viral vector

6.2. Adenoviral vectors

Adenoviruses are non-enveloped viruses responsible for causing respiratory problems, especially the common cold, intestinal infections, and eye infections in humans. The genetic material of adenovirus is in the form of linear double-stranded DNA, with a size of 36 kbp, enclosed in a nucleocapsid. The nucleocapsid contains various proteins with 3 major types: fiber, penton, and hexon. The hexon is a structural basis of adenovirus and composes most of the viral capsid. The adenovirus uses its fiber and penton-based proteins to bind to the coxsackie and adenovirus primary receptor (CAR) on the host cell, facilitating its internalization via endocytosis *(Fig 6-9)*. Once inside the host cell, the capsid of the virus is dismantled, and the viral genome is transported into the nucleus of the host cell (Barnett et al., 2002).

The genetic material of the adenoviruses is not incorporated into the host cell's genetic material. Instead, it remains in an episomal state in the host cell's nucleus, where it is transcribed.

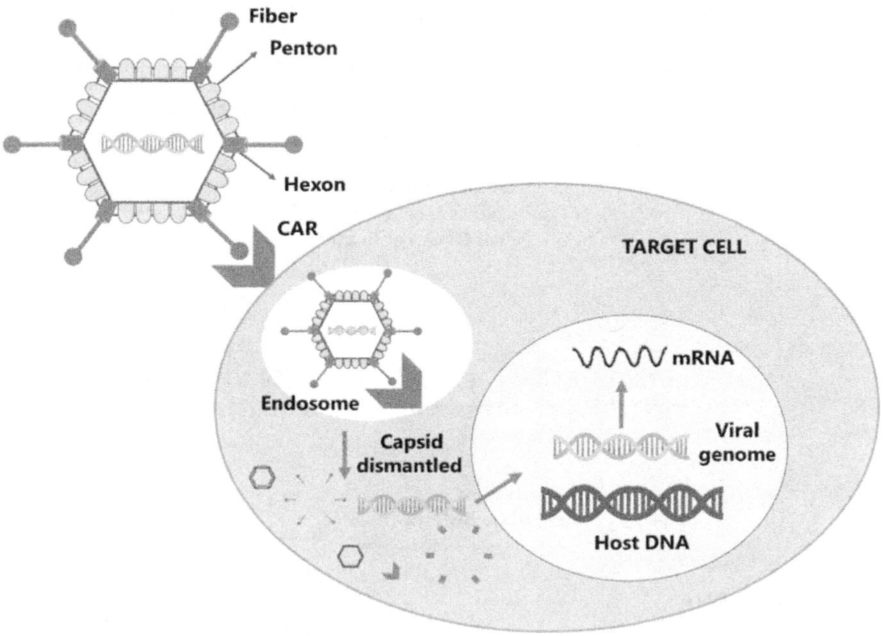

Fig 6-9: Mechanism of adenovirus infection

Genes of adenovirus are divided into early E1-E4 and late L1-L5 genes (Kovesdi, 1997; Yeh and Perricaudet, 1997). As soon as the viral DNA reaches the nucleus, early genes coding for the proteins required for viral genome replication, transcription, and translation are expressed. On the other hand, the late genes, which mainly code for structural or capsid proteins, are expressed later during the infectious cycle.

The first gene to be expressed is E1A. The transcription and translation of the proteins use the host cell machinery. The E1A protein then triggers the expression of other early genes E1B, E2, E3, and E4. Among these, E1B and E4 genes encode for viral DNA polymerase, responsible for viral DNA replication *(Fig 6-10)*. E2 encodes for viral RNA polymerase that forms an RNA intermediate called virus-associated RNA. The virus-associated (VA) RNA is not fully functional and can not be translated to form a protein. This viral associated RNA is then transcribed by host cellular RNA polymerase III to form mature viral RNA. Now, this mature viral RNA gets translated by the host ribosomes to synthesize viral proteins.

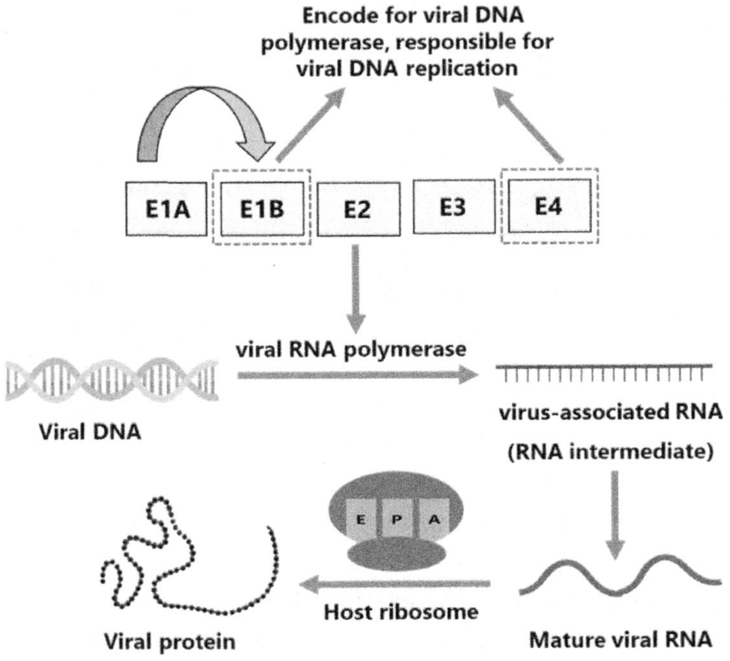

Fig 6-10: Expression of early genes of adenovirus

At either end of the genome are inverted terminal repeats abbreviated as ITRs. The ITRs contain the packaging signals for the packaging of the viral genome into the capsid. After the replication of adenoviral DNA, the structural and capsid proteins encoded by late genes are expressed, which then pack all the viral genetic material produced by DNA replication. After the assembly of viral particles, they are then released from the cell as a result of virus-induced cell lysis.

But before using the virus for gene delivery, it has to be modified so that the virus cannot cause disease in humans. Adenovirus is made replication-deficient after removing the coding regions of the viral genome. Now let's go through the development of adenoviral vectors. The development of these vectors can be categorized into first, second, and third-generation vectors (St George, 2003; Volpers and Kochanek, 2004; Douglas, 2007).

Adenoviral vectors

In the first generation vector, two genes E1 and E3 are deleted and are replaced by the transgene to be delivered to the target cell. The transgene contains a promoter. The transgene can be Cas9 or guide RNA. The vector also contains the Ψ sequence that is required for the packaging of viral particles. If the Ψ sequence is absent, then the DNA of interest or whatever you put in the vector will not get packaged into virus particles. In order to propagate the virus that lacks E1 expression, the vector is transfected into the packaging cell lines that are stable enough to express the viral E1A and E1B proteins *(Fig 6-11)*. Despite the deletion of E1 and E3 genes, there is evidence that these genes are still expressed in low levels in cells transduced with first-generation vectors. This can cause toxicity and immune reactions in the patients.

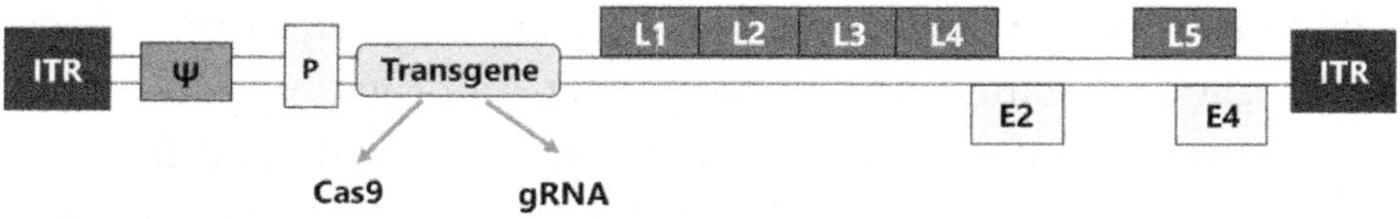

Fig 6-11: E1 and E3 genes are replaced by the transgene in the first generation adenoviral vector

In the second generation vectors, in addition to deletions of E1 and E3 genes, two more genes are deleted. These are the E2 and E4 genes. Because of the more deletions, larger DNA of interest can be inserted into the vector. In the first generation vector, DNA of around 8 kb size could be inserted while in the case of the second-generation vector,

DNA of around 12 kb size could be inserted in the vector. Then the adenoviral vector is transfected into the packaging cell line expressing E1, E2, and E4 proteins to form the virus particles that contain the transgene as their genome *(Fig 6-12)*. Because of the deletions of E1, E2, E3, and E4 genes, the adenoviral vectors show fewer immune reactions in the patients. However, still, there is a leaky expression of the viral genes.

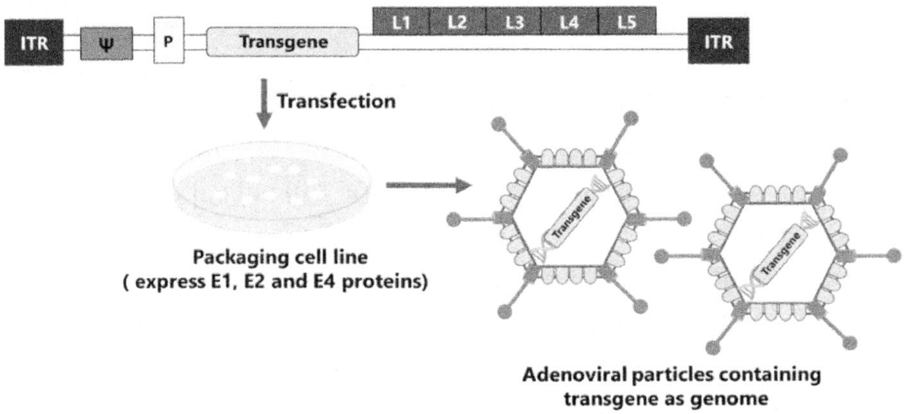

Fig 6-12: Second-generation adenoviral vector

The third-generation adenoviral vectors are also called helper-dependent adenovirus, abbreviated as HD-Ad or gutless adenovirus. In these vectors, all the viral coding sequences are deleted from the viral genome (Kochanek et al., 2001; Alba et al., 2005). In other words, the vector is devoid of all early and late genes. Therefore, the vectors show reduced toxicity and immunogenicity in the patients. But the genome retains the ITRs or inverted terminal repeat sequences and Ψ sequence. In between the ITRs, the transgene is inserted into the vector. Because the third generation vector lacks all the viral encoding genes, it requires co-infection with a helper adenovirus to provide the capsid proteins to encapsulate the transgene, thus making adenovirus containing transgene as its genome *(Fig 6-13)*.

To generate this, the third generation vector containing the transgene is transfected into the packaging cell lines that are infected with the helper virus.

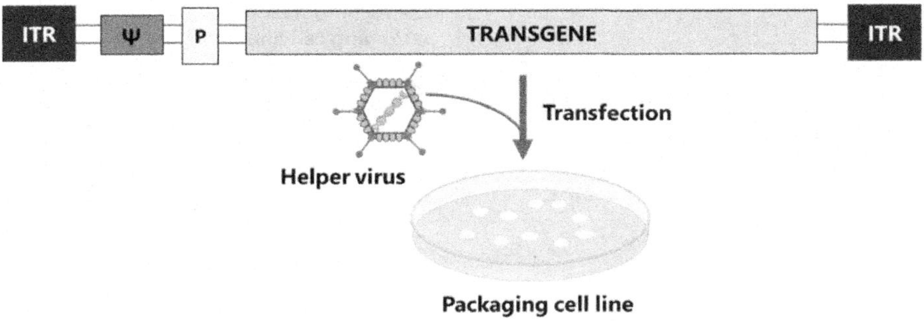

Fig 6-13: Third-generation/ Helper-dependent adenoviral vector

The genome of the helper virus has packaging sequences flanked by 2 loxP sites. But the ψ sequence is excised from the genome of the helper virus to prevent packaging of helper construct. Now, LoxP sites are the sites where cleavage occurs by Cre recombinase. When the helper virus infects the cells, the Cre recombinase produced by the cell lines cuts the packaging sequences from the genome of the helper virus. These packaging sequences code for the proteins required for the formation of capsid proteins *(Fig 6-14)*. And these capsid proteins expressed by the helper virus help in the packaging of the transgene to form viral particles.

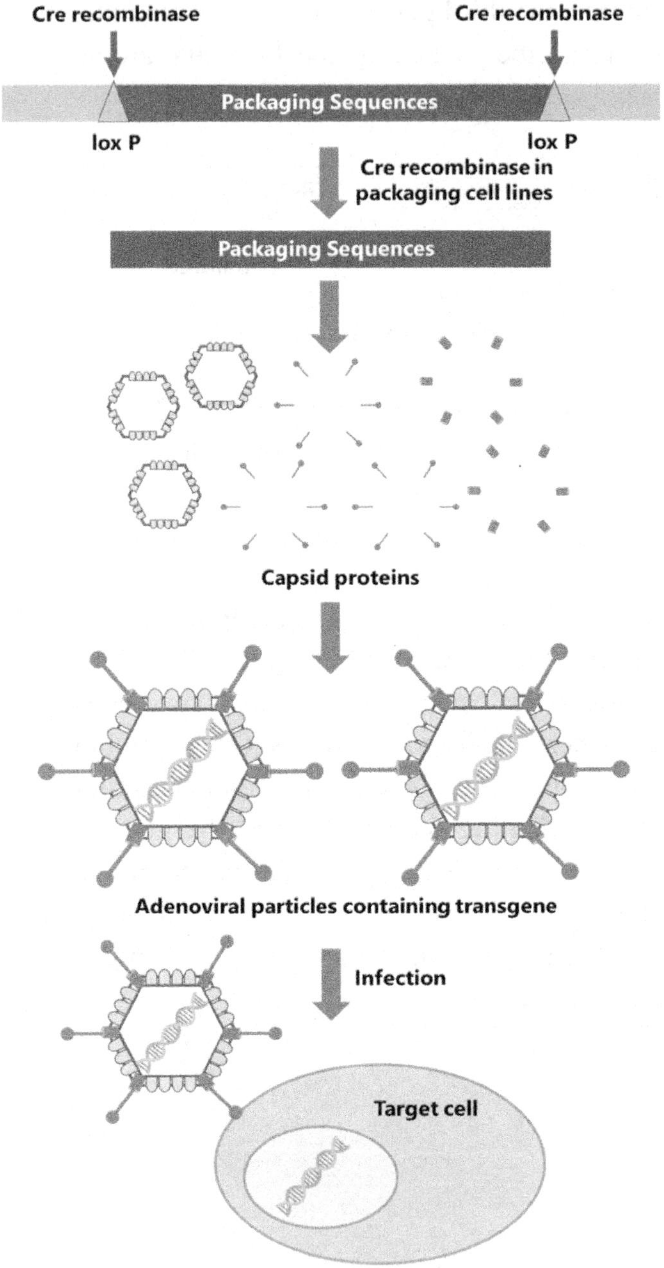

Fig 6-14: Capsid proteins expressed by the helper virus help in the packaging of the transgene to form adenoviral particles

Now, these adenoviral particles are allowed to infect the target cells of the patient. After binding to the target cells, the virus transfers the transgene into the target cells.

6.3. Components of sgRNA adenovirus CRISPR vector

The single gRNA adenovirus CRISPR vector consists of 5' and 3' ITR on the two sides of the viral vector.

Between the ITRs, the Cas9 gene and gRNA expression cassette are cloned. The gRNA itself contains a complementary sequence to bind to the target DNA. On the upstream, it is flanked by the U6 promoter sequence, which drives the high-level expression of gRNA. And on the downstream, the gRNA sequence is flanked by the scaffold RNA sequence, which helps gRNA to bind to the Cas9 enzyme and a terminator sequence that terminates its transcription *(Fig 6-15)*. On the other hand, the expression of the Cas9 gene is driven by a chicken beta-actin promoter (CBh). The vector also contains the Herpes simplex virus thymidine kinase polyadenylation (TK pA) signal that facilitates transcriptional termination of the Cas9 gene. There are two PacI restriction sites on the vector that can be used to linearize the vector and remove its backbone from the viral sequence, which is necessary for efficient packaging.

The vector is then transfected into the packaging cells containing the adenoviral genes necessary for virus production. In the packaging cells, the vector region that includes sequences encoding gRNA and Cas9 enzyme between the ITRs is packaged to form viral particles.

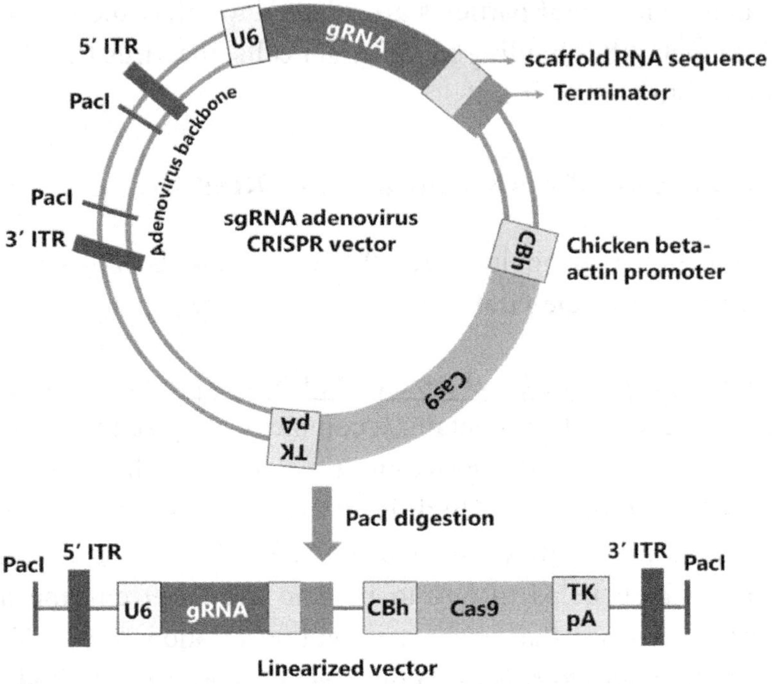

Fig 6-15: sgRNA adenovirus CRISPR vector

The viral particles are then allowed to infect the target cells of the patient. After binding to the target cells, the virus transfers the gRNA and Cas9 gene into the target cells *(Fig 6-16a)*. The gRNA binds to the target cells by its guide sequence, and the expressed Cas9 enzyme introduces a double-stranded break in the target cell. Using the cell's own repair machinery, the double-stranded break can be repaired either by a non-homologous end-joining pathway or by homologous-directed repair (HDR) pathway. For the homologous directed repair pathway, a vector containing donor template DNA with homology arms to the target DNA must also be inserted into the host cell along with the vector containing gRNA and Cas9 *(Fig 6-16b)*.

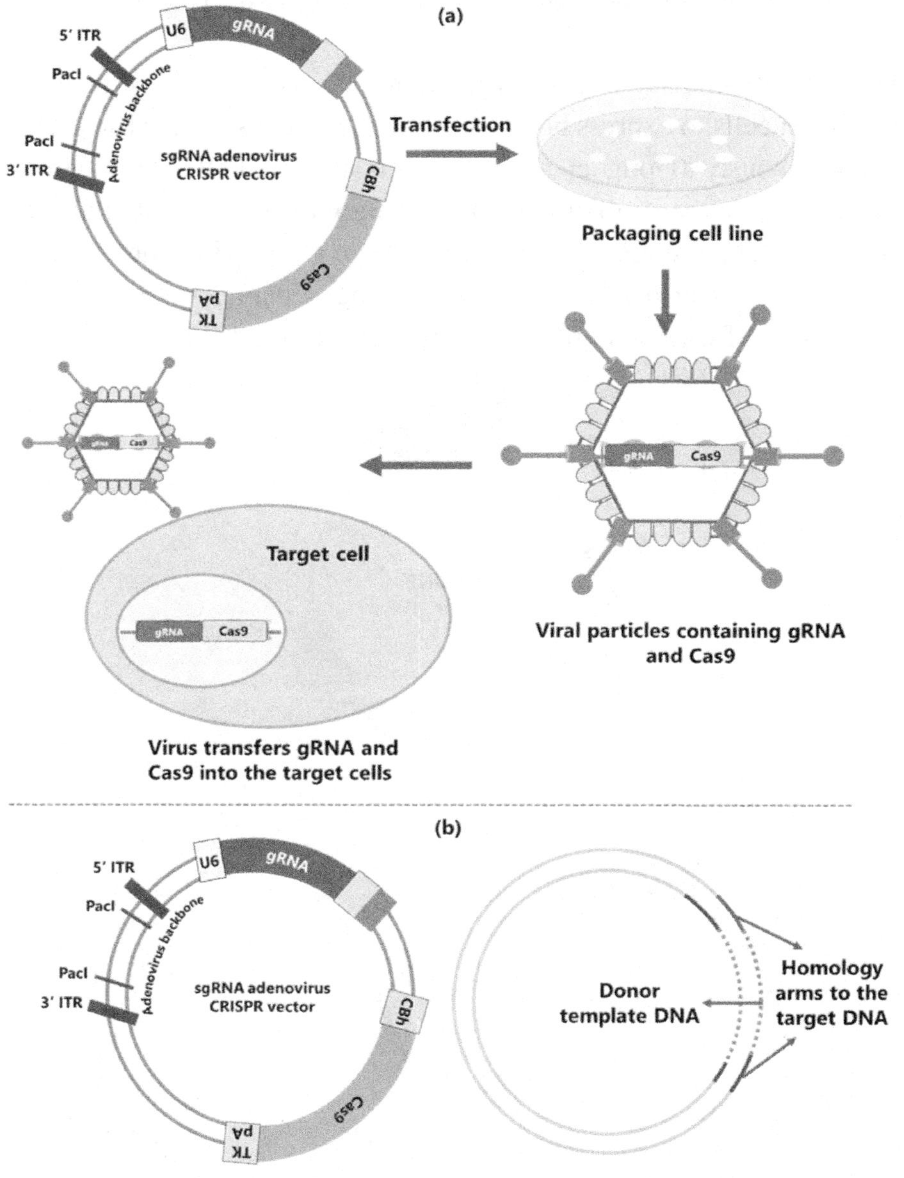

Fig 6-16: (a) Adenoviral vector transfers gRNA and Cas9 into target cells, and (b) vector containing donor template DNA is also inserted into the host to induce HDR pathway

6.4. All-in-one and separate adenoviral CRISPR vectors

For CRISPR-mediated gene targeting to be successful, it's crucial for the target cells to express both Cas9 and the target site-specific gRNA simultaneously (Gong et al., 2017; Jin et al., 2019). This can be achieved by either utilizing a single adenoviral vector that expresses both Cas9 and the gRNA sequence (known as an "all-in-one" vector, as shown in *Fig 6-17a*) or by employing separate vectors for the expression of Cas9 and gRNA (as shown in Fig *6-17b*).

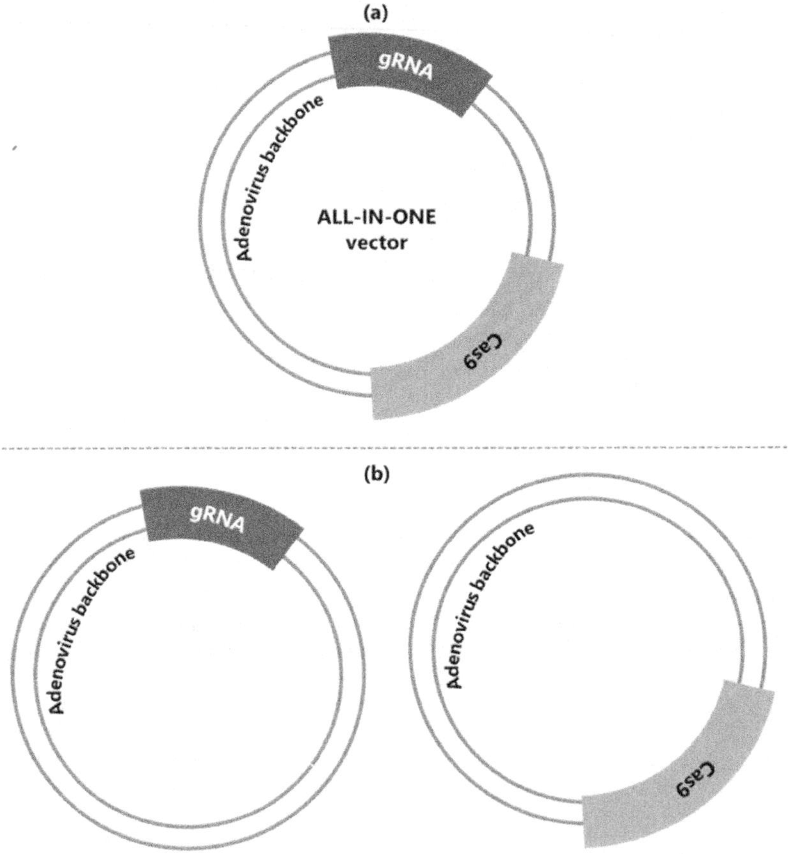

Fig 6-17: (a) All-in-one and (b) separate adenoviral CRISPR vectors

Using an all-in-one vector to express both Cas9 and gRNA offers the benefit of delivering all the necessary components for a CRISPR-mediated gene-editing system to the cell in a single vector. In contrast, when separate vectors are used to express Cas9 and gRNA, it becomes necessary to co-transduce the target cells with both vectors, which can pose a technical challenge as not all cells may be transduced with both gRNA and Cas9 vectors at the same time.

6.5. Single-gRNA and dual-gRNA adenoviral CRISPR vectors

All-in-one adenoviral CRISPR vector can be used for expressing either single-gRNA or dual-gRNAs.

The single-gRNA vector is widely utilized in CRISPR-based gene editing applications to target a specific gene, as illustrated in *Fig 6-18a*. However, for applications that involve simultaneous targeting of a pair of genomic sites, dual-gRNA vectors become necessary *(Fig 6-18b)*. For example, dual-gRNA vectors are often used by paired Cas9 nickase. The two gRNAs in the vector are designed to target opposite strands of the target DNA, and the paired Cas9 nickase generates single-strand cuts on each strand. This results in a double-stranded break (DSB), which can then be used to create various genetic modifications, such as deletions or insertions. The use of paired Cas9 nickase with dual-gRNA vectors offers improved specificity and reduced off-target effects compared to using a single Cas9 nuclease with a single gRNA.

Another scenario where two guide RNAs are required is in the case of generating deletion of a fragment between two DSBs.

The third scenario where two guide RNAs may be required is when the goal is to target two different genes simultaneously.

In these cases, when two guide RNAs are required, a dual gRNA vector is needed. In the single gRNA vector, the target site-specific gRNA sequence is driven by a single U6 promoter and is located between the two ITRs. In contrast, the dual gRNA vector utilizes two consecutive U6 promoters to drive the expression of two separate gRNA sequences, each specific to one of the two desired genomic target sites.

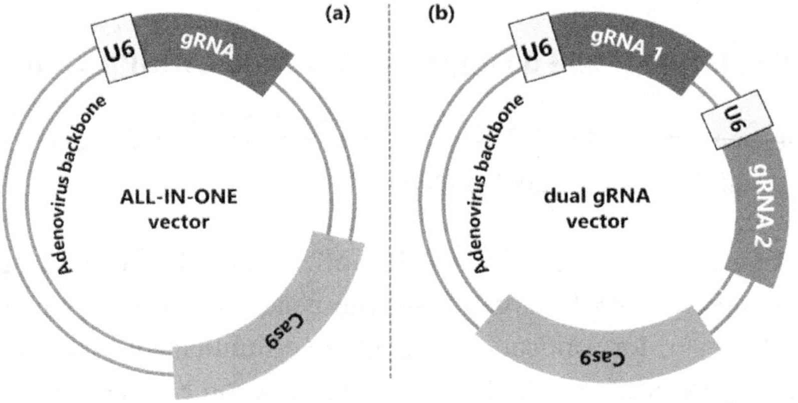

Fig 6-18: (a) single-gRNA and (b) dual-gRNA adenoviral CRISPR vectors

6.6. Adeno-associated viral (AAV) vectors

The second type of viral vectors used for CRISPR-mediated gene editing are adeno-associated viral vectors, abbreviated as AAV vectors. The adeno-associated virus was initially discovered as a contaminant of adenovirus preparations; hence the name adeno-associated virus or AAV. Also, AAV is dependent on co-infection with other viruses, mainly adenoviruses, in order to replicate. The genome of the adeno-associated virus is made up of single-stranded DNA, which is about 4.7 kb long. The genome contains only two genes, *rep* and *cap (Fig 6-19)*. The *rep* gene encodes for the proteins that help the AAV to replicate and integrate into the host genome. And the *cap* gene encodes the expression of adeno-associated viral capsid proteins. These coding sequences are flanked by inverted terminal repeats (ITRs) that are required for genome replication, packaging, and integration of the virus genome into the host cell genome.

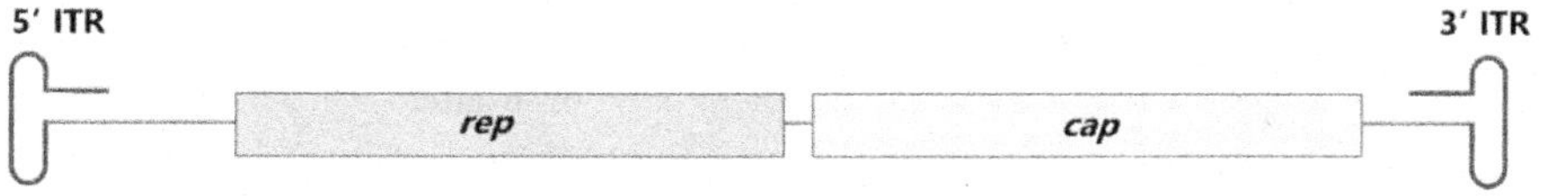

Fig 6-19: AAV genome

When AAV alone infects the host cells, its gene expression is autorepressed, and the genome of the virus is integrated into the 2 kb region of chromosome 19. This allows the virus to stay in a lysogenic or dormant state into the cell until the host cells are infected with the helper virus such as adenovirus. Once a helper virus infects the host cell, the AAV genome can replicate.

E1A, E1B, E2A, E4, and viral-associated (VA) genes of helper adenovirus are required for the replication, transcription, and translation to synthesize the AAV genome and capsid proteins *(Fig 6-20)*. The AAV genome gets assembled and packaged in the capsid proteins to form AAV particles which are then released from the cells along with the helper viruses.

For transforming the AAV into adeno-associated viral vector for gene editing purposes, the two native AAV genes, *rep* and *cap,* are removed and are replaced with a transgene that contains instructions and a promoter to produce a normal functional protein. The promoter and transgene are flanked by ITR signals that allow the transgene to be packaged inside the protein shell of the AAV vector.

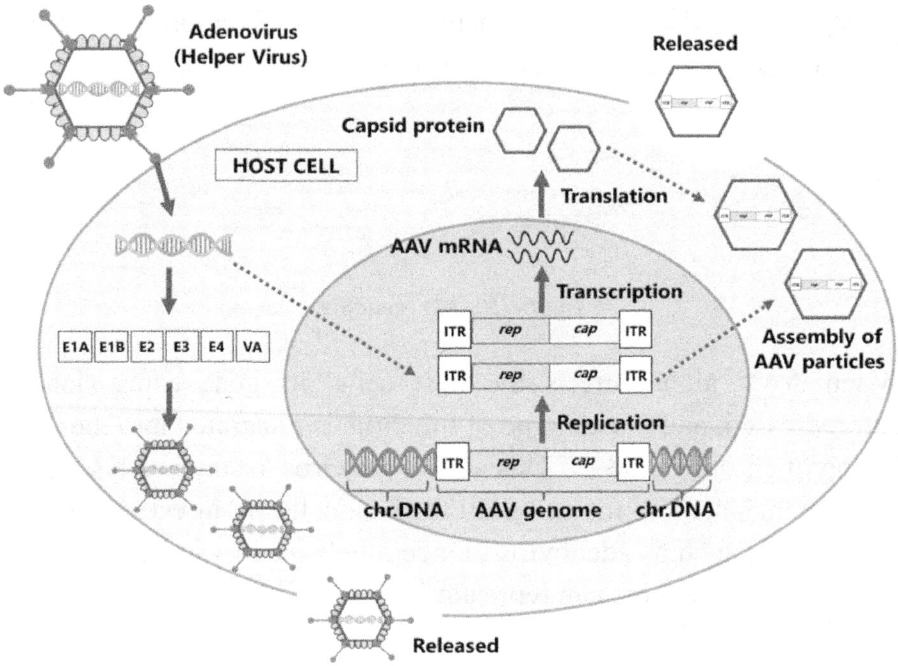

Fig 6-20: Assembly of AAV particles with the help of helper adenovirus

Then the vector is transfected into the packaging cell lines to generate an adeno-associated virus containing the transgene as its genome. At the same time, the packaging cells are also transfected with a plasmid that contains *rep* and *cap* genes. So this plasmid will express the proteins required for the replication of the viral genome and for packaging of it. AAV is dependent on co-infection with other viruses, mainly adenoviruses, in order to replicate and make virus particles. But you would not want to transfect the packaging cells with adenovirus, because adenovirus can cause the lysis of the cells.

Therefore the plasmids that contain adenoviral genes required by adeno-associated virus are transfected into the packaging cells. These genes are E1, E2, E4, and viral-associated (VA). So, for getting adeno-associated viral particles with genes of interest, there are 3 plasmids required: the vector with transgene construct, the plasmid that contains *rep* and *cap* genes, and the plasmid that contains E1, E2, E4, and viral-associated genes *(Fig 6-21)*. Even a few other modifications in the vector are done so that the vector does not get integrated into the host cell genome. Instead, it remains in the nucleus in an episomal state.

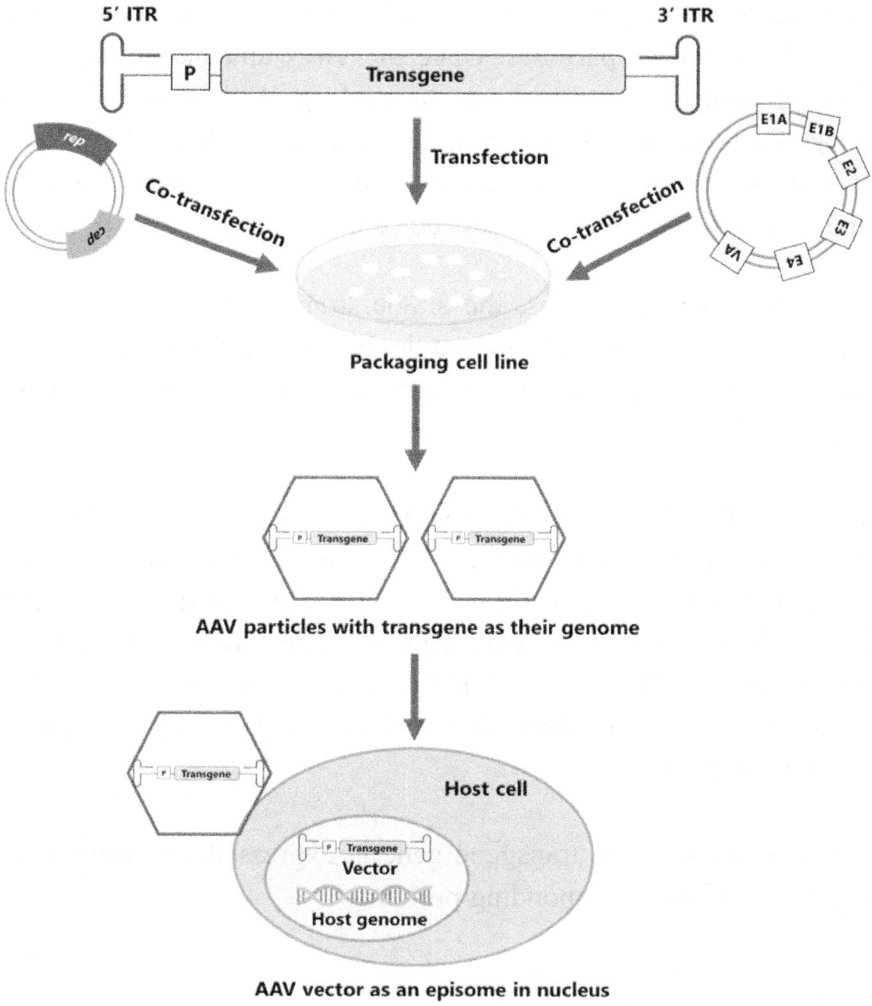

Fig 6-21: AAV vector does not get integrated into the genome of the host cell

Now the AAV particles carrying transgene are then delivered to the target cells. This can be accomplished by either injecting or intravenously administering the vector directly to a particular tissue in the body. Upon administration, the vector enters the target cells and delivers the transgene directly to the nucleus.

Alternatively, the patient's target cells can be isolated and then infected with AAV particles. Once the virus attaches to the target cells, it transfers the transgene to the nucleus. Without Rep proteins, transgenes flanked by ITR within recombinant AAV (rAAV) do not integrate into the host genome but instead form circular structures that persist as episomes in the transduced cells' nuclei.

Upon entering the nucleus, the single-stranded recombinant AAV genome undergoes conversion to a double-stranded form *(Fig 6-22)*. The second strand synthesis is initiated by the host cell DNA polymerase.

The ITRs of AAV contain self-complementary sequences, which are hairpin-like structures; therefore, the ITR at the 3' end of the genome acts as a primer for host cell DNA polymerase. The double-stranded genome then undergoes circularization, leading to the formation of episomal DNA. The episomal DNA is not integrated into the host cell's genome, thus, minimizing the risks of insertional mutagenesis and oncogene activation.

Finally, the functional transgene gene is expressed, resulting in the production of the corresponding protein.

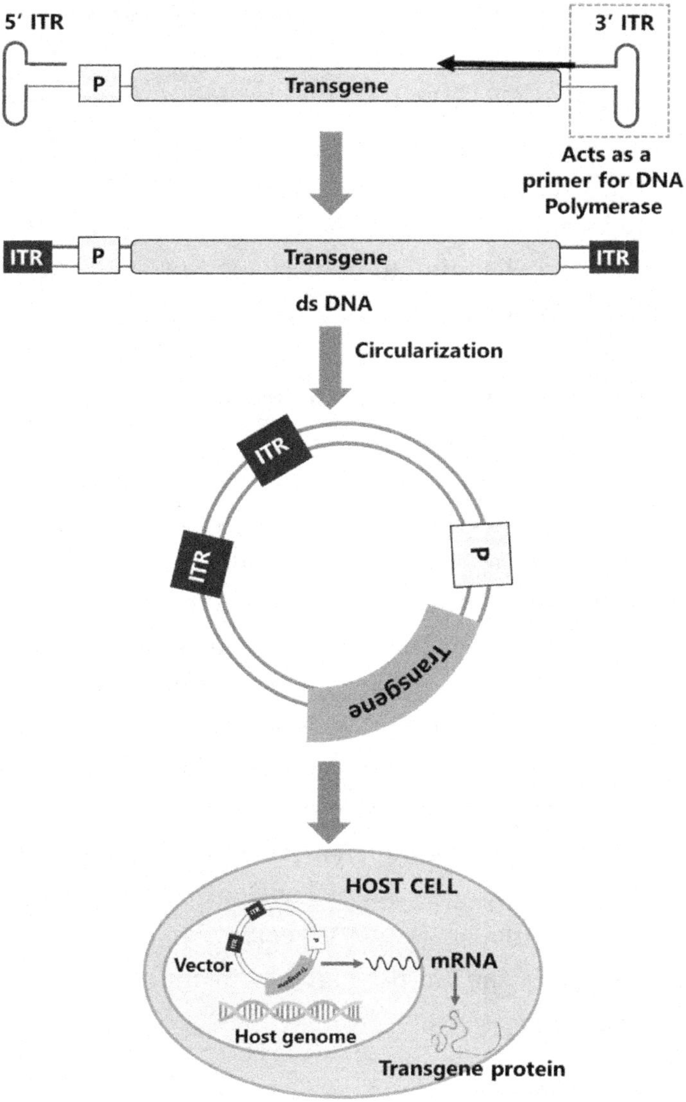

Fig 6-22: Circularization of AAV vector

6.7. AAV vectors for CRISPR-mediated gene editing

For AAV-mediated CRISPR-Cas9 gene editing, two transgenes are required that code for gRNA and Cas9 enzyme (Senís et al., 2014; Lau and Suh, 2017). Because of the limited size capacity of the AAV vector, i.e., 4.7kb, it has limited use for delivering large transgenes. Thus, it becomes challenging to co-package SpCas9 which is of size 4.1 kb, and sgRNA into an all-in-one AAV vector for genome editing. To circumvent this problem, dual AAV vector system is used, with one vector carrying the Cas9 with the promoter and the other vector carrying the sgRNA driven by the U6 promoter *(Fig 6-23a)*.

The double-stranded breaks produced by the Cas9 enzyme could be repaired either by a non-homologous end-joining pathway or by a homologous directed repair pathway. For the homologous directed repair pathway, donor template DNA with homology arms to the target DNA is inserted into the second vector along with the sgRNA.

Additionally, SpCas9 orthologs have smaller sizes compared to SpCas9. For instance, *Staphylococcus aureus*-derived Cas9 has a size of 3.16 kb, and *Campylobacter jejuni*-derived Cas9 has a size of 2.95 kb. Thus SpCas9 orthologues SaCas9 and CjCas9 can be used to pack Cas9 and sgRNA into a single AAV vector *(Fig 6-23b)*. Studies have shown that, CjCas9 can be packaged with multiple gRNAs into a single all-in-one AAV vector.

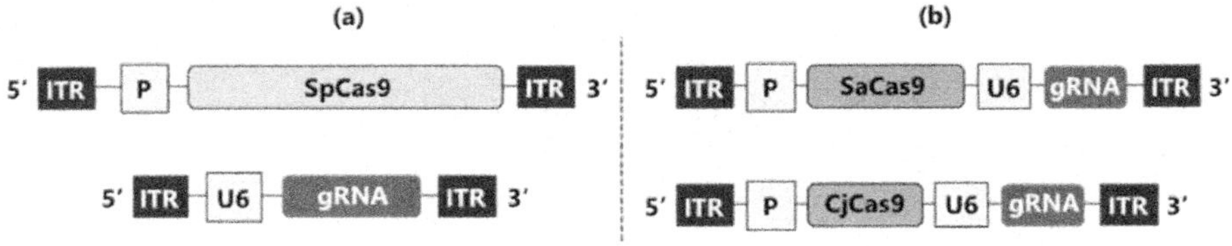

Fig 6-23: (a) Dual vector and (b) All-in-one vector

Adeno-associated viruses are a popular choice for gene delivery systems in research studies due to several features:

(i) Firstly, AAV can infect both dividing and non-dividing cells.

(ii) Further, the AAV elicits only a weak immune response and has never been shown to cause disease in humans, ensuring safety during gene delivery.

(iii) Moreover, over 12 different serotypes of AAV have been identified to date. A serotype is a distinct variation within a virus because of the differences in envelope proteins and interaction with different receptors of host cells. This means that different serotypes can bind to different receptors on various cell types, allowing them to infect cells from multiple tissue types.

As a result, AAVs can be used as excellent in vivo gene delivery systems to target specific tissues in the body (Xu et al., 2019; Wang et al., 2019; Tabebordbar et al., 2021). For instance, AAV serotype 2 has the preference to infect skeletal muscles, neurons, vascular smooth muscle cells, and hepatocytes; thus, AAV serotype 2 can be used as an efficient gene-editing system in these tissues.

AAV serotype 1 vector can be used to deliver the CRISPR-Cas system to vascular endothelial cells, as well as the retina, heart, and lungs. AAV-4 exhibits a preference for kidney and heart cells, AAV-5 exhibits a preference for astrocytes, AAV-6 can efficiently deliver transgenes to airway epithelial cells and hepatocytes, and other serotypes of AAV can be used as an efficient gene-editing system for various other cells.

To achieve tissue- or organ-specific gene editing with CRISPR, a promoter specific to the target tissue is used to drive the expression of the CRISPR transgene. For example, the thyroxine-binding globulin

TBG promoter is liver-specific and has been used to drive the expression of CRISPR for in vivo genome editing in mouse liver. Similarly, the myosin heavy polypeptide 6 Myh6 promoter is cardiac-specific and has been used for CRISPR-mediated gene editing in the heart. Other tissue-specific promoters can also be used to target other tissues or organs, depending on the research or therapeutic application.

6.8. Lentiviral vectors

The third type of viral vectors used for CRISPR-mediated gene editing purposes are lentiviral vectors, derived from human immunodeficiency type-1 (HIV-1) lentivirus. Lentivirus is a type of retrovirus that can infect both dividing and non-dividing cells. It is an RNA virus because its genome is made up of ssRNA, while its host's genetic material is in the form of DNA.

When a lentivirus infects a host cell, it transfers its RNA genetic material along with specific enzymes, namely reverse transcriptase and integrase, into the host cell. The virus then undergoes a process called reverse transcription, where it creates a DNA copy from its RNA molecule using the reverse transcriptase enzyme that it carries with it. After this, the produced DNA is incorporated into the host cell genome with the help of another enzyme carried in the virus called integrase.

Structurally, the lentivirus genome contains 3 major genes: *gag*, *pol*, and *env* genes, flanked by long terminal repeats (LTRs) on both the 5' and 3' ends *(Fig 6-24a)*. The *gag* gene encodes for the viral core proteins like matrix, capsid, and nucleocapsid. The *pol* gene encodes for the reverse transcriptase, integrase, and protease enzymes, which are crucial for reverse transcription and genomic integration. And the *env* gene encodes for the surface glycoproteins gp120 and gp141 that are responsible for binding to receptors on the surface of target cells and facilitating virus entry into the cell *(Fig 6-24b)*.

Within the lentivirus genome are two regulatory genes, tat and rev, which facilitate viral transcription. Additionally, the genome carries four accessory genes- *vif*, *vpr*, *vpu*, and *nef* that contribute to HIV virulence, although they are not as critical to viral replication in host cells. The genome also contains Ψ sequence, which is required for the packaging of viral particles. And the long terminal repeat sequences present on both ends of the genome help the reverse transcribed DNA to integrate into the host genome.

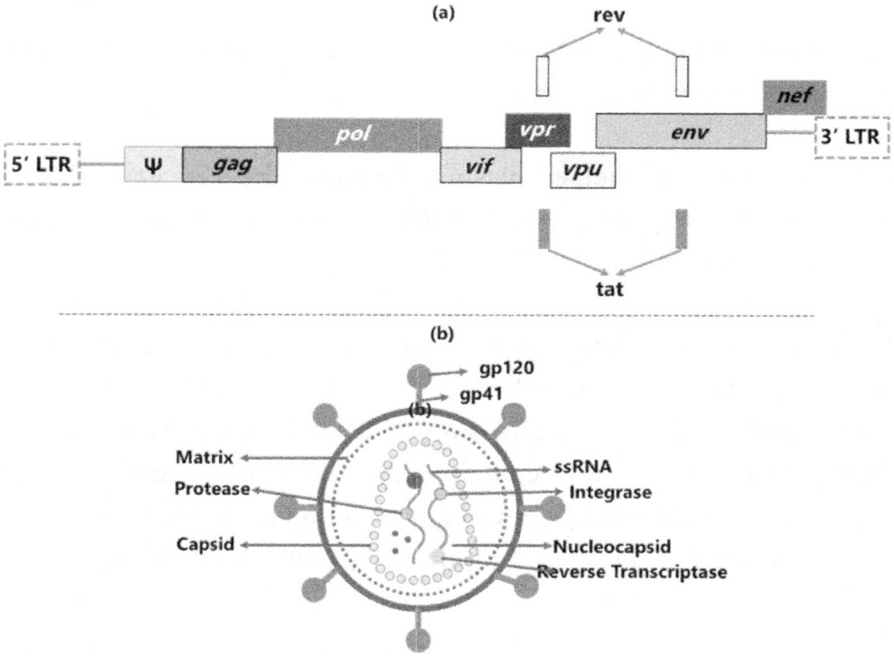

Fig 6-24: (a) Genes in the lentivirus genome (b) Structure of the lentivirus

In the past two decades, these properties of the genetic material of lentiviruses have turned them into an attractive system for use as vectors for transferring genes into mammalian cells (Sakuma et al.,

2012; Milone and O'Doherty, 2018). But before using the HIV viruses as vectors, its genome must be modified extensively to be safe for use in gene editing or gene therapy purposes.

To construct a safe and effective lentiviral vector, the accessory HIV virulence genes *vif*, *vpr*, *vpu,* and *nef* are removed entirely. And the other genes necessary for viral infection *gag*, *pol*, *env*, and *rev* are provided in trans, meaning these genes are expressed in different plasmids. For instance, 3 types of helper plasmids are used for the expression of different genes.

(i) Helper plasmid 1 contains *gag* and *pol* genes *(Fig 6-25a)*.

(ii) The second helper plasmid contains the 5' and 3' long terminal repeats and an *env* gene *(Fig 6-25b)*.

(iii) And the third helper plasmid contains the *rev* gene that is important for exporting viral mRNA from the nucleus to the cytoplasm during infection *(Fig 6-25c)*.

Each of these plasmids drives its gene expression from the well-studied promoters from other viruses such as cytomegalovirus, abbreviated as CMV, or respiratory syncytial virus, abbreviated as RSV. These plasmids also contain additional genetic elements, such as polyadenylation sites polyA, that are important for the stability of mRNA molecules transcribed from these plasmid constructs.

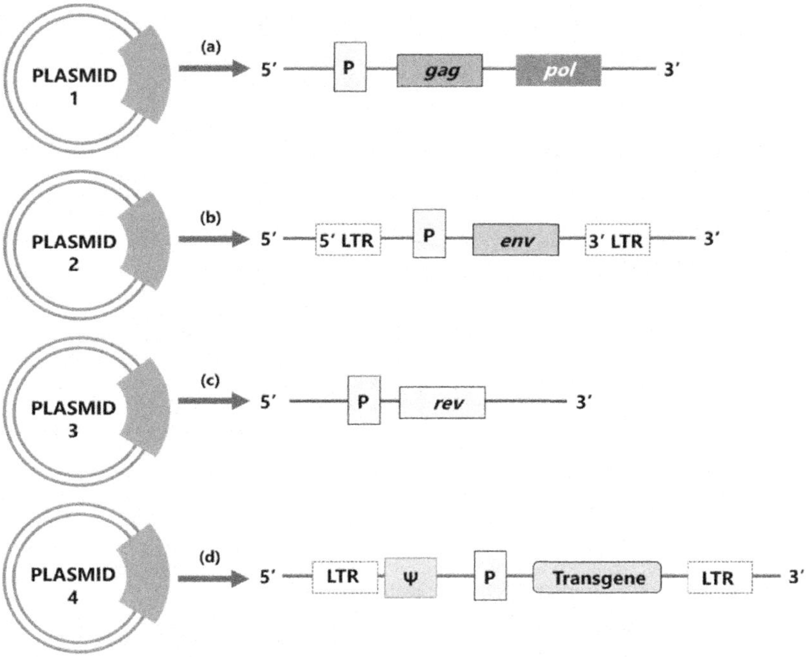

Fig 6-25: Four plasmids for lentiviral gene transfer system

The envelope glycoprotein of HIV, gp120, specifically binds to the CD4 receptor on the surface of CD4 T cells, which are a type of immune cell. This specificity of envelope glycoproteins narrows the host cell range of HIV vectors and limits the usefulness of HIV lentiviral vectors for gene editing and gene therapy purposes. To overcome this limitation, the *env* gene in the helper plasmid 3 is replaced with an envelope glycoprotein from another virus called the vesicular stomatitis virus, which is known as the VSV-G protein. The VSV-G protein interacts with phospholipids rather than specific receptor molecules on the target cell, allowing the lentiviral vector to infect a broader range of cell types.

Apart from 3 helper plasmids, there is a fourth plasmid that contains a transgene with the promoter, 5' and 3' long terminal repeat sequences for the genome integration into the host DNA and the packaging signal Ψ sequence *(Fig 6-25d)*. The insert capacity of the lentiviral vector is 8 kb of sequence.

The combination of 4 vectors is known as the lentiviral gene transfer system. When these plasmids are introduced into the packaging cell lines by a process called transfection, the transgene vector integrates into the cell line DNA and produces mRNA *(Fig 6-26)*.

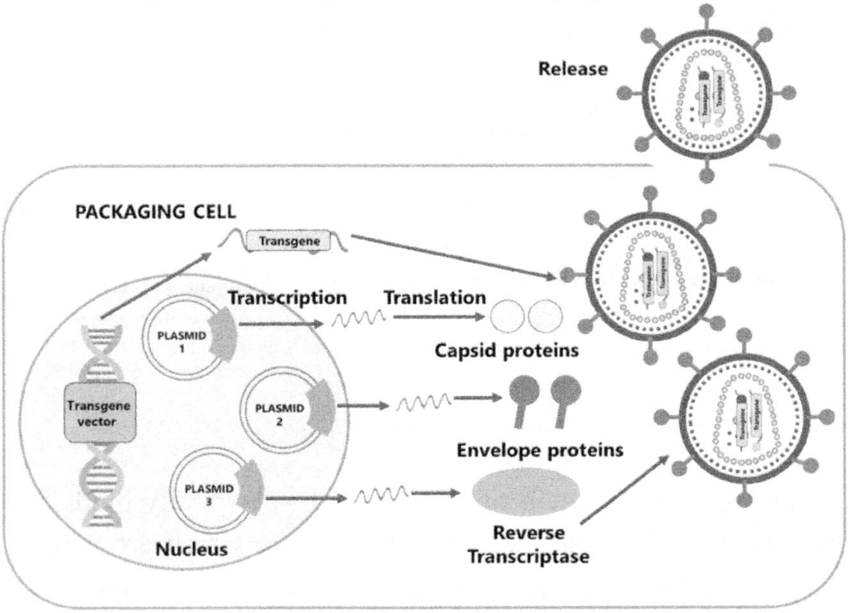

Fig 6-26: mRNAs containing the transgenes are packaged into the lentiviral particles in the packaging cells

The other three plasmids also produce mRNAs which are translated to form viral proteins, including the capsid, reverse transcriptase, and envelope proteins. These proteins assemble to form the viral particles, which package the mRNA containing the transgene. Once the lentiviral particles are released from the packaging cell lines and infect the target cells, the mRNA containing the transgene is converted into DNA by reverse transcriptase; the new DNA enters the host cell's nucleus and integrates into host chromosomal DNA *(Fig 6-27)*. The integrated DNA contains LTR and Ψ sequence, but none of the other components from the HIV genome. This means that the integrated DNA cannot form viral particles, as it lacks the other necessary viral components. However, the transgene can be expressed and produce functional proteins in the target cells.

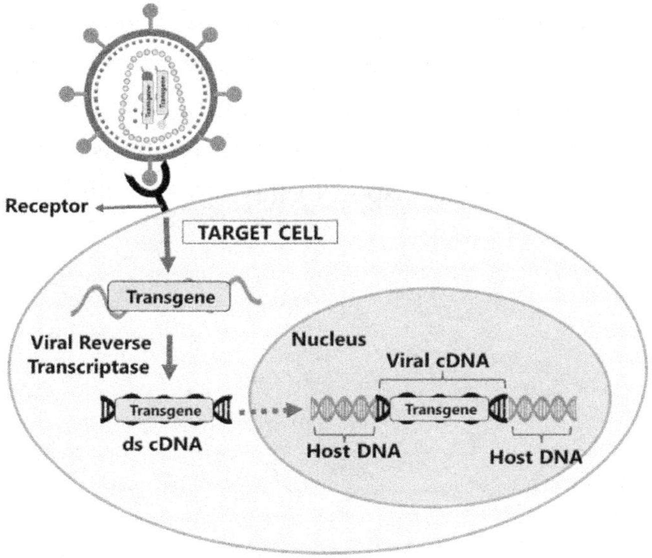

Fig 6-27: Integration of transgene into the host cell genome

6.9. Lentiviral vectors for CRISPR-mediated gene editing

For lentivirus-mediated CRISPR-Cas9 gene editing, two transgenes are required that code for gRNA and Cas9 enzyme. The two transgenes along with their promoters U6 are inserted into the lentiviral vector. When the viral particles containing transgenes gRNA and Cas9 as their genome are allowed to infect the target cells, gRNA, and Cas9 genes are expressed and perform gene editing at the desired position in the genome (Ryø et al., 2019; Dong and Kantor, 2021).

One of the challenges of using lentiviral vectors for gene editing is the potential for random integration of the transgene into the host genome, which can lead to unwanted effects such as disruption of endogenous gene function or activation of oncogenes.

To address this issue, non-integrating lentiviral vectors have been developed. These vectors are engineered to eliminate the activity of the integrase enzyme, which is responsible for integrating the transgene into the host genome. This is typically achieved through specific mutations in the integrase coding region, which disable its activity without affecting other important functions such as reverse transcription and mRNA transport (Vijayraghavan and Kantor, 2017; Ortinski et al., 2017).

Chapter 7: Non-viral methods for delivery of CRISPR components

Until now, we have understood that the viral vectors are crucial for transferring the CRISPR system components gRNA and Cas9 into the target cells. But there are certain drawbacks of using viral vectors, such as immunogenicity and cytotoxicity issues. The viral vectors can elicit inflammatory responses in the body.

An additional cause of concern over using viral gene transfer vectors is insertional mutagenesis. It means that the integration of viral DNA into the host chromosomes can disrupt the expression of some other functional gene or activate an oncogene leading to the malignant transformation of cells.

To overcome these drawbacks, non-viral approaches can be used for the transfer of desired transgenes (Li et al., 2018; Lino et al., 2018; Behr et al., 2021). Non-viral methods offer few advantages over viral vectors like reduced pathogenicity, low cost, and ease of production. Also, non-viral methods have significant safety advantages over viral vectors.

7.1. Cargo for CRISPR-Cas9-based gene editing therapy

For gene editing by CRISPR/Cas9, there can be various options to deliver guide RNA, Cas9, and template DNA into the target cell:

(i) Cas9 enzymes can be delivered in the form of purified Cas9 protein; or by plasmid DNA or mRNA encoding Cas9 protein. Using mRNA or plasmid DNA encoding Cas9 can avoid problems related to protein production and purification while reducing the cost. If mRNA or plasmid DNA are successfully delivered, Cas9 can be produced in target cells. The advantage of using mRNA over plasmid DNA is that it does not pose the risk of insertional mutagenesis in chromosomal

DNA. Additionally, unlike plasmid DNA, mRNAs do not need to enter the cell nucleus and can produce Cas9 protein in the cytoplasm by the ribosomes as quickly as ~6 hours post intracellular delivery.

(ii) gRNA can be delivered in the form of RNA or plasmid DNA encoding gRNA sequence. Plasmid DNA takes additional time to transcribe gRNA in the cell, causing a delayed expression of gRNA. If Cas9 nuclease is co-delivered along with plasmid DNA encoding gRNA, Cas9 nucleases may not survive in the cell by the time when gRNA is expressed. This timing issue will lead to a decreased genome editing efficiency. However, if a plasmid DNA is delivered to express both Cas9 and gRNA in target cells, the timing issue is generally not a concern.

(iii) DNA donor template is required to be delivered to mediate HDR-based gene editing in the target cells. It can be delivered in the form of plasmid DNA, usually when the insertion is a long DNA sequence. Alternatively, the DNA donor can be provided as a synthetic single-stranded DNA oligonucleotide when the insertion sequence is short.

Therefore, for gene editing by CRISPR/Cas9, three types of cargo can be delivered to the target cell:

- DNA plasmid encoding both the Cas9 protein and the guide RNA,
- mRNA for the translation of Cas9 protein and a guide RNA,
- or a ribonucleoprotein (RNP) complex composed of Cas9 protein and a guide RNA.

Non-viral delivery approaches can be divided into physical or chemical carrier-mediated methods.

Physical methods: These methods require directly injecting CRISPR cargo into the target cells by creating pores in the cell membrane of the target cells. A few of the physical methods are as follows:

7.2. Microinjection

In this method, the CRISPR cargo is directly injected into individual cells. This is done by using a microscope and a 0.5–5.0 μm diameter needle, which pierces the cell membrane, and then the CRISPR components are injected directly into the nucleus within the target cell (Chen et al., 2022). After which, the cell transcribes and translates the CRISPR components to cleave and modify the desired position in the genome. Using the microinjection technique, researchers have been able to successfully deliver CRISPR components to knock out four genes into rat zygotes, correct a cataract-causing mutation and Duchenne muscular dystrophy (DMD)-causing mutation in mice.

7.3. Electroporation-mediated gene delivery

Another method of introducing CRISPR/Cas components into the target cells is electroporation. Electroporation is a technique in which an electrical field is applied to the cell membrane of target cells in order to produce pores in their membrane (Isaka and Imai, 2007; Young and Dean, 2015; Che et al., 2021). Once the pores are formed on the cell membrane, the CRISPR cargo can be transferred into the cells. A cell membrane, also known as the plasma membrane, is a double layer of lipids and proteins that surrounds a cell and separates the cytoplasm from its surrounding environment. In a typical cell membrane, phospholipids are arrayed with their polar head groups pointing outward and their hydrophobic tail groups pointing inward, an arrangement that impedes the passage of polar molecules like DNA plasmids.

To create pores in the cell membrane, short pulses of high voltage are applied across the cell membrane. The application of electric pulses disturbs the membrane's phospholipid bilayer and results in the formation of temporary pores *(Fig 7-1)*. As a result, the CRISPR cargo can pass through the membrane. Under the controlled electric pulse, the membrane pore closes, returning the cell to its original state, except that the target cell now has the transgene.

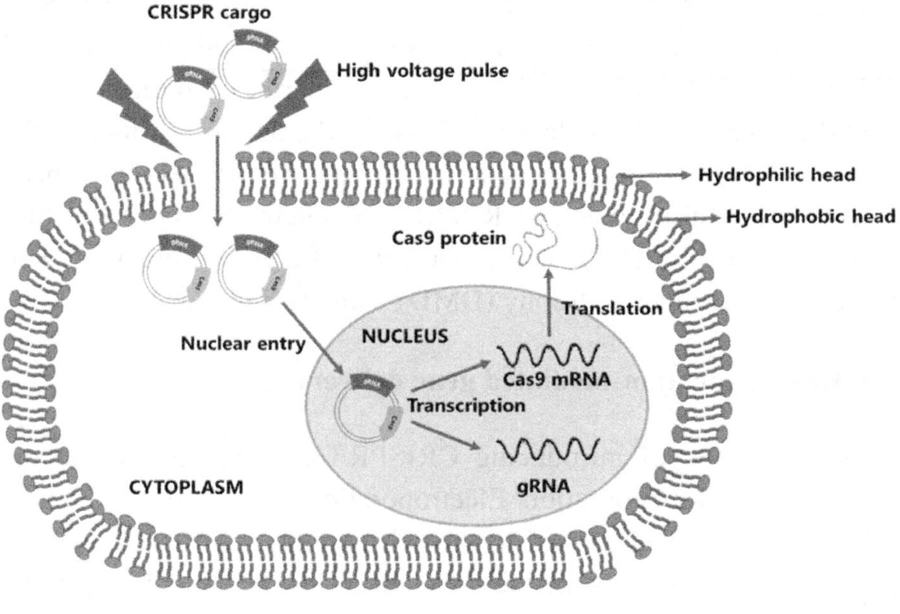

Fig 7-1: Electroporation method to deliver CRISPR cargo into the target cells

Following the insertion of CRISPR cargo comprising of DNA plasmid encoding transgenes Cas9 protein and the gRNA into the interior of the cell, the transgenes must enter the nucleus for productive gene expression to take place. If the plasmids carrying transgene sequences contain DNA nuclear localization sequences (NLS), then the specific proteins called nuclear localization signal-containing proteins bind to the plasmid. This, in turn, facilitates the entry of the plasmid through the nuclear pore complex into the nucleus. Alternatively, any plasmid, regardless of sequence, has the ability to enter the nucleus at the time of cell division because, at that time, the nuclear envelope breaks down. Electroporation is generally efficient and works across a broad range of cell types like lungs, heart, liver, kidney, skeletal muscle, and tumors.

Like microinjection, electroporation can also be used for in vitro and ex vivo studies. Electroporation has been used to deliver CRISPR cargo to correct a cataract-causing mutation in mouse spermatogonial stem cells; confer resistance to HIV infection by mutating the CCR5 gene in human cells; and cure latent herpesvirus infection by Cas9-based cleavage and destruction of latent viral genomes.

However, one disadvantage of using this method is that a high rate of cell death can occur following electroporation.

7.4. Gene gun

The use of particle bombardment, or the gene gun, is another method of delivering CRISPR/Cas components into the target cells (Sandhya et al., 2020; Miller et al., 2021). This method was first used as a gene transfer technique to plants but is now being used for both in vivo and in vitro gene transfer to animals and human cells. In this technique, the CRISPR cargo is first coated onto micro-particles, usually gold or silica particles. Gold particles are used because of their low toxicity and lack of chemical reactivity. On the other hand, silica particles are biodegradable; therefore, these two materials are used onto which the transgenes are coated. The CRISPR cargo-coated microparticles are then loaded into one side of the gene gun. A pressurized gas, usually helium, provides the force to penetrate the gene-coated microparticles into the cells *(Fig 7-2)*.

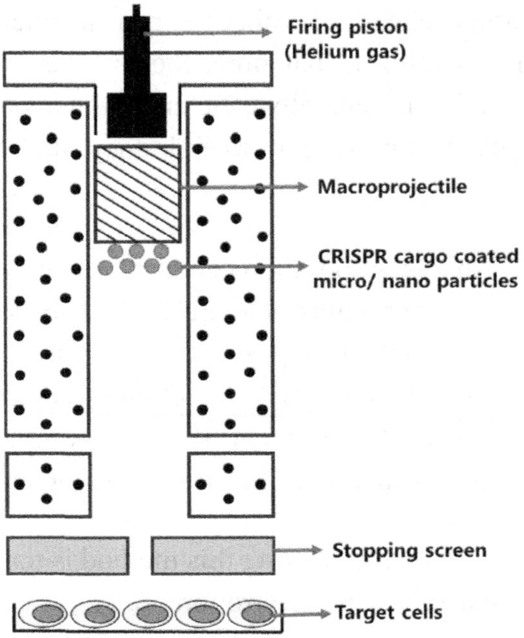

Fig 7-2: Gene gun method to deliver CRISPR cargo into the target cells

However, gene delivery by this method can cause some tissue damage. Tissue damage can be reduced by performing the gene transfer at an operating pressure of less than 200 psi. Also, the researchers have found that if nanoparticles are used instead of microparticles to coat the transgene, then these gene-coated nanoparticles create less cell damage, but the efficiency of gene transfer is similar to those of microparticles (Zhang et al., 2014).

Apart from the advantages of biosafety and low immunogenic reactions, the gene gun also has another major advantage. This method also allows the delivery of several transgenes encoding different gRNAs simultaneously to the same target tissues in case more than one gene is required to be edited. The gene gun can transfer the transgenes to a wide number of cells under in vitro conditions, i.e., after isolating the cells, the transgene is transferred into them, and then the modified cells are returned to the patient's body. The attractive candidates for in vivo gene gun-mediated gene transfer are skin and muscle cells.

7.5. Sonoporation

Sonoporation is the phenomenon that refers to the formation of small pores in the cell membranes of target cells by using ultrasound waves to transfer CRISPR cargo into the cells (Belling et al., 2020; Fajrial et al., 2020; Lo et al., 2021). In the medical field, many ultrasound diagnostic tests are routinely used. Now, ultrasonic energy is finding its path in gene editing and gene therapy. Sonoporation employs the use of injectable microbubbles. These microbubbles are composed of a gas-filled core- either air, nitrogen, or inert gas with high molecular weight, such as perfluorocarbon or sulfur hexafluoride. The core is surrounded by an outer shell consisting of biocompatible compounds such as lipids, proteins, or synthetic biopolymers. If sonoporation has to be used under in vivo conditions, then the specific proteins can be attached to the shell of microbubbles so that microbubbles get attached to specific target cells and then facilitate gene delivery. The size of microbubbles ranges between 1 and 10 μm. There are two methods by which these microbubbles can be paired with the CRISPR cargo.

(i) First, the CRISPR cargo can be incorporated into the shell of the microbubbles.

(ii) The second method is the co-incubation of microbubbles with the CRISPR cargo, which leads to the attachment of the CRISPR cargo to the shell of the microbubble by electrostatic interactions.

In both methods, the cationic microbubbles are required, meaning the bubbles should have a positive charge; only then can they either attach to the plasmid DNA or RNA, which is negatively charged, or incorporate the negatively charged DNA or RNA into its shell.

When these microbubbles are subjected to acoustic or sound frequencies, the microbubbles expand and contract in response to compression and the rarefaction of ultrasound. The expansion and contraction cause microbubbles to collapse, thus emitting shock waves that perturb the cell membrane by forming pores, and then the collapsed bubbles microinject the CRISPR/Cas9 components into the target cells *(Fig 7-3)*. Researchers have used sonoporation to deliver plasmid encoding Cas9 and gRNA to knock out the *c-erbB2* gene in human endometrial cancer (HEC)-1A cells. In vivo ultrasound-mediated gene delivery can be applied to the heart, blood vessels, lungs, kidneys, muscles, brain, and tumors.

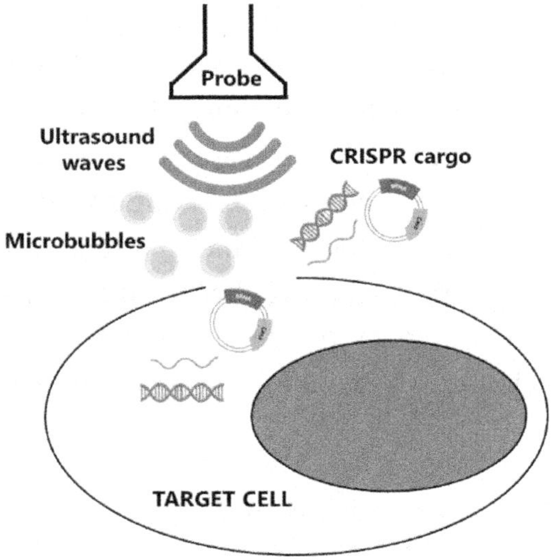

Fig 7-3: Sonoporation to deliver CRISPR cargo into the target cells

7.6. Optoporation or laser beam-mediated

Optoporation is another physical method of delivery of CRISPR cargo into the target cells with the help of a laser beam. In this technique, the emitted laser beam is concentrated at the target site. The concentrated laser beam causes local thermal changes at the cell membrane, due to which the cell membrane gets ablated, and transient pores are formed, hence facilitating the plasmid DNA encoding CRISPR components to get inserted into the target cell.

Few studies have revealed that nanoparticles can increase the efficiency of CRISPR cargo delivery by a laser beam. Because when the laser beam is irradiated, the nanoparticles absorb and scatter more light at the target site, thus causing enhanced thermal changes and leading to increased permeability of the target cell. Thus the efficiency of a transgene-containing cargo delivery into the target cells is increased. Gold nanoparticles and carbon nanoparticles can be used for this gene delivery technique.

7.7. Limitations of physical systems for gene delivery

Though the discussed nonviral techniques are much safer than the viral vectors in terms of inflammatory and immune responses, but there are a few limitations to using physical methods for gene delivery systems. From the moment of injection until the agent reaches targeted cells, genetic material encounters extracellular and intracellular barriers affecting therapeutic and gene editing results.

(i) First, the plasmid DNAs are unstable under physiological conditions, resulting in enzymatic degradation by endogenous nucleases. Nucleases are the enzymes that degrade the nucleic acids by cleaving the phosphodiester bond between the nucleotides of the nucleic acids DNA or RNA. The degraded nucleic acids are recognized as foreign by the body, and thus cleared from the body. To protect the transgene from getting degraded by nucleases, chemical modification of the transgene and a proper delivery method are required (Niidome and Huang, 2002; Jing et al., 2014).

(ii) The second limitation of physical systems is that DNAs are anionic hydrophilic polymers with negative charges. The negative charges on the nucleic acids impede them from binding to the cell membranes, which are also negatively charged because of negatively charged glycoproteins and proteoglycans. Even in case, the nucleic acid is uptaken by the endocytic pathway, degradation by lysosomal enzymes is another major issue.

(iii) The third problem with using physical methods is that these gene delivery methods can cause significant cell damage.

(iv) Also, the studies have shown that these methods have generally resulted in much lower levels of gene delivery and expression when compared to their viral counterparts

The success of gene therapy depends largely on the development of a vehicle or vector that can efficiently and effectively deliver genetic material to target cells and obtain sufficient levels of gene expression in vivo with minimal toxicity. Virus-derived vectors for gene therapy are efficient in gene delivery and transfer, but safety issues limit the use of viral vectors in gene therapy. Therefore, certain **nonviral chemical vectors such as lipids and polymers** have been designed that have the potential to overcome the barriers in the delivery of genetic material and enhance its transfer to the target cells.

There are few intracellular barriers that the vectors used to transfer transgene face. It has been reported that although more than 95% of cells in culture typically internalize vectors, but only a small fraction, less than 50% of the cells, express the transgene. The reason behind this is the inability of the vectors to release nucleic acids into the target cells. Successful gene therapy depends upon the ability of the vector to deliver the nucleic acids to the target sites inside the cells to obtain sufficient levels of gene expression. The escape of DNA/RNA from endosomal compartments is considered a major obstacle. Therefore, the lipid and nanoparticles vectors are designed that can facilitate the release of the transgene from the endosomal compartment, and thus, the transgene can express itself inside the cytoplasm. Current chemical vectors designed for gene delivery systems are cationic lipids, also known as lipoplexes, and polyplexes polymeric nanoparticles. Unlike the physical methods, which can deliver all three types of CRISPR cargo, the chemical vectors generally deliver CRISPR cargo comprised of plasmid DNA encoding both the Cas9 protein and the guide RNA into the target cells.

7.8. Cationic lipids/lipoplexes

Cationic lipids are very important non-viral lipid vectors used in gene editing and gene therapy purposes because they can be easily synthesized and modified (Parker et al., 2003; Chen et al., 2016). The cationic lipids contain positively charged hydrophilic heads and hydrophobic tails. The positively charged head group of the lipid is able to bind with negatively charged phosphate groups in the plasmid DNA encoding gRNA and Cas9.

Also, cationic lipids are highly soluble in an aqueous solution, forming positively charged micellar structures termed liposomes. The liposomes, after binding with the plasmid DNA, form a uniquely compacted structure called lipoplexes *(Fig 7-4a)*. Lipoplexes, due to their positive charge, interact with negatively charged glycoproteins and proteoglycans of the cell membrane, which may facilitate cellular uptake of the plasmid DNA.

After the interaction of lipoplexes with the cell membrane, endocytosis occurs, which internalizes the lipoplexes. A portion of the cell membrane invaginates to engulf the lipoplexes to be internalized, which subsequently buds off into the cell and are taken up into the endosomes *(Fig 7-4b)*.

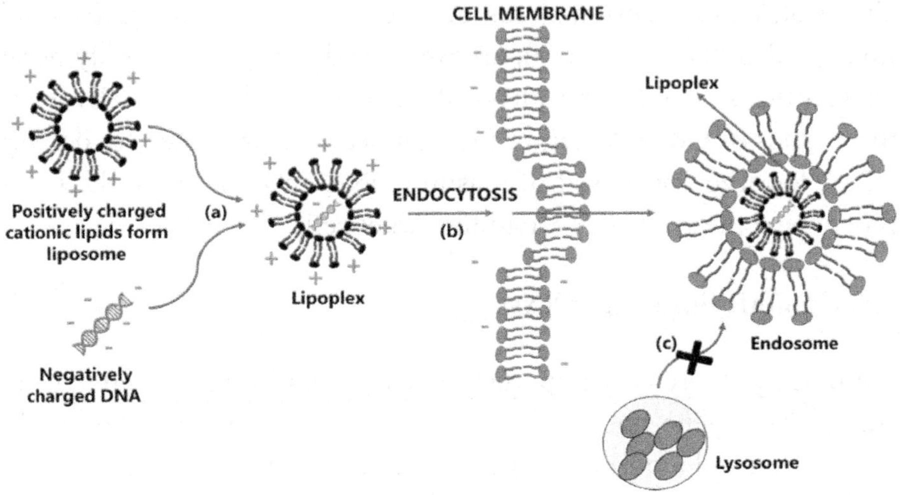

Fig 7-4: Endocytosis of lipoplexes

Endosomes containing the internalized lipoplexes have mild acidic pH. The natural pathway of endosomes is to undergo fusion with lysosomes. Lysosomes contain enzymes that can degrade the foreign substance that enters the cell. Thus definitely, fusion of the endosome with the lysosome is not expected to be productive for gene delivery because the lysosome enzymes can degrade the transgene *(Fig 7-4c)*.

Therefore, lipid vectors are to be designed in such a way that mediates the release of the plasmid DNA from the endosomal compartment into the cytoplasm. The lipoplexes mediate the release of the plasmid DNA in the cytoplasm by destabilizing the membrane of endosomes (Thapa, and Narain, 2016; Zhen and Li, 2020; Sousa et al., 2022).

Following endocytosis, the mixing of the liposomal cationic lipids and the endosomal anionic lipids makes the endosomal membrane destabilized. The destabilization of the endosomal membrane induces a flip-flop movement of negatively charged phospholipids from the cytoplasmic facing monolayer of the endosomal membrane into the inner monolayer *(Fig 7-5a)*.

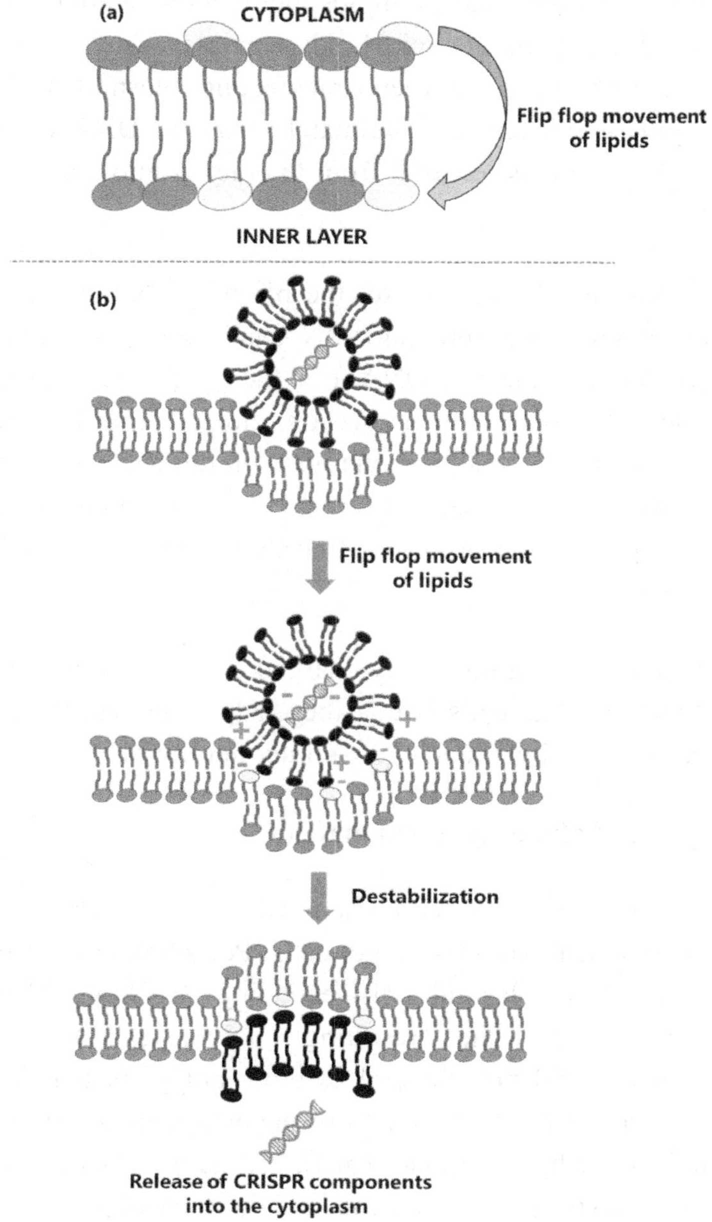

Fig 7-5: (a) Flip-flop movement of endosomal lipids and (b) release of CRISPR components into the cytoplasm

The negatively charged phospholipids then interact with a positively charged head group of cationic lipids (lipoplexes). Consequently, electrostatic interaction between transfecting plasmid DNA and cationic lipid is reduced, and ultimately plasmid DNA containing CRISPR components is released from lipoplexes into the cytoplasm *(Fig 7-5b)*.

Once released into the cytoplasm, the plasmid DNA moves to the nucleus to get transcribed into guide RNA and mRNA encoding Cas9. The transportation of plasmid DNA from the cytoplasm to the nucleus is favorable at the time of cell division, either during the breaking down of the nuclear envelope or via pores in the nuclear membrane. Import efficiency of the plasmid DNA via nuclear pore can also be enhanced by adding nuclear localization signal (NLS) sequence to the plasmid DNA.

The advantage of using positively charged lipids for transferring the plasmid DNA into the cells is that they help to protect the plasmid DNA against intracellular and extracellular nucleases.

7.9. Polyplexes Polymeric nanoparticles

Polyplexes are another non-viral chemical vectors for delivering the CRISPR components into the target cells. Polyplexes are formed by the self-assembly of plasmid DNA encoding gRNA and Cas9 with cationic polymers. The resulting polyplexes are usually nanoparticulate <100 nm. Several cationic polymers that have the ability to encapsulate DNA and form nanoparticles are poly-lysine (pLL) and polyethylene-imine (pEI). Targeting ligands can be incorporated into the polymers to mediate the entry of polyplexes into the target cells by binding the cognate receptors *(Fig 7-6)*. This can lead to increased levels of uptake of polyplexes into the receptor-positive cells, subsequently increasing transgene expression in the target cells. Polyplexes exhibit different mechanisms of releasing the plasmid DNA into the cytoplasm than lipoplexes.

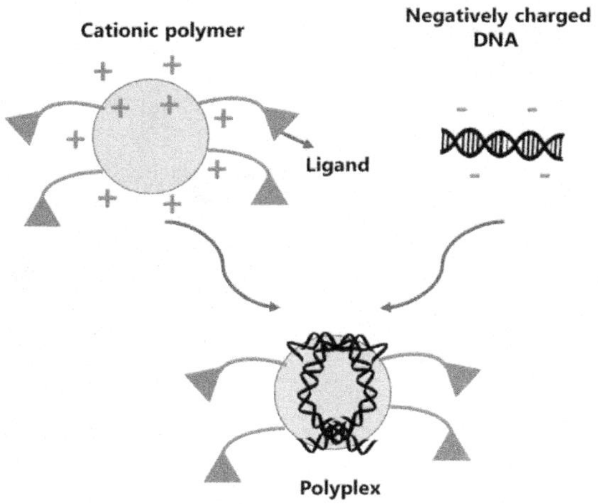

Fig 7-6: Formation of polyplex polymeric nanoparticle

For instance, when polyplexes made up of polymer polyethylene-imine undergo endocytosis and enter the endosome, which then eventually fuse with lysosomes. In the lysosomes, ATPase-mediated acidification occurs, which pumps protons (H+) into lysosomes, followed by a chloride influx to counterbalance the protons. Because of the high buffering capacity of polyethylene-imine and the increase of Cl− ion concentration, osmotic swelling occurs, resulting in the bursting of lysosomes in the cytoplasm. This is referred to as proton sponge *(Fig 7-7)*.

The released polyplexes then enter the cytoplasm, from which they are transported in the nucleus.

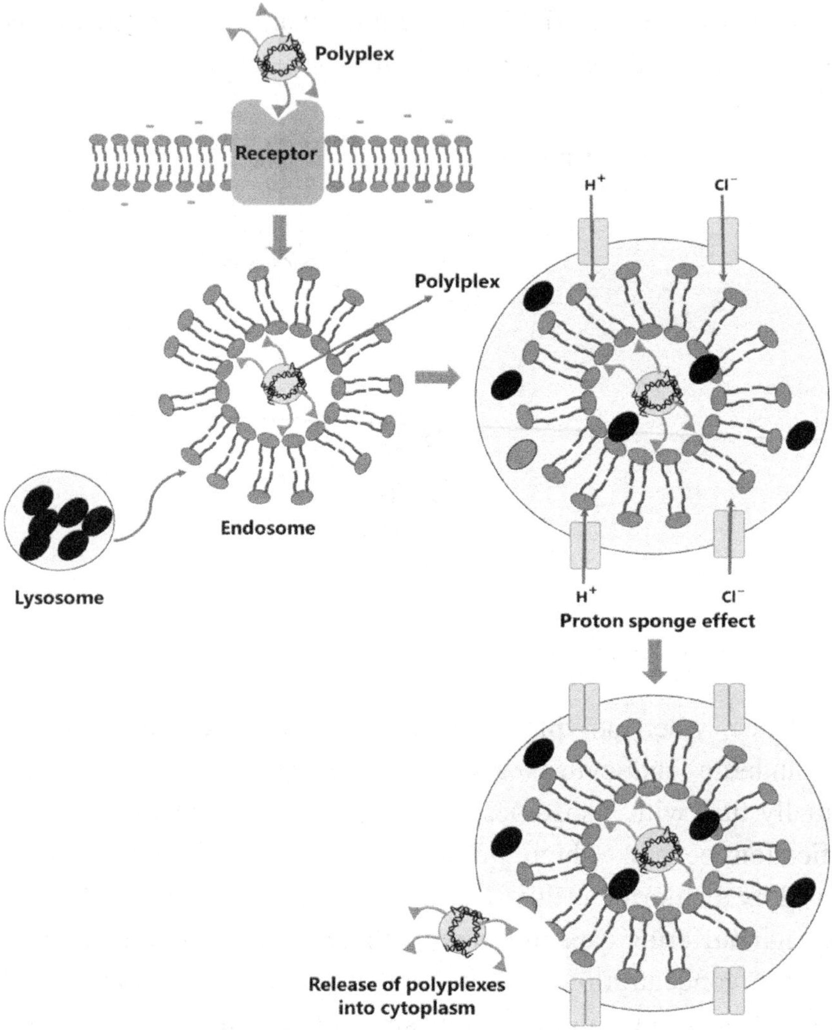

Fig 7-7: Release of polyplexes in the cytoplasm

In other polyplexes, which are made up of polymers like poly-lysine, the proton-sponge phenomenon does not occur. In these cases, they are required to be co-transfected with endosome-lytic agents to lyse the endosome. Thus when polyplexes enter the endosome, the endosome-lytic agents lyse the endosome, and the polyplexes enter the cytoplasm, where they dissociate to release the plasmid DNA encoding CRISPR components.

Then the plasmid DNA gets transported from the cytoplasm into the nucleus, where the CRISPR components gRNA and Cas9 are expressed and perform gene editing at the desired position in the genome.

One disadvantage of using synthetic polymers is their toxicity in humans. Therefore, researchers are exploring natural polymers, such as chitosan, dextran, and gelatin, to form polyplexes for gene therapy purposes. These polymers are biodegradable and exhibit low toxicity and immunogenic reactions in the body.

Apart from this, another important property of these nanoparticles is their ease of elimination, i.e., these nanoparticles do not accumulate inside the body after repeated administrations.

7.10. Magnetofection

Magnetofection is a highly efficient transfection method that utilizes a magnetic field to introduce the transgene into the target cells. Gene transfer uses magnetic nanoparticles made of iron oxide, which is fully biodegradable. The iron oxide nanoparticles are then coated with specific cationic lipids. We have already discussed cationic lipids are lipids that have a positive charge. The magnet nanoparticles coated with cationic lipids are then associated with the plasmid DNA containing the transgene through the electrostatic interaction. Next, this complex is delivered into the cell culture, and a magnet is placed underneath the petri dish containing the cell culture. Under the influence of an external magnetic field generated by a magnet, the magnetic particles associated with plasmid DNA containing the transgene are concentrated on the target cells (Hryhorowicz et al., 2019; Ghanbarlou et al., 2021). Then the target cells mediate the uptake of cationic lipids-coated nanoparticles associated with plasmid DNA through the process of endocytosis. The nanoparticles then fuse with the endosome, which contains acidic pH. The release of a plasmid-containing transgene into the cytoplasm from the endosome follows a similar principle as we discussed how cationic lipids mediate the release of the transgene into the cytoplasm.

The major advantage of chemical vectors and magnetofection-mediated gene transfer techniques is that membrane architecture and structure stay intact, in contrast to other physical transfection methods like electroporation, sonoporation, gene gun, etc., that damage the cell membrane in order to create pores in it. Magnetofection method can be used for both in vitro and in vivo gene therapy. Magnetofection has been broadly used in gene transfer for endothelial cells, tumors, kidneys, skin cells, etc.

However, several challenges remain for in vivo magnetofection, such as rapid systemic clearance of iron oxide particles because of the high accumulation of iron oxide in cells, especially with multi-dosing experiments, which can be toxic for cells.

Chapter 8: Therapeutic applications of CRISPR-Cas

8.1. CRISPR-Cas in adoptive T-cell Immunotherapy for cancer

Adoptive cell therapy, abbreviated as ACT, is a type of immunotherapy which seeks to harness immune cells, particularly T cells, for fighting cancer. Various adoptive cell therapies are:

(i) Tumor-infiltrating lymphocyte (TIL) therapy: It is one of the first types of ACT that is aimed at isolating naturally occurring T cells that have already infiltrated patients' tumors, then activating and expanding them in vitro. After the T cells have been activated in large numbers, they are reintroduced into the patient's body, where they can locate and eliminate cancerous cells *(Fig 8-1a)*.

(ii) Engineered TCR Therapy: The engineered T-cell receptor (TCR) therapy involves attaching new T cell receptors to the T cells' surfaces, enabling them to target specific cancer antigens. It is done by cloning the TCR genes of specific tumor-infiltrating lymphocytes and transferring them to the isolated T cells. The result is the engineered T cells with defined specificity for tumor cells *(Fig 8-1b)*.

However, the reliance of engineered TCR-T cells on major histocompatibility molecules (MHCs) to recognize tumor antigens is a major limitation of this therapy, as tumor cells are capable of downregulating MHC expression to evade immune surveillance.

(iii) CAR-T cell therapy: This technique involves equipping the patient's extracted T cells with a synthetic receptor known as chimeric antigen receptor, abbreviated as CAR. A key advantage of CARs is their ability to bind to cancer cells by targeting the naturally occurring antigens present on the surface of cancer cells without the need for antigen presentation by MHCs (Mollanoori et al., 2020; Al Saber et al., 2021).

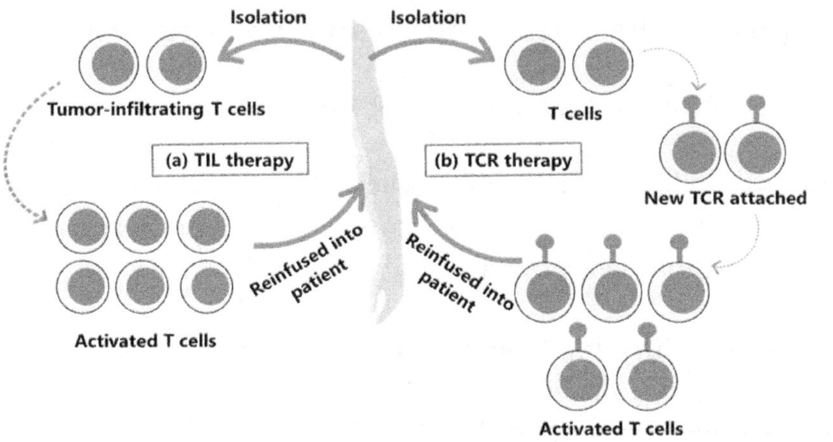

Fig 8-1: Adoptive cell therapies

Few FDA approved CAR-T cell therapies are:

- CD19-specific axicabtagene ciloleucel or Yescarta for the treatment of diffuse large B cell lymphoma (DLBCL)
- Tisagenlecleucel or Kymriah for the treatment of CD19+ diffuse large B-cell lymphoma and relapsed or refractory acute lymphoblastic leukemia (ALL)
- Brexucabtagene autoleucel or Tecartus for treating mantle cell lymphoma (MCL) and relapsed or refractory B-cell precursor acute lymphoblastic leukemia (ALL)
- Lisocabtagene maraleucel or Breyanzi for treating relapsed and refractory large B cell lymphoma.
- Idecabtagene vicleucel or Abecma for treating relapsed or refractory multiple myeloma.

Chimeric Antigen Receptor in CAR-T cells: Chimeric antigen receptor or CAR consists of 3 parts:

(i) an extracellular antigen recognition domain of the single-chain Fragment variant (scFv), derived from a monoclonal antibody molecule specific to a unique antigen present on the target tumor cells. The scFv is produced by fusing variable domains of the IgG antibody's heavy chain (V_H) and the light chain (V_L) through a short flexible peptide linker. Variable domains are chosen because they are responsible for forming an antigen-binding site. The scFv identifies the target antigen expressed on the surface of tumor cells without the need for antigen presentation by MHCs *(Fig 8-2)*.

(ii) a transmembrane domain, which is connected to the scFv through a flexible linker. The main function of the transmembrane domain is to anchor the CAR in the T cell membrane.

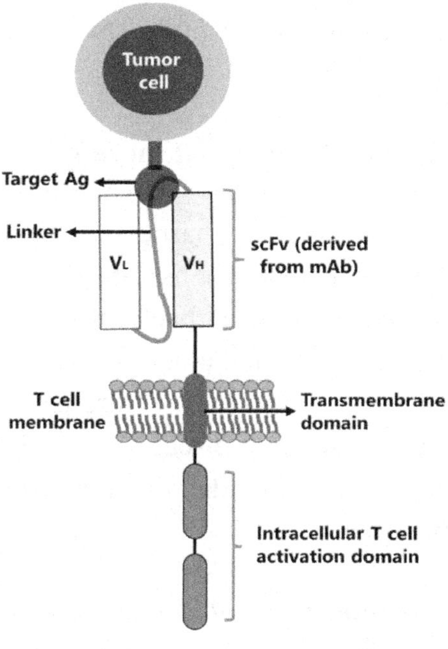

Fig 8-2: Structure of chimeric antigen receptor (CAR)

(iii) and an intracellular T cell activation domain that includes costimulatory molecules. After the CAR scFv binds to the tumor antigen, the intracellular activation domain initiates T cell proliferation and activation signaling, ultimately leading to the cytotoxicity of the tumor cells.

The gene editing tools such as Zinc finger nucleases (ZFN), transcription activator-like effector nucleases (TALEN), and CRISPR/Cas9 can be used to precisely insert the CAR construct into the correct position in the genome of the T cells so that it is expressed efficiently at sufficient levels to generate engineered CAR-T cells. However, as already discussed in previous chapters, CRISPR-Cas9 is easier to construct and a more efficient gene editing tool with few off-target effects. In addition, the previous studies also suggest CRISPR-Cas9 is a better tool for gene editing in T cells and can edit multiple genes simultaneously using several guide RNAs (Ghaffari et al., 2021; Quazi, 2022).

Autologous vs. allogenic CAR-T cell therapy

Autologous CAR-T cell therapy harnesses the power of T cells isolated from the cancer patient. The isolated T cells are modified to target cancer cells and then infused back into the same patient. This process takes several weeks to manufacture CAR-T cells, and a delay in treatment availability can have dire consequences, particularly in patients with highly proliferative diseases. Even in some conditions, lymphocyte repertoire is depleted due to which isolating T cells from the patient is not possible.

The challenges of autologous CAR T therapy can be overcomed by engineering T cells derived from healthy donors/volunteers instead of patients. These are also called "off-the-shelf" CAR T cells. Off-the-shelf or allogenic CAR-T cells can be made in advance and administered immediately when the patient needs. Unlike autologous therapies that directly treat one patient, allogeneic therapies can treat multiple patients.

But there are certain issues with allogenic CAR-T cell therapy as the patient's own immune system may recognize the transferred CAR-T cells as foreign, causing serious life-threatening reactions. This rejection is caused by the interaction of T cell receptors (TCR) and human leukocyte antigen class I (HLA I) that are expressed on both the CAR-T and patient's T cells.

The endogenous TCR on transferred CAR-T cells can identify alloantigens in HLA mismatched recipients, thereby leading to Graft-versus-host disease (GVHD), thus impeding the anti-tumor activity of CAR-T cells. On the other hand, if the patient's immune system detects foreign HLA molecules on the CAR-T cells, it can reject them. This can also interfere with the effectiveness of CAR-T cell therapy.

CRISPR in gene editing of "off the shell" allogenic CAR-T cells

There are different ways to avoid GVHD when designing allogeneic CAR-T cells, the most widely used strategy being the generation of CAR-T cells that abrogate the expression of endogenous TCRs. It can be done using genome editing tools CRISPR-Cas9 (Ureña-Bailénet al., 2020; Razeghian et al., 2021; Dimitri et al., 2022).

TCR is a membrane-bound protein complex that consists of α and β chains and is expressed on the surface of all T cells. The HLA class I molecules appear ubiquitously on the surface of cells throughout the body and consist of an α chain that is stabilized by β2-microglobulin (β2M). The allogeneic rejection arises when the TCR on transferred CAR-T cells recognizes the HLA I complex on the recipient's cells as foreign and attacks them. Similarly, the TCRs of the recipient's T cells recognize the HLA complex on donor CAR-T cells as foreign and attack them.

To remove these rejection barriers, disrupting the α or/and β chains of allogenic CAR-T cells TCR through CRISPR-Cas9 can prevent them from recognizing and attacking the recipient T cells. Similarly, the disruption of β2M through CRISPR-Cas9 can be used to prevent maturation and expression of HLA class I molecules on CAR-T cell surface. This will prevent the patient's own T cells from recognizing the donor allogenic CAR-T cells as foreign and attacking them.

CRISPR in disruption of Immune checkpoints

In addition to disrupting endogenous TCR and β2-microglobulin expression in CAR-T cells, disruption of immune checkpoint inhibitor genes is also required. In normal conditions, after cellular immunity is activated and has successfully neutralized the target, the immune checkpoints regulate T-cell over-activation to prevent them from attacking the cells indiscriminately. However, cancer cells use the similar ways to protect themselves from being attacked by T cells by stimulating immune checkpoints. When the checkpoint on the cancer cells and partner proteins on T cells bind together, they send an "off" signal to the T cells and prevent them from destroying the cancer cells. Cytotoxic T-lymphocyte-associated antigen 4 (CTLA-4) and PD-1 are key immune checkpoints. For instance, CTLA-4 expressed by activated T cells, after binding to the B7 molecule present on antigen-presenting/cancer cells, mediates immunosuppression by transmitting inhibitory signals to T-cells. PD-1 is another checkpoint protein present on T cells that usually acts as an "off switch" by attaching to a protein PD-L1 present on the other cells. The PD1-PDL1 interaction signals the activated T cells from not attacking those cells in the body and leaving them. Some cancer cells stimulate the expression of significant amounts of PD-L1, which helps them evade immune attacks. To counter this, drugs called therapeutic monoclonal antibodies that target PD-1 (nivolumab and pembrolizumab) and CTLA-4 (e.g., ipilimumab) are administered with CAR T-cell therapy for cancer treatment. Unfortunately, systemic blockade of immune checkpoints can also cause immune-related adverse effects throughout the body.

CRISPR-Cas9 technology can overcome this challenge by disrupting single or multiple immune checkpoint genes of allogenic CAR T-cells. Thus, avoiding the need to administrate antibodies that target immune checkpoints systemically (Xia et al., 2019; Wu and Cao, 2019; Ou et al., 2021). According to studies, CAR-T cells with disrupted immune checkpoint genes show stronger responses to PD-L1 and B7-expressing cancer cells, resulting in increased anti-tumor activity and prevention of cancer relapse.

An additional challenge faced by CAR-T cells is the **immunosuppressive tumor microenvironment** (TME) after infiltrating the solid tumor. TME is composed of various chemicals like transforming growth factor-beta (TGF-β), produced by TME and cancerous cells. TGF-β suppresses T-cell function by inhibiting T-cell activation, proliferation, and effector mechanisms. By knocking out the TGF-β receptor II gene in CAR-T cells using CRISPR-Cas9 technology, tumor eradication can become more efficient.

Thus, multiple genes are required to be knocked out to boost the therapeutic potential of CAR-T cells, and disrupting multiple genes is easier to do with CRISPR-Cas technology than with other gene editing tools like ZFNs and TALENs.

8.2. CRISPR-Cas9 as antiviral therapy

CRISPR-Cas9 tool can also act as antiviral therapy in treating viral infections like:

Human Immunodeficiency Virus or HIV: HIV belongs to the retrovirus family of viruses and stores its genetic information using single-stranded RNA. The outer shell of the virus contains 'glycoproteins' gp120 and gp41, which allow HIV to lock onto the CD4 receptor and another co-receptor, either CCR5 or CXCR4, present on the surface of CD4 T cells. Following fusion with the cell membrane, HIV genome RNA enters the T cell, where it is reverse transcribed into proviral DNA using the viral reverse transcriptase enzyme. And then, the integrase enzyme of the virus integrates the viral DNA into the DNA of the host cell. The integrated viral DNA provides the instructions to the host transcription machinery to drive transcription of HIV RNA, producing enough progeny virions that bud off from the cell, leading to binding and infecting other CD4 T cells.

Helper CD4 T cells play an essential role in the adaptive immune system by activating and recruiting other immune cells, including cytotoxic T cells, and promoting the formation of antibodies by B cells. Thus, by infecting and killing the helper T cells, HIV prevents the immune system from working properly and can lead to the development of acquired immune deficiency syndrome or AIDS, leaving the host vulnerable to other pathogens and cancerous cells.

Active antiretroviral therapy (ART), a combination of three or more drugs, can keep HIV replication under control but cannot eradicate the integrated HIV DNA in infected CD4+ T cells. An integrated HIV genome will not form viral particles and will not circulate in bloodstream as long as the ART effect is there. Thus, HIV even with long-term ART, HIV cannot be entirely eliminated from the body. ART can also have various side effects and must be taken for life to keep HIV infection under control.

(i) However CRISPR-Cas9 system can provide a permanent effective antiviral therapy to eliminate HIV infection completely by targeting different regions of the viral genome (Saayman et al., 2015; Xiao et al., 2019; Ghorbani et al., 2021; Kang et al., 2022). In vivo delivery of Cas9 and multiple gRNAs using vectors like Adenovirus-associated virus vector serotype 9 (AAV9) or lentiviral vectors, the gene editing tool can result in complete HIV eradication by excising the entire integrated viral DNA from the host DNA. Thus, impeding the chances of viral gene reactivation and virus replication in HIV-1 infected T cells.

CRISPR-based one-time treatment to cut HIV DNA from human cells **is under clinical trials**.

(ii) CRISPR-Cas9 tool can also be used to prevent new HIV infections by creating mutations in the genes encoding co-receptors CXCR4 and CXCR5 required for HIV entry into the CD4 T cells. According to studies, mutated co-receptors showed no obvious effects on cell viability but can have a significant effect in preventing helper T cells from HIV-infection when compared to unedited T cells.

Hepatitis B virus: Hepatitis B virus (HBV) is a dsDNA virus that infects hepatocytes causing acute or chronic infection of the liver, including cirrhosis and hepatocellular carcinoma. Upon infection, the partial dsDNA genome of HBV is converted into an episomal stable covalently closed circular DNA (cccDNA) form using the host DNA repair machinery. The cccDNA associates with host transcription factors and viral proteins to enable the viral gene expression and viral replication. The cccDNA molecule pose a major roadblock to HBV cure because current antiviral therapies can suppress HBV replication, but fail to disrupt cccDNA and integrated HBV DNA, which leads to the relapse of the disease after the cessation of the treatment.

In vitro and in vivo studies have shown that AAV vectors containing SaCas9 and multiple gRNAs can cleave the HBV cccDNA and integrated HBV DNA inside the infected cells, thus demonstrating the potential cure for HBV infection (Seeger and Sohn, 2014; Kennedy et al., 2015; Ramanan et al., 2015).

Herpes Virus: Of the more than 100 known species of herpesviruses, 9 are known to infect humans: herpes simplex virus types 1 and 2, which cause oral cold sores and genital herpes, respectively; varicella-zoster virus which causes chickenpox and shingles; cytomegalovirus which causes mononucleosis, Epstein-Barr virus which causes infectious mononucleosis and Burkitt lymphoma, human herpesvirus 6 (variants A and B), human herpesvirus 7, and Kaposi's sarcoma virus or human herpesvirus 8 which causes Kaposi's sarcoma in AIDS patients and a B virus.

Herpesviruses possess a double-stranded linear DNA genome. All herpesviruses have a latent period in their life cycle, during which the virus is "hidden," genes expressed during the lytic cycle are suppressed and do not produce virus progeny. The ability to establish a latent infection represents a key challenge in treating herpesvirus infections. Most antiviral therapies suppress the virus replication but are unable to eradicate the latent infection.

Genome editing using CRISPR-Cas9 technology may serve as a promising therapeutic strategy against the latent infection of the herpesvirus (van Diemen and Lebbink, 2017; Chen et al., 2018). Delivery of Cas9 and multiple gRNAs using appropriate vectors can disrupt the viral genome directly, thus eradicating the latent herpesviruses from the infected cells.

Other viruses: Other studies have also revealed the therapeutic applications of CRISPR-Cas9 in the treatment of numerous other viral diseases, including papillomavirus, various other oncogenic viruses responsible for causing cancer in humans, influenza virus, Severe Acute Respiratory Syndrome Coronavirus 2 (SARS-CoV-2), respiratory syncytial virus (RSV), human neurotropic polyomavirus, etc.

8.3. CRISPR-Cas9 in the treatment of genetic diseases

In the context of genetic diseases, CRISPR has the potential to be used to correct the underlying genetic defect that causes the disease. For example, in diseases caused by a single mutation in a specific gene, such as cystic fibrosis or sickle cell anemia, CRISPR could be used to correct the mutation in affected cells, potentially providing a cure for the disease.

However, using CRISPR to treat genetic diseases is still in the early stages of development and there are many technical and ethical challenges that need to be addressed. One challenge is ensuring that the CRISPR system only modifies the targeted gene and does not have unintended effects on other parts of the genome. Another challenge is delivering the CRISPR system to the appropriate cells in the body.

Despite these challenges, there have been some promising results in animal studies and early-stage clinical trials using CRISPR to treat genetic diseases. With continued research and development, CRISPR has the potential to revolutionize the treatment of genetic diseases.

Here are some examples of how CRISPR has been used in the treatment of genetic diseases:

Sickle cell anemia: Sickle cell anemia is a genetic disease caused by a mutation in the *HBB* gene, which codes for the beta-globin subunit of hemoglobin. The mutation is a single nucleotide substitution, where adenine (A) is replaced by thymine (T) at position 17 of the gene. This leads to substituting the amino acid glutamate with valine at position 6 of the beta-globin protein. The resulting mutated and abnormal hemoglobin S produced causes red blood cells to become rigid and sickle-shaped rather than the normal round and flexible shape. It interferes with oxygen binding and release. As a result, sickle cells are less efficient at delivering oxygen to tissues than normal red blood cells. Sickle-shaped red blood cells can also get stuck in small blood vessels, blocking blood flow and leading to tissue damage, pain, organ damage, and various other health problems.

CRISPR-mediated gene therapy for sickle cell anemia involves extracting the blood stem cells from patient's bone marrow. The extracted stem cells are then treated with CRISPR to edit the genetic mutation causing sickle cell disease (Dever et al., 2016; Cai et al.,

2018; Frangoul et al., 2021; Park and Bao, 2021). The corrected stem cells are then multiplied in the laboratory to create a large enough population of edited cells. Once a sufficient number of corrected stem cells are obtained, they are infused back into the patient's bloodstream through a vein with the hope that they will migrate to the bone marrow and produce healthy red blood cells. Thus, alleviating the symptoms of sickle cell anemia.

One of the major advantages of using CRISPR to treat sickle cell anemia is that the therapy can be developed using the patient's own blood stem cells, which avoids the risk of immune rejection.

In traditional gene therapy approaches, viral vectors are used to deliver the corrected gene into the patient's cells. However, this can lead to an immune response against the vector, which can limit the effectiveness of the therapy and potentially cause harmful side effects. But by using CRISPR to edit the patient's own stem cells, the risk of immune rejection is significantly reduced. The edited cells are genetically identical to the patient's own cells, which minimizes the risk of the immune system attacking them as foreign invaders.

Each patient's genetic mutation is unique, and using CRISPR to edit the patient's own cells allows for a more personalized approach to treatment, which could potentially lead to more effective and targeted therapies.

However, there are still several challenges that need to be addressed before CRISPR-based gene therapy can become a routine treatment for sickle cell anemia. One challenge is ensuring the edited cells are safe and gene editing machinery does not cause unintended mutations or other harmful effects. Another challenge is finding an efficient and effective way to deliver the edited cells to the patient's bone marrow, where they can repopulate the blood cell supply.

Despite these challenges, the success of preclinical studies in animal models suggests that CRISPR-based gene therapy offers a promising approach to treating genetic diseases like sickle cell anemia.

FDA has also approved several clinical trials for CRISPR-mediated sickle cell therapy in humans, and some patients have already received this treatment (Cyranoski, 2016; Ledford, 2020; Cereseto et al., 2021; Rautela et al., 2021). In early 2019, a patient named Victoria Gray became the first person in the United States to receive an experimental CRISPR-based therapy for sickle cell disease. The treatment, which was developed by CRISPR Therapeutics and Vertex Pharmaceuticals, involved using CRISPR to edit the patient's stem cells to produce healthy red blood cells.

Following the treatment, Ms. Gray experienced some side effects, including pain and nausea, but she also reported feeling significantly better overall. In the months that followed, her condition continued to improve, with fewer episodes of pain and fewer hospitalizations.

Ms. Gray's case is still considered an early-stage success, and more research is needed to determine the long-term safety and efficacy of CRISPR-based therapies for sickle cell disease. However, her experience suggests that this approach could be a viable treatment option for people with this debilitating condition.

Like sickle cell anemia, CRISPR can be used for the treatment of many other genetic diseases like cancer, Beta Thalassemia, neurodegenerative disorders, Huntington's disease, Duchenne Muscular Dystrophy, inherited retinal diseases, etc.

It's worth noting that while CRISPR-based therapies hold tremendous promise for treating genetic diseases, they are still in the early stages of development and face many challenges before they can become widely available. However, with continued research and clinical trials, we may see more and more patients benefitting from this exciting technology in the years to come.

Chapter 9: Ethical issues regarding CRISPR-mediated gene editing

CRISPR-mediated gene editing can be classified into two types: germline editing and somatic cell editing. Germline editing involves editing genes in the reproductive cells, such as sperm and egg cells, which means that any changes made would be inherited by future generations. Somatic cell editing involves editing genes in non-reproductive cells, such as cells in the liver or lungs, which means that any changes made would not be passed on to future generations.

Both types of gene editing raise ethical concerns, but germline editing is considered to be more controversial due to its potential long-term impact on future generations.

Here are some of the ethical concerns of gene editing:

(i) Safety: Gene editing techniques, such as CRISPR/Cas9, can cause undesired changes in the genome, in which the Cas9-gRNA complex can cut DNA at unintended locations in the genome, potentially causing mutations and altering genes that were not desired to be edited (Doudna and Charpentier, 2014; Cribbs and Perera, 2017; Shinwari et al., 2018; Brokowski and Adli, 2019). Another safety concern is the possibility of long-term effects on human health and the environment. While CRISPR has the potential to treat genetic diseases, there is still much that is unknown about the long-term effects of these treatments. It is still unclear how the edited genes will interact with other genes in the body and what will be the effects of these interactions over time. For example, editing the *CCR5* gene has been shown to be effective in preventing HIV infection, as the CCR5 protein is a co-receptor that HIV uses to enter immune cells. So editing this gene can potentially make cells resistant to HIV infection. But there are some concerns that deleting the *CCR5* gene could have unintended effects on the immune system, as this gene plays a role in immune function beyond its role as an HIV co-receptor. For example, a study published in Nature in

2019 found that deleting the *CCR5* gene in mice led to increased susceptibility to West Nile virus and reduced survival rates.

It is critical to understand that the consequences of altering the human genome in the germline are not just limited to the individual but can also affect future generations (Lanphier et al., 2015; Tang et al., 2017; Lander et al., 2019; Schleidgen et al., 2020). Undoubtedly, gene editing could potentially eliminate genetic diseases from a family's lineage, but it could also introduce unintended mutations or genetic changes that could have negative consequences for future generations.

Therefore, it is crucial to ensure that the safety of gene editing is thoroughly evaluated before it is used in humans. It is essential to take a cautious approach to gene editing and carefully consider the potential risks and consequences. As we continue to explore the possibilities of gene editing, it is essential to keep in mind that any decision to alter the germline DNA could have significant implications for future generations and the course of human history.

(ii) Equity: Equity is a concern because access to gene editing technology may not be available to everyone equally. There is a risk that technology could become a tool for the wealthy, leading to greater disparities between the rich and the poor. Gene editing could also exacerbate existing health disparities. For example, certain genetic diseases are more prevalent in certain populations, such as sickle cell anemia in people of African descent. The cost of gene editing technology is likely to be very high initially, and it may take time before the technology becomes more widely available and affordable. This means that if gene editing technology is only accessible to those who can afford it, it will prevent those in poorer regions or people from certain populations with a higher incidence of certain genetic diseases from having access to gene editing technology. Thus, widening the health disparities between different populations.

(iii) Genetic enhancements: Once we fully understand the genetic factors determining human health and performance, genetic enhancements will become another ethical concern. The gene editing of human embryos could potentially be used to enhance certain traits, such as intelligence or athletic ability. This raises concern that genetic enhancement could be used to create "designer babies," in which parents select certain genetic traits for their offspring. This raises serious ethical questions about the extent to which parents should be allowed to manipulate the genetic makeup of their children and whether it is ethical to manipulate genomes for non-medical purposes (Ishii, 2017; Enriquez, 2018; Baylis, 2019; Al Shakaki, 2022).

Additionally, the potential for genetic enhancements to create a societal divide between those who have access to gene editing technology and those who do not is a significant concern. If only certain segments of the population have access to genetic enhancements, it could lead to a new form of discrimination based on genetic traits. This would exacerbate existing societal inequalities based on factors such as socioeconomic status or race.

(iv) Consent: Gene editing also raises questions about informed consent before undergoing gene editing procedures. Patients need to fully understand the risks and benefits of gene editing and give their informed consent before undergoing the procedure. There is also a question of how consent can be obtained for future generations, who may not have the ability to give their own consent.

(v) Eugenics: CRISPR-mediated gene editing, a cutting-edge technology, also raises concerns about its potential for eugenics. Eugenics is a social and scientific movement that emerged in the early 20th century, aiming to improve the genetic quality of human populations through selective breeding and forced sterilization of individuals deemed "unfit" or "undesirable" due to disabilities, mental illness, or low intelligence. This movement was widely discredited and condemned after it was used by the Nazi regime in Germany to justify the genocide of millions of people, including Jews, homosexuals, and individuals with disabilities.

Nevertheless, germline CRISPR editing uses methods different from eugenic measures in the past, but there are concerns that eugenic ideologies and practices may reemerge. Some have argued that gene editing could be used for eugenic purposes, leading to the creation of a genetically homogenous society in which only certain traits or characteristics are valued. It may disallow people with genetic differences from coming into the world while simultaneously claiming to "improve" the human race by eliminating certain traits in future generations (Krishan et al., 2016; de Araujo, 2017; Garland-Thomson, 2020). This could result in discrimination and exclusion of individuals who do not meet certain genetic standards, leading to social inequalities and a lack of diversity if certain traits are edited out of the gene pool.

(vi) Moral and religious objections: Some people also have moral or religious objections to the use of human embryos for gene editing, arguing that it is playing with God or interfering with nature. These objections must be taken into account when considering the ethical implications of gene editing.

In summary, gene editing is a powerful technology that has the potential to improve human health, but it must be used responsibly and ethically. The ethical concerns raised by gene editing must be carefully considered and addressed before its widespread use in humans.

To address these concerns, certain steps are required to be taken:

(i) Educating the public: Effective communication between researchers and the public is critical to ensuring that everyone is on the same page when it comes to the use of gene editing technology. Scientists should make an effort to explain the science behind gene editing in a way that is accessible to the general public while also being transparent about the benefits, limitations, and potential risks associated with this technology. This could involve educational materials, public lectures, or engaging with the media to provide accurate and accessible information about gene editing technology.

At the same time, it's important for the public to be actively engaged in the discussion around gene editing. Providing feedback and input to researchers and policymakers can help ensure that the development and use of gene editing technology is aligned with the ethical values and priorities of the broader community. Public engagement can also help identify concerns and issues that scientists and policymakers may not have fully considered, enabling a more comprehensive and responsible approach to developing and using gene editing technology.

By working together and fostering open communication between scientists and the public, it can be ensured that gene editing technology is used in a responsible and ethical way that benefits society as a whole.

In fact, some researchers have already begun taking steps to engage with the public about gene editing. For example, the Broad Institute of MIT and Harvard launched a public dialogue initiative called the Human Gene Editing Initiative, which aims to encourage discussion and debate about the ethics of gene editing technology. The initiative has also published various resources, including background materials on gene editing technology, summaries of key ethical and policy issues, and reports on public engagement activities.

By engaging with the public in this way, researchers can help ensure that gene editing technology is developed and used in a way that reflects the values and priorities of the broader community.

(ii) Guidelines and regulations: Various guidelines and regulations have been issued by governments, international organizations, and scientific societies to provide guidance on the responsible use of this technology.

In the United States, the National Institutes of Health (NIH) released a policy in 2015 stating that it would not fund research that involves editing the genomes of human embryos. However, research on somatic cells (non-germline cells) is allowed.

In 2017, the National Academy of Sciences (NAS) and the National Academy of Medicine (NAM) released a report that provided recommendations for the use of gene editing technologies in humans, including the importance of informed consent, transparency, and continued monitoring of the technology.

The World Health Organization (WHO) established an Expert Advisory Committee on Developing Global Standards for Governance and Oversight of Human Genome Editing, which issued a report in 2019 that called for a global registry to track research on human genome editing and emphasized the importance of transparency and public engagement. The report also recommended that any use of genome editing in human embryos should only be done for the treatment or prevention of serious genetic diseases and not for genetic enhancement or non-medical purposes.

Scientific societies, such as the American Society of Gene and Cell Therapy and the International Society for Stem Cell Research (ISSCR), have issued guidelines and recommendations for the responsible use of genome editing technologies in research. The guidelines also stated that any use of genome editing in human embryos should be subject to rigorous oversight and conducted only for compelling reasons and under strict regulatory and ethical standards. The ISSCR also recommended that germline editing should not be used for clinical purposes until safety and efficacy concerns have been addressed.

It's important to note that guidelines and regulations related to CRISPR gene editing are still evolving and subject to change as new information and technologies become available. While the specific recommendations and requirements may vary depending on the country or organization, the overall goal is to balance the potential benefits of these technologies with the need to ensure safety, protect human dignity, and respect ethical principles.

Chapter 10: CRISPR-based gene drive

10.1. What are gene drives?

Gene drives are self-propagating genetic elements that rapidly promote the inheritance and spread of the desired gene variant throughout a population. In the traditional Mendelian inheritance—each allele has a 50% chance of being passed on to an offspring *(Fig 10-1a)*. But through gene drives, a desired genetic variant has a >50% chance of being passed onto the offspring and thus can be spread through a population faster *(Fig 10-1b)*.

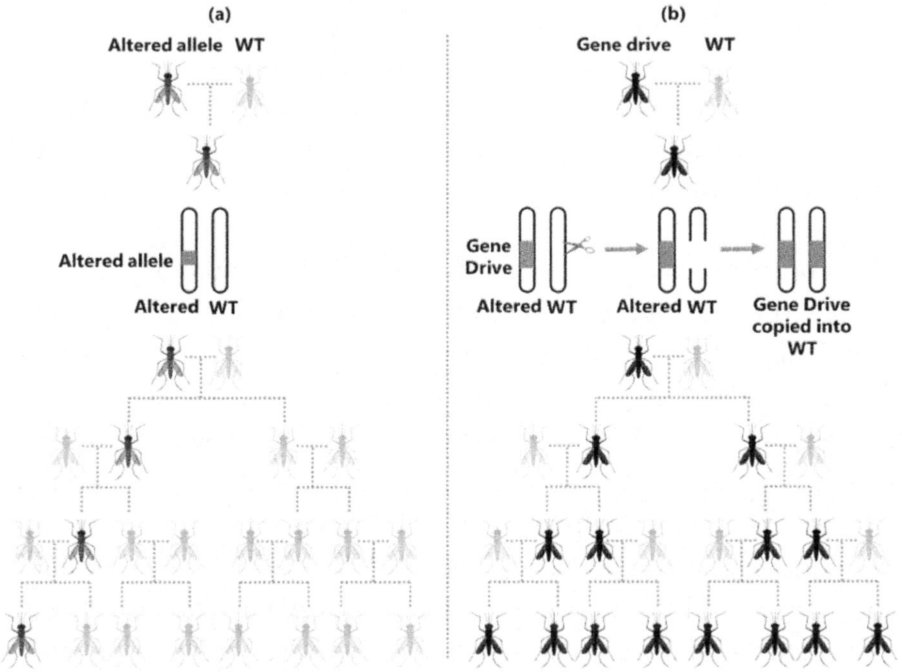

Fig 10-1: (a) Traditional Mendelian inheritance and (b) Gene drive Inheritance

Gene drives contain molecular tool that targets other versions of the target gene, cut it out, and replaces it with DNA using the gene drive as a template, resulting in the homozygous presence of this modification. This means the gene variant will be present in both copies of the chromosome pair and is inherited in 100% of the offsprings. Over successive generations, that gene variant spreads through a population until most individuals possess it.

This strategy can be used to spread a disadvantageous trait, such as sterility, to an organism. Thus, it presents potential solutions for various issues, including eradicating or altering disease vectors (such as mosquitoes), controlling invasive species of insects or mammals, and combating pesticide resistance.

10.2. Integration of gene drive technology with CRISPR

CRISPR gene drive includes the desired version of the target gene to be propagated and the additional DNA sequence for expressing Cas9 and guide RNA.

The process of releasing a gene drive into a wild population starts with constructing transgenic organisms in the laboratory. A plasmid containing sequences of Cas9, gRNA, and desired allele, which are flanked by DNA sequences homologous to the gRNA target site, is expressed in the target embryo *(Fig 10-2)*. The wild-type (WT) sequence on one chromosome is replaced with the gene drive, containing CRISPR components (guide RNA and Cas9 genes) and the cargo gene (desired allele). A guide RNA binds Cas9 and directs it to bind and cleave DNA at complementary sites 20 nucleotides in length. The double-stranded breaks generated are repaired either by the error-prone non-homologous end-joining or by homology-directed repair (HDR). HDR-mediated gene repair results in the copying of the gene drive cassette from the plasmid into the genomic cut site generating heterozygotes (Esvelt et al., 2014; Dury et al., 2017; Noble et al., 2017; Tanaka et al., 2017).

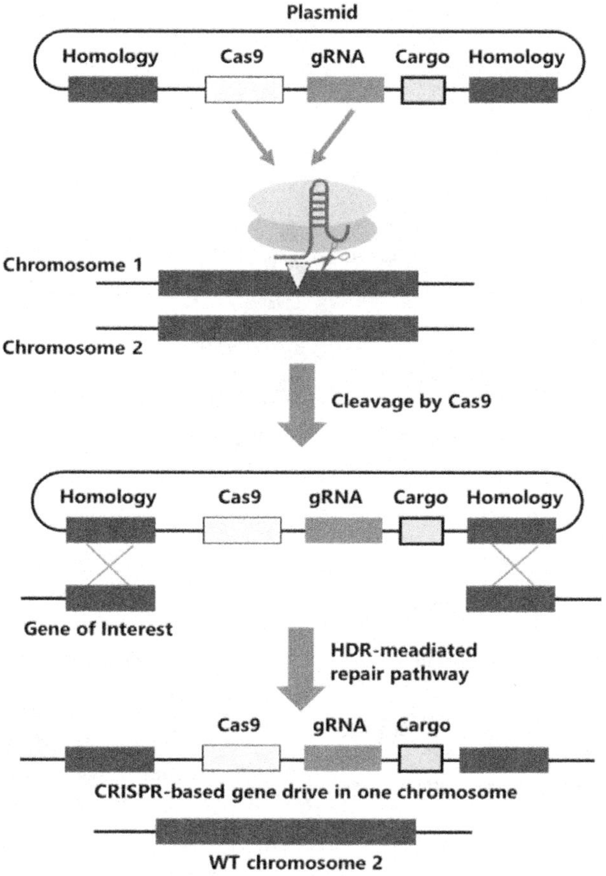

Fig 10-2: Insertion of CRISPR-based gene drive into the genome

Because a gene drive includes instructions to build more CRISPR proteins, it constantly edits the genome. Once inserted into the genome, the Cas9 + gRNA cassette is expressed and mediates cleavage of the homologous chromosome in the germline and is then copied into the DNA break by HDR pathway using gene drive-containing chromosome as the template, thus completely replacing the wild-type DNA sequence at this position in the genome *(Fig 10-3)*. Thus, resulting in the homozygous transgenic organism containing gene drive in both chromosomes.

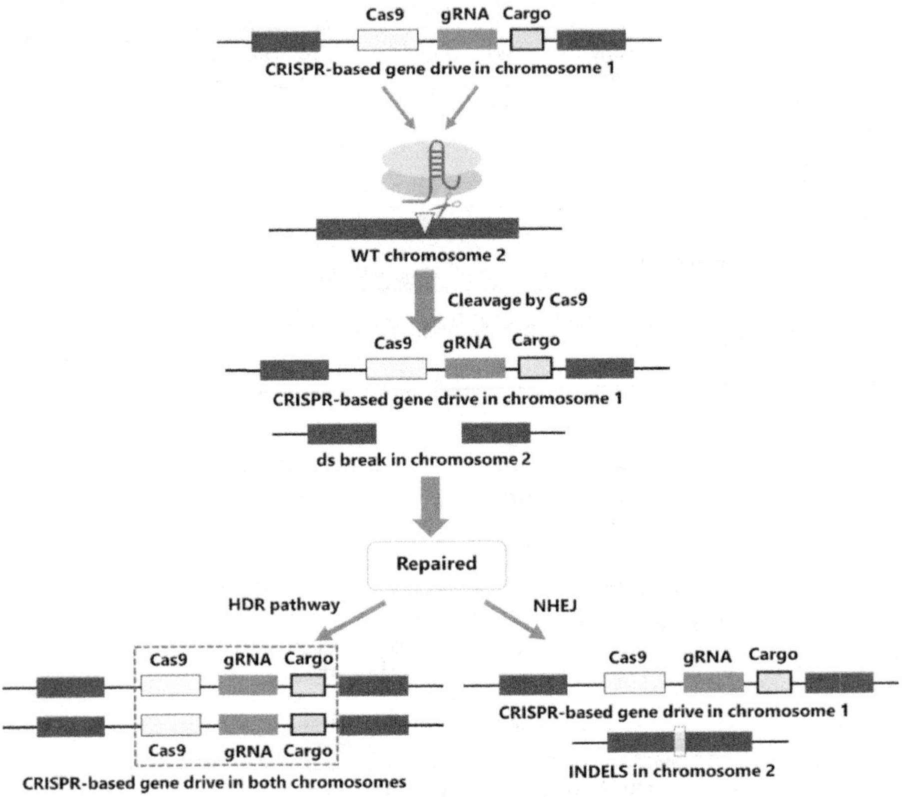

Fig 10-3: Insertion of CRISPR-based gene drives into both chromosomes by HDR-mediated repair pathway

The transgenic organism is then released into the wild to breed with wild-type individuals. During the early development, gene drive (obtained from the transgenic parent) will cut the partner chromosome (obtained from WT parent) by Cas9 and gRNA and gets copied into the WT chromosome. Thus, the heterozygous allele in the embryo gets converted into the homozygous allele with a gene drive construct *(Fig 10-4)*. By using gene drive technology, the desired trait sequence is transmitted to offspring, resulting in the replacement of the natural version of the target gene with two copies of the gene drive. This guarantees that the change is passed on to nearly all of the offspring, rather than the typical 50% in traditional Mendelian inheritance, allowing it to spread rapidly through the population over time.

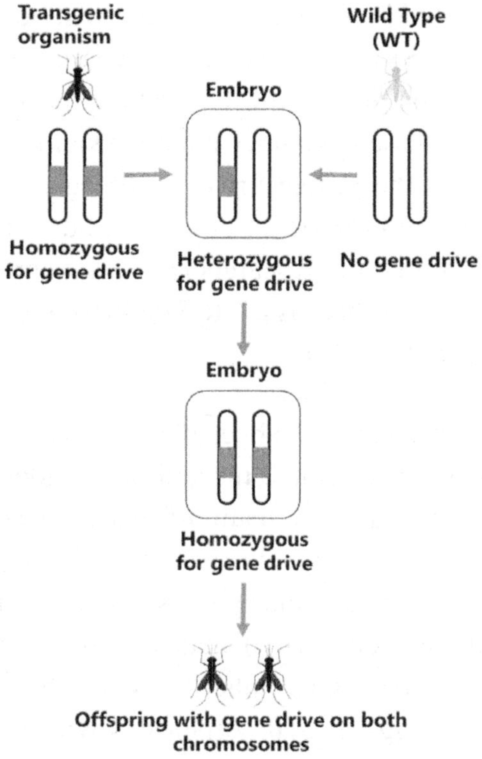

Fig 10-4: Gene drive with desired trait sequence is passed on to offspring

10.3. Applications of CRISPR drives

CRISPR gene drives can address some real-world problems, like:

(i) Eradicating malaria: One of the most common uses of gene drives is centered on eradicating vector-borne diseases like malaria, a disease that impacts millions of individuals worldwide every year. It is caused by *Plasmodium*, a parasite that multiplies within and spreads through female *Anopheles* mosquitoes, which serve as the vectors for transmission. Considerable effort has been devoted to creating vaccines that attack the malaria parasite. Nonetheless, due to the parasite's varying life and replication phases within mosquitoes and humans, it is improbable that such a vaccine will be accessible in the near future. Thus, the central focus on preventing malaria spread is through using insecticides to kill and reduce the breeding sites for mosquitoes. But, malaria rates have not been significantly reduced in recent years, one of the primary reasons being the development of insecticide-resistant mosquitoes. Complete eradication of malaria is far from sight. Therefore, researchers have been pursuing other possibilities, and one of them is a CRISPR-based gene drive (Burt et al., 2018; Liebert et al., 2021).

Several gene drive systems have been developed to control mosquito-borne diseases using gene-edited mosquitoes. Under the guidance of Professor Andrea Crisanti, a team of researchers from Imperial College London developed a gene drive that disrupts a fertility gene called *doublesex* in female mosquitoes. When a transgenic male mosquito (containing gene drive) mates with an unaltered female, Cas9 snips the *doublesex* gene inherited from the female mosquito inside the fertilized egg. Female offsprings with homozygous gene drives and disrupted *doublesex* gene cannot bite human skin, becomes infertile, and can not lay eggs (Simoni et al., 2020). Unable to bite, those female mosquitoes can't transmit malaria-causing parasites from their bodies to humans. And the inability of female mosquitoes to lay eggs will lead to the collapse of insect populations after a few generations. The gene drive can reach 100% prevalence within 8-12 generations, to the point of total population collapse within the laboratory environment. The introduction of these genetically modified mosquitoes into the natural environment could result in the complete eradication of the mosquito species, thus increasing the potential for malaria eradication. Thus, this technique has a high probability of aiding in the eradication of malaria.

(ii) Controlling invasive species: The rise in global trade has resulted in the transfer of goods and commodities across the planet, which has led to the migration of species from their original habitats to new areas, where they may become invasive. The introduction of invasive species can lead to adverse economic or environmental effects, as well as harm to human health. About 80% of recorded extinctions have occurred on islands, and 40% of the world's endangered and threatened species are island dwellers. The leading causes of these extinctions are the invasive rodents such as mice and rats that have successfully invaded nearly every landmass across the globe. They are particularly detrimental to island ecosystems, where they cause significant damage to endemic flora and fauna. For instance, black rats in New Zealand have eradicated native plants and negatively affected several insects, snails, spiders, reptiles, and bird species. Current methods to control invasive rodent populations via poisoning, trapping, or habitat management have produced considerable gains but also have significant limitations. The use of anticoagluant toxicants, currently the main method to control populations, often results in the killing of non-target species, and the slow internal bleeding induced by the toxin raises major animal welfare concerns. Moreover, toxicants also pose a risk to human health on inhabited islands. Overall, the negative impacts of invasive rodents result in the loss of biodiversity.

Gene drives can provide a solution to eradicate invasive rodents by inducing female infertility. A rat genome could be modified to contain a gene variant responsible for female fertility, and introducing some of these modified rats in an island. They would mate with wild rats, and their female offspring would inherit that variant gene responsible for inducing infertility. The male offspring will mate with other female wild rats, and all of their offspring will inherit that variant gene. As a result, gene drive spreads from one generation to another, and with time, the entire female population would become infertile, and the rats would die out.

(iii) Protecting endangered species: While gene drives are most commonly used to control or decrease the population of target organisms, they may also be used to help save species that are in danger of becoming extinct. Through modification gene drives, a protective gene could be spread. For instance, frogs and other amphibians worldwide are under severe threat from pathogenic chytrid fungus. A gene preventing fungal infections could potentially save many frogs and other amphibians from extinction.

Adding useful genes to endangered plants, such as drought tolerance or disease-resistance genes, can potentially ensure their long-term survival.

10.4. Safety, Ethics, and Effectiveness of CRISPR gene drive

The application of gene drive technology to eradicate a species can have direct effects such as the loss of a species, or indirect effects that could lead to unforeseen impacts on human and environmental health. Such consequences raise a host of ethical concerns related to the use of gene drives for species eradication.

(i) Significance of a species: Biodiversity, also known as biological diversity, encompasses the diversity of life in all its forms, ranging from genes and species to populations and entire ecosystems. The gene drive technique, if employed, may result in the extinction of a species, thereby leading to a decline in biodiversity. Despite this, the conservation of biodiversity is a critical objective of numerous organizations and governments, such as the World Wildlife Fund and the European Commission. Biodiversity is highly valued as it plays a crucial role in providing food, purifying water, stabilizing the earth's climate, and has aesthetic benefits, among other things. However, this raises the question of whether all species hold value and merit protection.

Some organisms, such as viruses, bacteria, and other pathogens, are often excluded from conservation programs despite their high extinction rates. While efforts have been made to eradicate certain disease-causing pathogens that pose a threat to humans, such as the smallpox virus, the guinea worm parasite, and the SARS-CoV-2 virus, the elimination of the malaria-causing *Plasmodium* parasite is not yet feasible. As a result, preventing malaria by targeting mosquitoes has been the primary focus. The importance of the conservation of mosquito populations has yet to be clearly defined.

The public's response to eradicating a species is primarily determined by the organism's moral status and the potential consequences that may arise from its elimination. Mosquitoes, as a species, are often considered to have little or no moral value. However, there is a distinction between killing a few mosquitoes with insecticides and deliberately using gene drive technology to eliminate an entire species of mosquitoes.

In terms of measuring biodiversity, it is recognized that malaria is caused by a small subset of mosquitoes, specifically 30-40 *Anopheles* species, out of over 3500 mosquito species. Thus, eradicating thirty of these *Anopheles* species would lead to only a 1% reduction in the overall biodiversity of mosquitoes. Additionally, the gene drive technique targeting the *doublesex* gene was utilized on the *Anopheles gambiae* species, which means that eliminating this species would have minimal effect on the total biodiversity of mosquitoes (Hammond and Galizi, 2017; Annoni and Pievani , 2021; Wise and Borry, 2022).

The selective protection or eradication of certain species might lead to programs for the elimination of other species with a lower status, which is a cause for concern. For example, while eradicating a mosquito species that transmits malaria may seem justifiable, it could provide a basis for the eradication of other mosquito species responsible for transmitting diseases like dengue, yellow fever, or Zika. Moreover, allowing the eradication of certain species for the sake of human lives may open the door to further justifications for the elimination of other species.

The use of gene drives to control invasive mammalian species is another controversial area that is being explored. Some groups have proposed using gene drives to eradicate invasive rodent species from islands, where they cause substantial ecological harm. A different group has suggested the use of gene drive technology to manage the large population of feral cats that hunt and pose a threat to indigenous and at-risk species in the area.

Therefore, it is crucial to consider the role of a species and explore alternative options for eradicating diseases caused by that species. The elimination of a species should only be considered if it saves a significant number of human lives and does not harm the ecosystem, or **if it helps protect the ecosystem**. It is crucial to carefully evaluate these factors before resorting to gene drives to eliminate a species, as eradication should be the last option, not an initial approach.

(ii) Ecological consequences: Eradication of an entire species can also have ecological consequences.

Consequences of species eradication: Insects, including mosquitoes, not only serve as a food source for other species but also play a crucial role in pollinating food crops and plant species. Therefore, assessing the losses a species' elimination may cause is essential. Though it is uncertain how much *Anopheles gambiae* mosquitoes contribute to pollination, studies suggest they do not occupy a significant role in the food cycle. However, determining losses can be difficult due to the intricate interrelationships within an ecosystem. It is also essential to consider whether other species can replace the functions of the extinct species to mitigate the negative effects of its eradication. Even though most researchers believe that eradicating *A. gambiae* mosquitoes would not significantly affect the ecosystem, field trials are crucial to understanding their role in the food web, pollination, and the ecosystem before releasing genetically modified mosquitoes.

Consequences of gene drives: One more concern is the potential spread of gene drives to unintended species. This could happen if crossbreeding or horizontal gene transfer takes place between the genetically modified *A. gambiae* and closely related species of mosquitoes. It is crucial to implement control strategies to prevent the gene drives from spreading to non-target species, as it could lead to the extinction or significant alteration of such species, endangering the food webs and ecosystems that depend on them.

Furthermore, there are situations where the Cas9 enzyme cuts the target site, but instead of being repaired by HDR-mediated pathway, it is repaired through a non-homologous end-joining (NHEJ) pathway. This process may produce **gene drive resistant** insertion/deletion (indel) alleles. Additionally, the Cas-9 enzyme and gRNA may unintentionally cause mutations in non-targeted areas of the DNA, leading to unpredictable effects.

Another point of concern is that a gene drive created for the purpose of managing invasive mammalian species may potentially spread beyond its intended geographical area and cause legal complications (Champer et al., 2016; Marshall and Akbari, 2018; Noble et al., 2018). For instance, the elimination of invasive black rats in New Zealand would be advantageous, but if a gene drive rat escapes to another location, intentionally or not, the consequences could be highly detrimental.

Other possibilities must also be taken into account, such as the accumulation of mutations in the parasite to enable it to find a new vector organism in case its original vector is eliminated.

Achieving a balance between maximizing the benefits and minimizing the risks of gene editing technology in public health is challenging. All interventions come with a certain level of risk, and therefore a risk assessment must be conducted. However, the ethical question that arises is whether it is acceptable to prioritize public health over the environment. The negative effects of environmental damage on public health should also be taken into consideration. Hence, the decision about the acceptable level of impact must be made carefully, considering ethical concerns.

Chapter 11: CRISPR in agriculture

With precise genome editing, CRISPR has the potential to resolve major challenges to crop improvement.

11.1. Traditional Selective breeding

Selective plant breeding of crops began thousands of years ago, even when anyone knew about DNA. It is the process of handpicking the best and most desirable traits from different plants and mating them together to create offspring with those desirable traits. The continuous accumulation of spontaneous mutations in the genome is the foundation of genetic variation for species evolution. Mutations cause changes in the DNA of an organism that may result in the natural emergence of plant variants with desired and beneficial traits like higher yield, nutritional content, shelf life, shortened growing seasons, taste less bitter, increased resistance to diseases and pests, can withstand harsh climate conditions, etc. The crops with desired characteristics are selected and bred together. Pollens with the genes for the desired trait from plants of one crop variety are transferred to the flowers of another variety with other desirable traits. From the offspring produced, the farmers again choose those with the best combination of desired traits and breed them together. The process is repeated over many generations to produce the offspring plants that inherit those desirable characteristics. For example, let's say a farmer wants to grow a crop that produces bigger and sweeter apples. They would start by selecting the apple tree that produces the sweetest apples and then breed it with another tree that produces bigger apples but maybe not as sweet. By cross-breeding these two trees, the farmer can create offspring that inherit the desired size and sweetness traits from both parents.

Our food crops today are, in fact, very different from their wild counterparts. Using selective breeding, farmers have converted wild plants into the crops we rely on today. For example, corn is descended from a wild grass called teosinte, which has small, hard kernels that are not edible. But it has been selectively bred by farmers over thousands of years to produce the much larger, sweeter, and tender corn we enjoy today. Similarly, broccoli, cauliflower, and cabbage all descended from the same wild mustard plant, but through selective breeding, farmers have created distinct varieties with different shapes and flavors. Selective breeding has also been used to create new varieties of other crops, such as wheat, rice, soybeans, and potatoes. Without selective breeding, our modern food supply would look very different, and we would not have access to the wide variety of crops that we have today.

11.2. Mutagenesis/Mutation Breeding

In the beginning of the 20th century, plant breeders and researchers learned that with a process called mutagenesis, they could make mutations happen faster than would occur naturally (Oladosu et al., 2016; Jankowicz-Cieslak et al., 2017; Shcherban, 2022). Physical methods like UV, X-ray, gamma radiation, fast neutron, or chemicals like methyl-N-nitrosourea, hydrogen fluoride, sodium azide, methyl methanesulfonate, or ethyl methanesulfonate can be used to induce mutations in the plant's DNA. Plant seeds, seedlings, whole plants, or only parts of the plant— pollen, spores, stem of a plant, etc. can be irradiated or given chemical treatment. If the cell's own repair mechanism does not eliminate the resulting mutations, a heritable mutation is generated. During this process, plant breeders grow, screen, and evaluate each plant from each seed produced. The plants with promising mutations are bred further with the wild plants until the breeders arrive at a substantially improved plant variant that possesses new and desirable characteristics.

More than 3000 desirable crops (including wheat, barley, soybeans, rice, tomatoes, grapefruit, and many fruits) have been developed using radiation mutagenesis in the last 100 years.

Limitations: But there are various limitations of traditional selective breeding and mutation breeding.

(i) Because the mutations induced are random, some may occur in functional genes, making them non-functional. Therefore, a great deal of effort is required to separate undesirable traits from desirable ones, which is sometimes not economically practical. Selecting desirable alleles and their combinations is a long-term process that involves crossing with wild genotypes and cultivating necessary ones for several generations.

(ii) During breeding, DNA from the selected parents may recombine randomly, and desirable traits, such as pest resistance, may get bundled with undesirable traits, such as lower yield or poor quality.

(iii) Additionally, traditional breeding techniques are time-consuming, often take decades to produce new viable crop varieties, and are labor-intensive.

11.3. Transgenic plants

The second method for obtaining desired traits lies with genetic engineering and transgenesis. The advantage, compared to induced mutagenesis, is that it allows for obtaining a desired trait through the introduction of a foreign gene (called a transgene) into an organism, which significantly reduces the time required to obtain a genetically modified organism/plant (GMO).

A transgenic crop variety is developed by:

(i) Identifying and isolating a desired gene that confers the trait of interest, for example, a gene responsible for disease or pest resistance, herbicide tolerance, or increased yield.

(ii) The desired gene is cloned into the carrier or vector plasmid.

(iii) The vector containing desired DNA is then introduced into the target plant cells using various techniques such as Agrobacterium-mediated transformation or gene gun.

(iv) After delivering, the foreign gene integrates into the plant's DNA and is expressed, resulting in the desired trait.

The modified plants obtained usually contain some DNA sequences from other organisms, sometimes referred to as “foreign DNA” that confer them the desired characteristics. An organism/plant with DNA from another organism is considered **transgenic**. This technique has widened the range of the modification of plants because transgenes could provide those traits to a recipient plant that could not be obtained by mutagenesis, like the synthesis of pharmaceuticals, insecticides, and other agents in plants (Schuler et al., 1998; Groot and Dicke, 2002; Prins et al., 2008; Ahmad et al., 2012). For instance, insect-resistant plants contain genes from the bacteria *Bacillus thuringiensis* (“Bt genes”) that, when expressed, produce a toxin. The toxin or natural insecticide protects crops from pests but is not harmful to humans. When growing Bt crops, farmers no longer need to use synthetic pesticides, which can be costly, toxic to inhale or handle, and pollute nearby soil and water.

Other examples of transgenic plants are Golden Rice which has a high content of β-carotene, transgenic soybean, etc.

Regulations: Although years of scientific research show that transgenic crops currently available are safe to eat and cause no harm to human health (Ubalua, 2009; Rani and Usha, 2013; Tarafdar et al., 2014). But they have struggled to gain popularity because of certain concerns:

(i) GMOs may present a certain danger for ecosystems, e. g., how can we be certain that the transgenes in herbicide-resistant plants will not be transferred to weeds by pollens while cross-hybridization?

(ii) The long-term consequences of transgenesis remain unclear; if a transgene gets integrated into different regions of a genome and affects other genes' expression?

(iii) There is also a risk that a transgenic plant can affect non-target organisms, such as plants possessing BT-toxin genes can kill non-hazardous insects.

(iv) **Resistance development:** Pests can develop resistance to the toxin produced by transgenic plants, making them less effective over time. This has already been seen in some transgenic crops that produce insecticides, such as Bt cotton, where some pests have developed resistance to the toxin.

(v) Some studies have raised concerns that consuming transgenic crops could have negative health effects on humans or animals, although most studies have found no evidence of harm. For example, some researchers have raised concerns that using antibiotic resistance genes in transgenic crops could lead to the development of antibiotic-resistant bacteria in humans. Another concern is the potential for allergic reactions in people who consume transgenic crops. For example, if a gene from a common allergen is introduced into a crop, it could potentially cause an allergic reaction in people who consume that crop. Despite these concerns, the **vast majority of studies have found no evidence of harm from consuming transgenic crops**.

To address these concerns, many countries regulate transgenic plants more strictly than non-transgenic plants. Therefore, transgenic plant producers must spend time and money on research to show the public and government that their products are safe for the ecosystem.

(vi) Apart from environmental and health-risk issues, there are several **Intellectual property issues**. The companies that develop transgenic crops often hold patents on the technology used to create them. They have exclusive rights to produce and distribute the seeds. This means that farmers who want to plant transgenic crops must purchase the seeds from the company that holds the patent, and they may be required to sign a contract that limits their ability to save seeds for future planting or to share the technology with others.

11.4. Genome editing of plants by CRISP/Cas9

With the recent advents in genome editing approaches like ZFNs, TALENs, and CRISPR-Cas9 technology, it is possible to precisely generate targeted mutations in the target gene. Whereas, as in the case of conventional mutagenesis, the random and high load of off-target mutations are generated in the plant's DNA. Thus, if already known which gene is required to be mutated to achieve an improvement, genome editing can massively speed up the process of producing plants with desired traits in the first generation.

Also, genome editing plant breeding technologies differ from transgenic methods in various ways: Genome editing technologies are used for precise genetic manipulation without introducing exogenous DNA such as antibiotic/virus-resistant genes. Thus, eliminating the fear that foreign DNA may be present in the target crop. For example, many viruses need plant proteins to grow, multiply, and spread. Researchers can use genome editing tools to break or change the plant proteins a particular virus likes to take advantage of. As a result, the viruses won't be able to spread within the edited plants, and thus the plants will become virus resistant. On the other hand, transgenic crop production requires the insertion of foreign DNA to confer virus resistance in the target plant.

Therefore, regulators place less stringent regulations on these non-transgenic, genome-edited plants because they have DNA sequences that could occur naturally in their wild counterparts.

Out of all genome editing tools, CRISPR/Cas9 technique has several advantages over ZFNs and TALENs. It is easy to use, less laborious to make gene editing components, more efficient, cost-effective, has fewer off-target effects, and has the ability to target multiple genes simultaneously, which makes them suitable for editing genomes of numerous polyploid species of crops (Demirci et al., 2018; Wada et al., 2020; Saraswat and Ranjan, 2021; Camerlengo et al., 2022).

11.5. Transgene-free genome editing in plants

The CRISPR-Cas9 system can be used to generate transgene-free genome edited plants. To make the plants completely transgenic-free, the transforming agent is not a vector or plasmid but a ready-to-use complex including Cas9 and a gRNA (Woo et al., 2015; Chandrasekaran et al., 2016; Svitashev et al., 2016; Gu and Zhang, 2021). Here is a general overview of the steps involved:

(i) Design and synthesize guide RNAs (gRNAs): The first step is to design and synthesize gRNAs that will target the specific DNA sequence to be edited. The gRNAs are typically designed to be specific to the target sequence to avoid off-target effects.

(ii) Delivery of CRISPR-Cas9 components: The CRISPR-Cas9 components, including the Cas9 protein and the gRNA, need to be delivered into the plant cells. This can be done using various methods, such as Agrobacterium-mediated transformation, particle bombardment, or electroporation. After the CRISPR-Cas9 components have been delivered, the Cas9 protein cuts the DNA at the target site, and the plant's natural repair mechanisms attempt to fix the break. This repair process can result in one of two outcomes: the introduction of small insertions or deletions (INDELs) at the target site or the precise replacement of the target DNA sequence with a new one using the donor template DNA. The donor DNA template can be designed to be similar to the plant's native DNA sequence, thus avoiding the introduction of foreign DNA.

(iii) Screening for edited plants: Once the plant cells have been transformed with the CRISPR-Cas9 components and donor DNA template, they must be screened to identify those that have undergone successful genome editing. This can be done using various methods, such as PCR amplification and sequencing.

(iv) Regeneration of edited plants: The edited cells are then selected and induced to regenerate into whole plants.

(v) Selection of transgene-free plants: Finally, the plants need to be screened to ensure that they are transgene-free. This can be done using various methods, such as PCR amplification and sequencing, to confirm that there are no foreign DNA sequences in the genome.

By following these steps, it is possible to generate transgene-free genome edited plants using the CRISPR-Cas9 system. This approach has the advantage of avoiding the introduction of foreign DNA into the genome, which can simplify the regulatory process and potentially reduce public concern about genetically modified plants/organisms. However, it is still a relatively new technology and further research is needed to fully optimize and validate the approach.

11.6. Applications of CRISPR-Cas9 in crop improvement

CRISPR-Cas9 has revolutionized the field of crop improvement by providing a precise and efficient tool for genome editing. Here are some examples of how CRISPR-Cas9 has been used in crop improvement:

(i) Disease resistance: CRISPR-Cas9 has been used to create crops that are resistant to various diseases caused by viruses, bacteria, and fungi (Ahmad et al., 2020; Tripathi et al., 2020; Tyagi et al., 2021). For example, researchers have used CRISPR-Cas9 to edit the genome of rice plants to make them resistant to bacterial blight, a disease caused by the bacterium *Xanthomonas oryzae* that can devastate rice crops. CRISPR-Cas9 has also been used to develop wheat that is resistant to powdery mildew, a fungal disease that affects the leaves and stems of plants.

Similarly, CRISPR-Cas9 has been used to create a disease-resistant version of the Cavendish banana, which is the most commonly consumed banana variety worldwide. The Cavendish banana is susceptible to a fungal disease known as Panama disease, which has devastated banana crops in the past. By using CRISPR to edit the banana's genome, researchers have created a variety that is resistant to Panama disease, offering a promising solution to this ongoing problem.

(ii) Drought and salt tolerance: Drought and high salinity are major challenges to agricultural productivity worldwide, and climate change is expected to exacerbate these challenges in the future (Zhang et al., 2019; Santosh Kumar et al., 2020; Li et al., 2022). To address this problem, CRISPR-Cas9 can be used to edit genes involved in drought and salt tolerance, leading to crops that can better withstand these harsh conditions. Researchers have used CRISPR-Cas9 to edit the genome of maize, wheat, and rice plants to increase their tolerance to drought, resulting in increased yield in areas with limited water availability. For example, using CRISPR-Cas9 to modify the *OsDST* gene in rice, researchers have been able to create rice plants that are more tolerant to drought conditions. The gene modification leads to an increase in the accumulation of the stress hormone abscisic acid (ABA) in the plant, which triggers a response that helps the plant to conserve water and survive under drought stress.

Similarly, in wheat, researchers used CRISPR-Cas9 to modify a gene called *TaHKT1;5*, which regulates sodium uptake in the plant. The modification resulted in wheat plants that are better able to tolerate high levels of salt in the soil.

(iii) Improved shelf life: CRISPR-Cas9 has been used to edit genes involved in fruit ripening, leading to crops with improved shelf life (Yu et al., 2017; Chakravorty et al., 2022). One example of this is the CRISPR-Cas9 modification of the tomato genome to delay fruit ripening and improve its shelf life, resulting in tomatoes that can be transported over longer distances without spoiling. The researchers targeted genes involved in the production of ethylene, a plant hormone that triggers fruit ripening, and were able to create tomato plants that ripen more slowly and stay fresh for longer periods of time.

(iv) Improved crop quality: CRISPR-Cas9 has been used to edit genes involved in crop quality, leading to crops with improved taste, aroma, and nutritional value. For example, researchers have used CRISPR-Cas9 to edit the genome of wheat plants to reduce the levels of gluten, making the wheat more tolerable for people with gluten sensitivity.

(v) Improved yield: CRISPR-Cas9 can be used to edit genes that control key plant growth and development pathways, leading to improved yield. For example, scientists have used CRISPR-Cas9 to increase the number of grains per wheat spike, resulting in higher yield.

Another example is the use of CRISPR-Cas9 to modify genes involved in the production of hormones that regulate plant growth and development, such as gibberellins and auxins. By modifying these genes, scientists have been able to create crops that produce more fruit or grain per plant, resulting in increased yield.

(vi) Nutritional enhancement: CRISPR-Cas9 can be used to enhance the nutritional value of crops by editing genes involved in the synthesis and accumulation of nutrients. For example, scientists have used CRISPR-Cas9 to develop rice that produces higher levels of beta-carotene, a precursor of vitamin A, which can help to address vitamin A deficiency and improve public health.

Similarly, CRISPR-Cas9 can be used to increase the levels of other important nutrients, such as iron and zinc in crops like wheat, maize, and potatoes.

Furthermore, CRISPR-Cas9 can be used to modify the composition of plant oils and fats, making them more healthy for human consumption. For example, scientists have used CRISPR-Cas9 to modify the fatty acid composition of soybean oil, resulting in a healthier oil with lower levels of saturated fats.

(vii) Environmental sustainability: CRISPR-Cas9 can be used to develop crops with reduced environmental impact by editing genes that control plant traits related to nitrogen use efficiency, water use efficiency, and carbon sequestration. For example, scientists have used CRISPR-Cas9 to develop rice that uses nitrogen more efficiently, reducing the need for nitrogen fertilizers. Nitrogen fertilizers are widely used in agriculture to promote plant growth and increase yield, but they can also lead to environmental problems such as nitrogen runoff into waterways and greenhouse gas emissions.

These are just a few examples of the many ways that CRISPR-Cas9 is being used to improve crops. As the technology continues to advance, we will likely see even more applications of genome editing in crop improvement.

Chapter 12: CRISPR in animals

CRISPR has various applications for animals as well.

12.1. Models for human diseases

CRISPR gene editing technology has been used in various animals to generate disease models that can help researchers better understand and develop treatments for human diseases (Shrock and Güell, 2017; Zarei et al., 2019; Khalaf et al., 2020).

Mouse models are one of the most widely used animal models for studying human diseases using CRISPR gene editing. This is because mice share many genetic and physiological similarities with humans. This involves modifying their genomes to mimic the genetic mutations or variations that are associated with specific human diseases. These animal models can then be used to study the underlying mechanisms of the diseases, test potential treatments, and identify new targets for drug development (Birling et al. 2017; Lu et al., 2019; Sharma et al., 2021; Lim et al., 2022).

For example, researchers have used CRISPR to create mouse models of cystic fibrosis, a genetic disorder that affects the lungs and other organs. By introducing mutations in the mouse genome that mimic the human genetic mutations that cause cystic fibrosis, researchers were able to create mice that develop lung diseases similar to that seen in human patients. These mice can then be used to test potential therapies and identify new drug targets for cystic fibrosis. Mouse models have been used to study a wide range of human diseases, including but not limited to Huntington's disease, sickle cell disease, muscular dystrophy, retinitis pigmentosa, cancer, diabetes, and various other genetic diseases. The mouse models provide valuable tools for studying disease mechanisms and testing potential therapies.

While mouse and rat models are commonly used for studying a wide range of human diseases, larger animals such as pigs and monkeys may be preferred for certain disease models due to their physiological and anatomical similarities to humans. One example is the study of cardiovascular diseases such as heart disease and stroke. Pigs have a cardiovascular system that is more similar to humans than mice, making them better models for studying these diseases (Perleberg et al., 2018; Hou et al., 2022). Furthermore, pigs have the advantages of early sexual maturity, a short reproductive cycle, and a high number of offspring per litter. One study used CRISPR gene editing to create pigs with a mutation associated with human familial hypercholesterolemia, a condition that causes high cholesterol levels and an increased risk of heart disease (Yao et al., 2016; Huang et al., 2017; Zhao et al., 2019). The pig models were able to replicate key features of the disease, making them useful for studying its underlying mechanisms and testing potential treatments.

Another example is the study of neurological disorders such as Parkinson's disease and Alzheimer's disease. Monkeys are more similar to humans in terms of brain structure and function, making them better models for studying these diseases. Researchers have also used macaques as a model for studying autism spectrum disorders (ASDs) because they share many similarities with humans in terms of brain structure, social behavior, and communication (Park et al., 2019; Aida and Feng, 2020). For example, researchers used CRISPR/Cas9 to delete the *SHANK3* gene associated with autism in fertilized macaque eggs, which were then implanted into surrogate mothers. The resulting offspring showed several behavioral and neurological abnormalities that were similar to those seen in humans with ASDs, such as reduced social interaction and repetitive behaviors (Zhao et al., 2018; Tu et al., 2019).

In general, larger animal models such as pigs and monkeys can provide a more realistic and clinically relevant model for certain human diseases than mouse and rat models. However, they also come with additional ethical and logistical considerations, such as higher costs, longer gestation periods, and greater regulatory oversight. Therefore, the choice of animal model ultimately depends on the specific disease being studied and the scientific questions being addressed. But with all animal research, it is important to ensure that the animals are treated humanely and that their use is justified by the potential benefits to human health.

Overall, the use of CRISPR to create animal models of human diseases is a promising area of research that has the potential to greatly advance our understanding and treatment of a wide range of diseases. By creating animal models of human diseases using CRISPR, researchers can gain valuable insights into the underlying mechanisms of these diseases, and test potential treatments in a controlled and ethical manner. These animal models also provide a valuable tool for drug development and testing, as they can help to identify new drug targets and test the safety and efficacy of potential therapies before they are tested in human patients.

12.2. Xenotransplantation

CRISPR has the potential to revolutionize the field of organ creation and transplantation. There are currently many challenges associated with organ transplantation, including a shortage of donor organs and the risk of rejection by the recipient's immune system. CRISPR offers a hopeful solution to these challenges by enabling the creation of organs that are more compatible with human recipients.

One of the most promising applications of CRISPR in organ creation is Xenotransplantation, i.e., the transplantation of organs from one species to another. CRISPR has the ability to edit the genomes of other species to make their organs more compatible with human recipients (Ryczek et al., 2021). Pigs are a particularly optimistic source of organs for xenotransplantation because their organs are anatomically and physiologically similar to human organs. However, pig organs contain antigens that can trigger an immune response in humans called as hyperacute rejection (HAR) process, which can occur within minutes to hours after transplantation. HAR is a type of rejection that is mediated by antibodies naturally present in the human recipient, which bind to specific epitopes on the porcine endothelial cells. This binding triggers the activation of the complement system, leading to the destruction of the graft's vascular system and subsequent transplant failure. This immune response is characterized by damage to the vessels, swelling, bleeding, and clotting.

The main xenoantigen involved in HAR is the galactose-α1,3-galactose (α-Gal) epitope, which is produced by the porcine *GGTA1* gene. This enzyme is found in pigs, but not in humans, and triggers the production of antibodies in humans. Additional antigens like, Neu5Gc and β4GALNT2 also significantly influence HAR. With CRISPR, it may be possible to edit the pig genome to inactivate these porcine genes. This could make pig organs suitable for transplantation in humans without triggering an immune response that leads to transplant failure (Zhang et al., 2018; Cowan et al., 2019; Wang et al., 2019).

Another challenge of pig organ xenotransplantation is the risk of infection with porcine endogenous retroviruses (PERVs), which are present in the DNA of all pigs. While PERVs are not harmful to pigs, there is a possibility that they could infect human recipient cells, potentially leading to disease. Scientists are working to develop methods to eliminate PERVs from pig cells using CRISPR before transplantation (Niu et al., 2017; Ross and Coates, 2018).

In addition to pig-to-human transplantation, CRISPR can also be used to create organs from scratch by editing the genomes of stem cells. For example, scientists have used CRISPR to edit the genomes of stem cells to differentiate them into functional liver cells, which could eventually be used to create a liver for transplantation.

While there is still much research to be done in this area, CRISPR has the potential to overcome many of the barriers that have made organ transplantation difficult in the past, improving the lives of countless individuals around the world. However, it is important to proceed with caution and ensure that any use of CRISPR in organ creation is safe, ethical, and well-regulated.

12.3. For livestock

CRISPR has several applications in livestock, including improving animal health, productivity, and disease resistance (Jabbar et al., 2021; Singh and Ali , 2021; Yunes et al., 2021; Mehra et al., 2022). One of the main uses of CRISPR in livestock is to create animals that are resistant to diseases, such as African Swine Fever or Foot-and-Mouth disease. For example, researchers could edit the genes of pigs to delete or modify a receptor that allows the African Swine Fever virus to enter their cells, making them resistant to the disease. Similarly, scientists could edit the genes of cattle to make them more resistant to Foot-and-Mouth disease.

Another application of CRISPR in livestock is to improve animal productivity. For example, CRISPR can be used to modify genes that regulate the cow's mammary gland development, resulting in increased milk production. CRISPR can also be used to modify genes that regulate the composition of the milk, resulting in milk with higher protein or fat content and reduced levels of allergens or lactose. These modifications could have important benefits for dairy farmers, who could improve the efficiency of their operations and increase their profits by producing more milk of higher quality. Additionally, by producing milk with specific properties, farmers could meet the growing demand for specialized dairy products, such as high-protein or lactose-free milk.

Similarly, researchers can use CRISPR to modify genes (e.g., Myostatin gene) that regulate the animal's muscle development, resulting in leaner meat.

However, the use of CRISPR in livestock also raises ethical concerns. Some argue that it is unethical to genetically modify animals for human benefit, particularly if it causes suffering or reduces their quality of life. Additionally, there are concerns about the potential impact on the environment and natural ecosystems if genetically modified animals escape into the wild.

There are also concerns about the long-term effects of genetic modification on animal health and the potential for unintended consequences. It is possible that editing the genes of livestock could have unforeseen effects on their physiology or behavior, potentially leading to negative impacts on their health and well-being. For example, creating animals with increased productivity could lead to health problems or reduced lifespans. Mutations in the myostatin gene have also been associated with muscle disorders in some species.

It is therefore important to carefully evaluate the potential risks and benefits of using CRISPR to modify the genes in livestock and to ensure that any research or applications are conducted in an ethical and responsible manner. Additionally, regulatory frameworks must be in place to ensure that genetically modified livestock and their products are safe for human consumption and do not pose any environmental risks.

Chapter 13: Prime Editing

Undoubtedly CRISPR-Cas technology has changed the way scientists approach gene editing, offering various advantages that were never seen before. Scientists are able to edit human, animal, and plant genomes with this technique. CRISPR-Cas technology has become an indispensable tool in many research laboratories, allowing scientists to more easily create animal models of genetic diseases. But CRISPR-Cas technology is not foolproof. Its practical use in curing human disease has been limited by challenges with delivery and precision because of off-target effects that alter DNA at unintended loci in the genome. As previously mentioned, CRISPR-Cas9 genome editing depends on DNA repair mechanisms, including non-homologous end joining (NHEJ) or homology-directed repair (HDR), to fix DNA breaks. However, this repair process can lead to the creation of undesired and random insertions or deletions (INDELs) in unintended areas of DNA, which may disrupt cell function.

13.1. Introduction to prime editing

Prime editing is a new genome-editing technique that can correct specific nucleotides within the genome with greater precision and efficiency while limiting the negative effects of the CRISPR-Cas system.

In October 2019, David Liu and colleagues published the development and applications of this new technique in Nature journal (Anzalone et al., 2019). Prime editing is similar to CRISPR technology in some aspects. But what sets prime editing apart from traditional CRISPR is that it enables targeted editing without generating double-stranded DNA breaks. We have already studied in CRISPR technology that Cas9 generates double-stranded breaks in the target DNA, and then the targeted insertions are achieved by adding a donor DNA template. But in the case of prime editing, targeted insertions can be achieved with great precision without the need for donor DNA templates. In short, prime editing is an all-in-one solution with small improvements to the CRISPR process that impacts the final result, i.e., gene editing with greater precision.

13.2. Components of prime gene editing

Similar to CRISPR, prime editing requires the presence of a Cas endonuclease and a single guide (sg) RNA. However, as the goal of prime editing is to edit DNA sequences without generating a double-stranded break, therefore both components are slightly modified. Instead of traditional Cas9, the prime editing method utilizes Cas9 nickase, which is a variant of Cas9 endonuclease (Lin et al., 2020; Liu et al., 2020; Marzec et al., 2020). As discussed in the previous chapters, the Cas9 enzyme contains two nuclease domains that can cleave DNA sequences, a RuvC domain that cleaves the non-complementary or coding strand of the target DNA and an HNH domain that cleaves the target DNA strand complementary to the guide RNA. The introduction of an H840A substitution in Cas9 in which the amino acid histidine at position 840 is replaced by alanine inactivates the HNH domain. With only the RuvC functioning domain, the catalytically impaired Cas9 introduces a single-strand nick rather than generating double-strand breaks hence the name Cas9 nickase. Also, Cas9 nickase is fused to a reverse transcriptase enzyme *(Fig 13-1a)*. The function of the reverse transcriptase enzyme is to synthesize DNA from a single-stranded RNA template. This Cas9 fusion is referred to as a prime editor or PE.

The second component, single guide RNA, is also modified in the prime editing method. In the prime editing technique, the sgRNA is fused with a primer binding sequence, simply written as PBS. The sgRNA also carries the template RNA sequence for the change that we want to make to our genomic target DNA *(Fig 13-1b)*. These sequences are added at the 3'end of the single gRNA. This fused single guide RNA is called prime editing guide RNA or simply referred to as pegRNA (Li et al., 2022; Nelson et al., 2022).

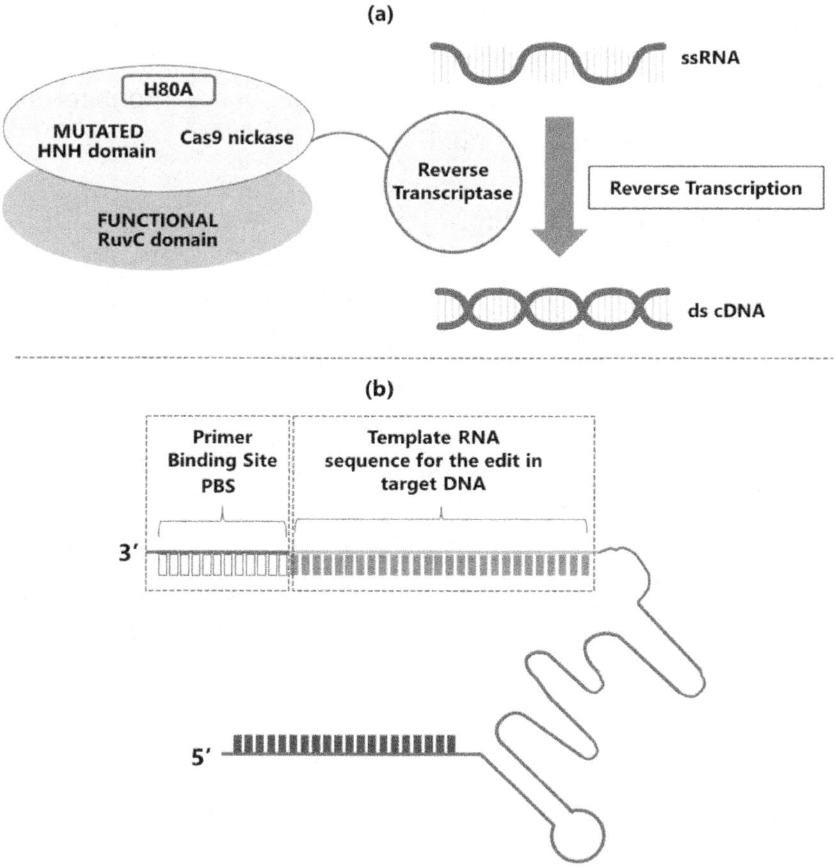

Fig 13-1: Components of prime gene editing (a) prime editor and (b) pegRNA

Simply put, prime editing guide RNA (pegRNA), is capable of performing 2 tasks simultaneously. The first is, identifying the target site with the help of a single gRNA, and the second is, providing the new template or genetic information to replace the target DNA nucleotides. Thus the prime editing method is able to mediate targeted insertions, deletions, and base-to-base conversions without the need for double-strand breaks (DSBs) or donor DNA templates. Because of the additional sequences, pegRNA is substantially larger than standard sgRNAs commonly used for CRISPR gene editing. Together, the Prime editor and pegRNA complex is used to mediate genome editing within the cell. The genome editing complex also contains an additional single guide RNA (sgRNA) that directs the Cas9 nickase to nick the unedited DNA strand in order to incorporate the edited sequences in this strand.

13.3. How does prime editing work?

(i) The Prime editor and pegRNA complex bind to the target DNA. As a reminder, all CRISPR target sites consist of a PAM sequence in the genomic DNA next to 20 base sequence that matches the protospacer in the guide RNA. Firstly the Cas9 nickase nicks only one strand of DNA, i.e., the non-target strand or the PAM strand. Additionally, it cuts at a very specific location - 3 bases away from the PAM sequence *(Fig 13-2)*. The first base after the nick is called +1. The prime editor can change any sequence after this point. So all the sequence after the +1, including PAM, is susceptible to prime editing.

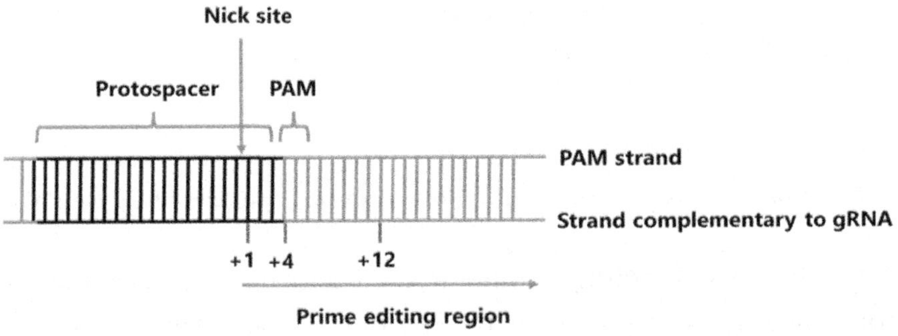

Fig 13-2: Components of prime gene editing

(ii) After nicking the PAM strand, the DNA opens up, and generates a 3' flap exposing a 3' hydroxyl group. Now the primer binding sequence or PBS, located on the 3' end of pegRNA, pairs up with the protospacer of the 3' DNA flap *(Fig 13-3a)*. Generally, the primer binding sequence is 12 bases long. We already know that all the reverse transcription reactions require a primer and a reverse transcriptase enzyme. In this case, the 3' overhang acts as a primer for 5'-3' reverse transcription, and the template for editing DNA present at the 3'end of the peg RNA is reverse transcribed into cDNA using the reverse transcriptase fused with Cas9 nickase *(Fig 13-3b)*. Now the edited DNA formed by reverse transcription is incorporated into the PAM DNA. In the *Fig*, you can see 3 bases to be edited in black, followed by the short sequence that matches the rest of the genomic DNA. A key thing to know is that the reverse transcriptase can not go any further into the pegRNA because it is bound up by Cas9. Therefore the reverse transcriptase enzyme can not copy the entire pegRNA sequence into the genomic DNA.

(iii) So now we have a 3' flap that contains a newly synthesized or edited sequence *(Fig 13-4a)*. 3' single-stranded strands are relatively stable and, therefore, will not be digested by the host enzymes. In *Fig*, you can see that in the genome, there is a 3' flap and a nick in the genome. This nick creates an instability in the genome, so ds DNA will unzip to some level to create a flap equilibrium in which a 3' flap will be incorporated into the genome, exposing a non-edited 5' flap *(Fig 13-4b)*. Unlike 3' flaps, 5' flaps are unstable and get digested and excised by the host enzymes like 5' exonucleases *(Fig 13-4c)*. This process allows 3' flap ligation and creates a heteroduplex DNA composed of one edited strand and one unedited strand *(Fig 13-4d)*. The reannealed double-stranded DNA contains nucleotide mismatches at the location where editing took place.

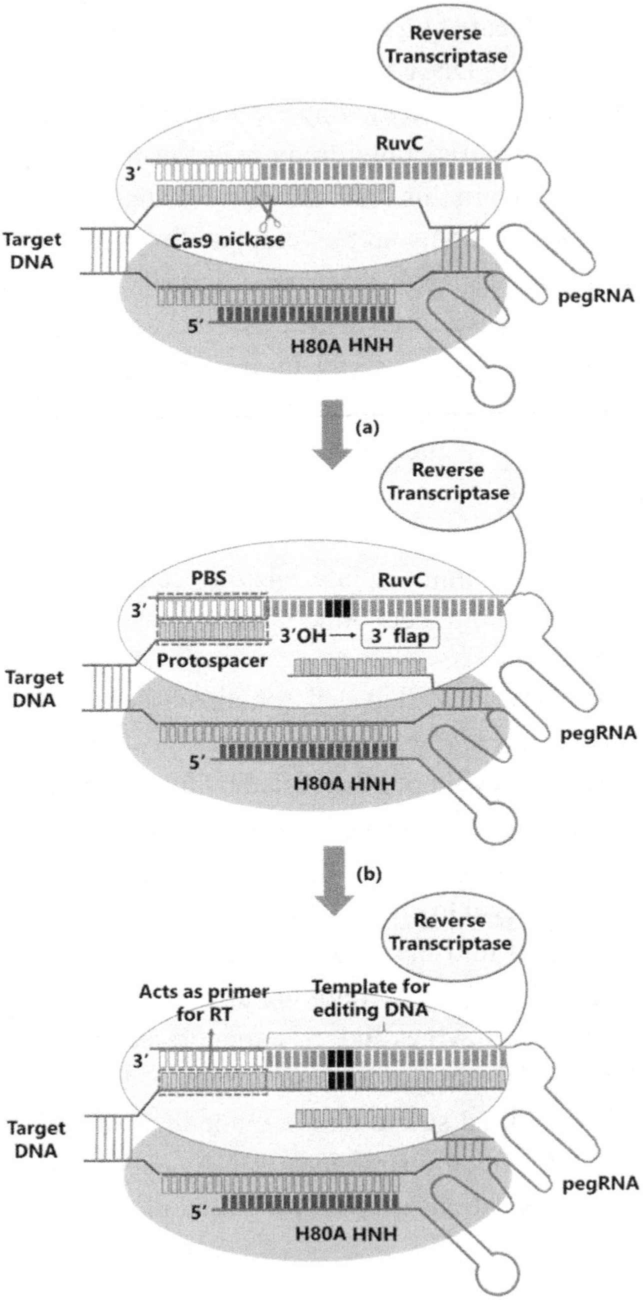

Fig 13-3: Mechanism of prime gene editing

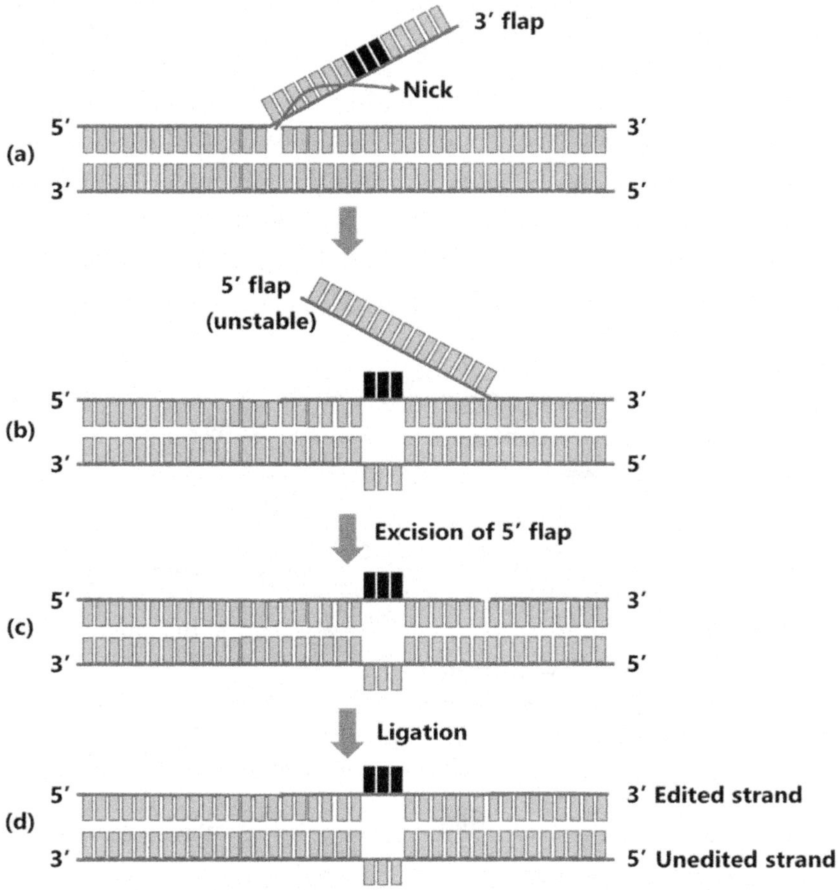

Fig 13-4: Heteroduplex DNA composed of one edited strand and one unedited strand

(iv) In order to correct the mismatches, the cells exploit the intrinsic mismatch repair mechanism, in which information in the edited strand is copied into the complementary strand, thus permanently installing the edit in the genome. To facilitate this step, there is an additional gRNA added to the reaction to create a nick in the original non-edited strand. This gRNA is designed to match the edited sequence introduced by the pegRNA but not to the original non-edited strand *(Fig 13-5a)*. Therefore it directs the Cas9 nickase to nick the non-edited strand rather than nicking the edited strand *(Fig 13-5b)*. By adding a nick into the non-edited strand, the cell repair process biases to remove mismatch sequences of the nicked strand and synthesize a new strand complementary to the opposite strand *(Fig 13-5c)*. In this case, the opposite strand is the edited strand which acts as a template to repair the nick, thus completing the edit on both strands. Therefore, edited sequences are permanently installed into the target gene. Thus in this way, the gene editing is done by the prime editing technique.

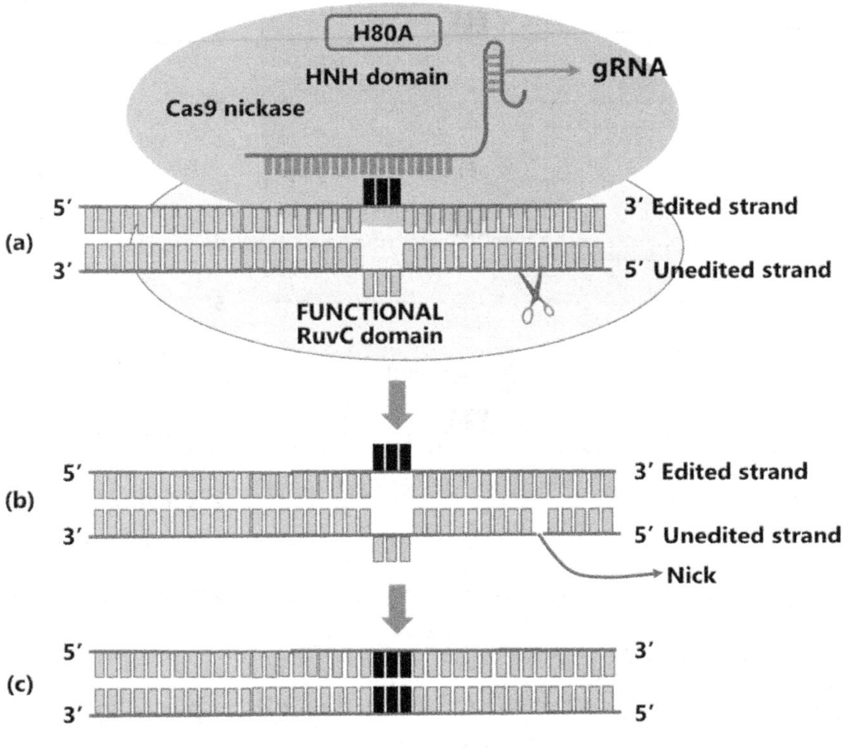

Fig 13-5: Edited strand acts as a template to repair the nick, thus completing the edit on both strands

13.4. Advantages & Limitations of Prime editing technique

Advantages: The prime editing tool offers advantages over CRISPR-Cas gene editing technology. CRISPR-Cas9 system generates double-stranded breaks in the genome. The edits rely on DNA repair mechanisms such as non-homologous end joining (NHEJ) or homology-directed repair (HDR) to fix DNA breaks, which is notoriously error-prone and can generate unwanted, random insertions or deletions (INDELs) in the unintended regions of DNA which in turn can interrupt cell function. And undesirable changes in the genome have been a significant concern in taking therapeutic gene editing to the clinic.

On the other hand, the prime editing system is comparatively precise and employs DNA mismatch repair. The prime system introduces single-stranded DNA breaks instead of the double-stranded DNA breaks observed in CRISPR technology.

Limitations: Even though prime editing is a major breakthrough in genome editing, the technology has some downsides as well (Chen and Liu, 2022; Hillary and Ceasar, 2022).

A major limitation of prime editing is the size of the prime editing construct, which is fairly large, making it inefficient in making larger genome alterations. Larger genetic alterations would require a longer reverse transcriptase template, which could hinder the efficient delivery of pegRNA into the target cells. Using Cas9 orthologs that are smaller in size could reduce the size of the prime editor and, thus, facilitate its delivery into the target cell.

Chapter 14: CRISPR-mediated base editing

The ability to precisely edit genomic DNA using the CRISPR-Cas system has revolutionized the field of genome engineering. Like prime editing, base editing is a CRISPR-based precise genome editing technology that can fix point mutations in the DNA by precisely editing it without generating double-strand breaks (Komor et al., 2016; Gaudelli et al., 2017; Rees et al., 2018; Anzalone et al., 2020). A point mutation is when a single DNA base has been substituted, inserted, or deleted. For instance, base T gets substituted with base C. There are more than 30,000 point mutations associated with genetic diseases in humans. Thus, base editing can be a promising therapeutic strategy for genetic diseases caused by point mutations, including many blood disorders.

14.1. How does base editing work?

For base editing, two major classes of base editors have been developed: cytidine base editors or CBEs that allow conversion of CG to TA base pair, and adenine base editors or ABEs convert AT to GC base pair. Cytidine base editors are formed by fusing Cas9 nickase with a cytidine deaminase (Kim et al., 2017; Kantor et al., 2020; Huang et al., 2021). On the other hand, Adenine base editors (ABEs) are composed of a Cas9 nickase fused to *E.coli* tRNA adenosine deaminase enzyme TadA to edit adenine in DNA (Jiang et al., 2020; Ma et al., 2020; Chen et al., 2022). Base editors are guided by a sgRNA to the target DNA sequence.

(i) Now, after the Cas9 nickase binds to the target DNA, the deaminase, such as a cytidine deaminase, recognizes the cytosine base at the target DNA, removes its amino group, and converts it into a Uracil or U base *(Fig 14-1a)*.

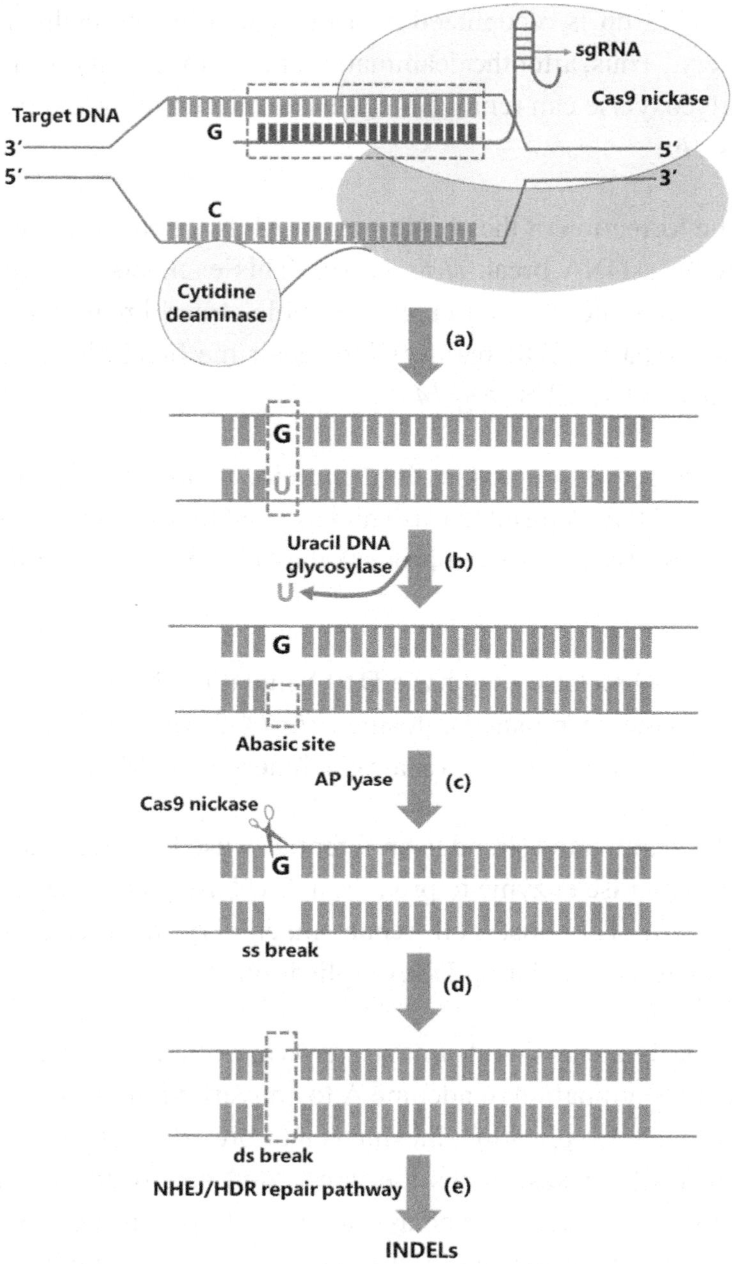

Fig 14-1: Generation of INDELS if the Uracil DNA glycosylase and AP lyase enzymes are functional

(ii) The G:U bp is recognized as a mismatch by the cellular repair machinery. Thus, after the deamination, i.e., C to U conversion, uracil DNA glycosylase can remove the U base, thus creating an abasic site *(Fig 14-1b)*.

Upon the formation of the abasic site, an AP lyase can remove the site forming an ssDNA break *(Fig 14-1c)*. This excision in conjunction with the Cas9 nickase can create a double-strand break, which then initiates NHEJ or HDR pathways to repair the break, leading to the generation of INDELS *(Fig 14-1d)*.

Thus to stop this process, a third protein, uracil DNA glycosylase inhibitor or UGI, is fused to Cas9 nickase that blocks base excision by uracil DNA glycosylase and protects the U base from being removed *(Fig 14-2a)*.

(iii) The Cas9 nickase used has a D10A mutation that nicks the strand opposite to the deaminated cytosine *(Fig 14-2b and c)*. The U base is interpreted as base T by the repair machinery *(Fig 14-2c)*.

Thus, the deaminated strand is preferentially used as a template by the DNA polymerase enzyme to produce a A:U base pair at the target site *(Fig 14-2d)*. The host cell repair machinery then converts this intermediate to A:T during DNA replication *(Fig 14-2e)*.

(iv) Similarly adenosine deaminase enzyme of adenine base editors, causes the deamination of adenine A to convert it into inosine I. In the context of DNA replication, inosine is interpreted as guanine G. Thus, after the Cas9 nickase nicks the non-edited strand, the polymerase enzyme uses the edited strand as a template to produce C:I base pair. The host cell repair machinery then converts this intermediate to C:G at the target site during DNA replication *(Fig 14-3)*.

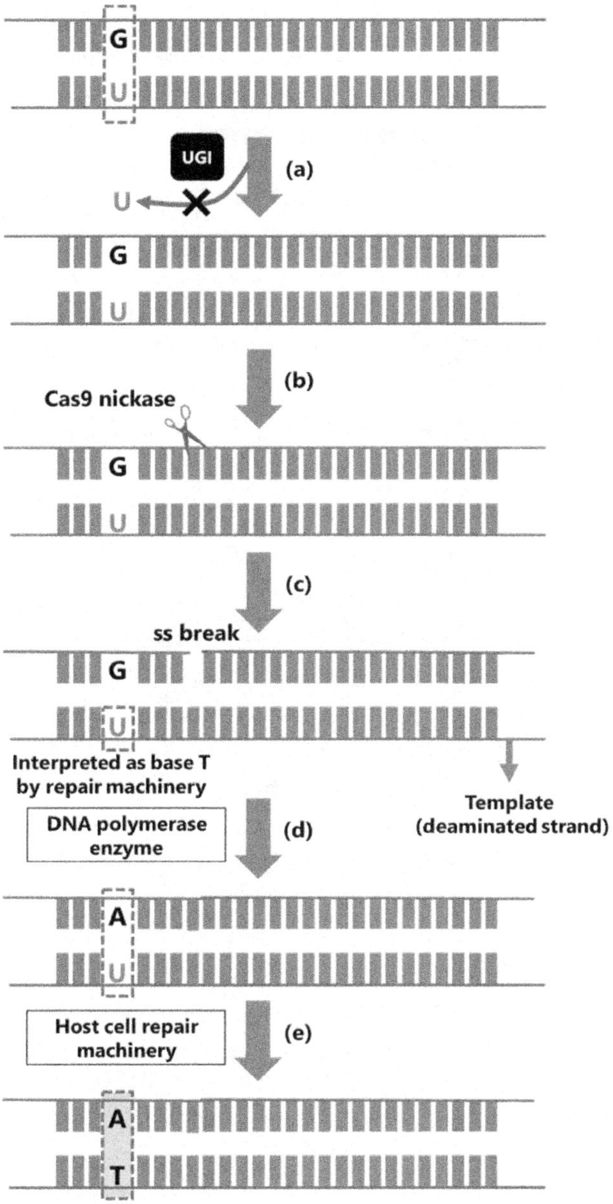

Fig 14-2: Conversion of U:G to T:A base pair by cytidine base editor or CBE

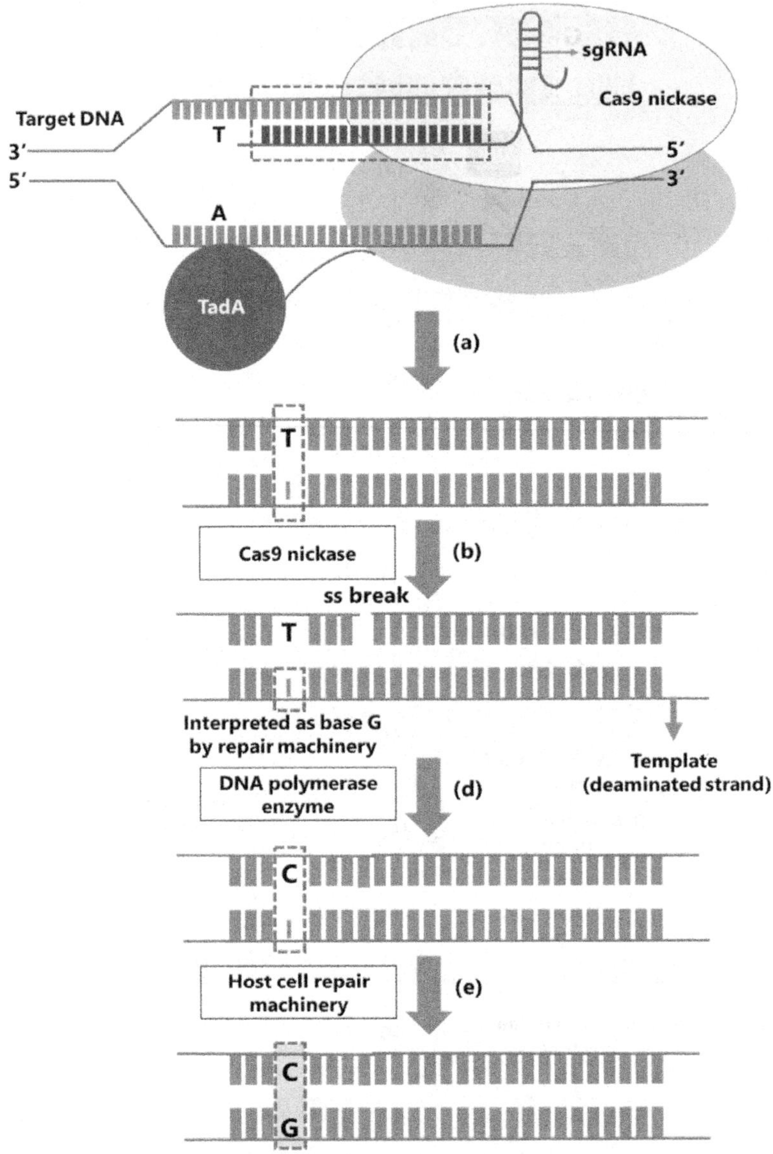

Fig 14-3: Conversion of T:A to C:G base pair by adenine base editor or ABE

14.2. Advantages and limitations of base editing

Precision editing systems like base editing can exhibit significantly lower unwanted INDEL rates, thus decreasing the chances of disrupting the function of the target gene. Hence, a precise base editor system can be a potential therapeutic approach for the targeted insertion of beneficial mutations.

Researchers have demonstrated the potential use of this system by correcting Alzheimer's Disease-associated mutations in APOE4 in mouse astrocytes and TP53 mutations in a mouse breast cancer line.

Precision editing systems also have an important role in agriculture. Efficient base editing in crops can provide a new method to select desirable crop traits in wheat, maize, etc (Monsur et al., 2020; Li et al., 2021; Molla et al., 2021).

Comparisons between the frequency of off-target mutations generated using base editing versus cleavage with active Cas9 nuclease suggest that base editing may have greater specificity. To further reduce non-specific mutations, a high-fidelity SpCas9 (HF-Cas9) can be used. Additionally, engineered SpCas9 nucleases with varied PAM specificities can be used to allow the base editor system to target a wide range of genes.

Practice Test 2

1. What is the main disadvantage of using natural restriction enzymes for genome editing?
(a) Their recognition sequences are too short
(b) Their recognition sequences are too long
(c) Their recognition sequences are not specific enough
(d) None of the above

2. What is the function of the zinc finger repeats in Zinc Finger nuclease?
(a) To cleave DNA
(b) To recognize and bind to specific DNA sequences
(c) To regulate gene expression
(d) All of the above

3. What is the DNA cleavage domain of Zinc Finger nuclease derived from?
(a) FokI restriction endonuclease enzyme
(b) EcoRI restriction endonuclease enzyme
(c) Zinc finger transcription factor
(d) None of the above

4. How many ZF motifs are required to recognize 12 nucleotides within the target DNA?
(a) 1
(b) 2
(c) 3
(d) 4

5. What is the main difference between Non-Homologous End Joining and Homology-directed repair (HDR) pathway?
(a) Non-Homologous End Joining results in the generation of small insertions or deletions of nucleotides, while HDR utilizes the presence of a template DNA for repairing double-stranded breaks.
(b) Non-Homologous End Joining is error-free, while HDR is prone to errors.
(c) Non-Homologous End Joining is faster than HDR, while HDR is more efficient.
(d) Non-Homologous End Joining is used for gene activation, while HDR is used for gene inactivation.

6. Each TALE repeat in an array binds to how many bases of target DNA?

(a) 1
(b) 2
(c) 3
(d) None of the above

7. How does the TALE protein recognize a specific target DNA sequence?
(a) By recognizing and binding to the desired target RNA
(b) By recognizing and binding to the desired target DNA sequence through its variable pair of amino acids at positions 12 and 13
(c) By recognizing and binding to the desired target DNA sequence through its tandem repeated segments
(d) By recognizing and binding to the desired target DNA sequence through its DNA cleavage domain

8. What does CRISPR stand for?
(a) Clustered Regularly Interspaced Short Palindromic Repeats
(b) Clustered Reporter Interspaced Specific Palindromic Rearrangement
(c) cDNA Regularly Intergenic Single Phenotypic Rearrangement
(d) Clustered Regularly Interacting Short Palindromic Repeats

9. CRISPR/Cas system provides adaptive immunity against viruses in what types of organisms?
(a) Bacteria
(b) Archaea
(c) Eukaryotes
(d) Both a and b

10. How does the methyltransferase enzyme protect bacterial DNA from being destroyed by restriction enzymes?
(a) By cleaving viral DNA
(b) By cleaving bacterial DNA at different sites than the restriction enzyme
(c) By causing methylation of bacterial DNA
(d) By binding to the viral DNA and preventing it from replicating

11. What is abortive infection?
(a) An altruistic mechanism by which an infected bacterium kills itself, thereby eliminating phage propagation
(b) An aggressive mechanism by which an infected bacterium actively fights off the phage attack
(c) A mechanism by which the bacterium prevents phage attachment to the cell surface

(d) A mechanism by which the bacterium repairs its own DNA after a phage attack

12. Which of the following is true about the CRISPR/Cas system and adaptive immunity in humans?
(a) They have no similarities
(b) They both generate a memory of past infections to mount a more efficient response upon recurrent infections
(c) They both attack viruses with enzymes
(d) They are used by bacteria to produce energy

13. What are the characteristics of the repeat sequences in CRISPR?
(a) Short and non-palindromic
(b) Long and palindromic
(c) Short and palindromic
(d) Long and non-palindromic

14. Which of the following components of CRISPR function as a memory bank in bacteria?
(a) Repetitive sequences
(b) Cas proteins
(c) Spacers
(d) tracrRNA

15. Which Cas proteins are responsible for excising and integrating the protospacer from the viral DNA during the adaptation stage?
(a) Cas1and Cas2
(b) Cas1 and Cas9
(c) Cas3 and Cas9
(d) Cas2 and Cas9

16. Which domains in the Cas9 protein are responsible for cleaving the target DNA in Type II CRISPR-Cas system?
(a) HNH and RuvC
(b) REC1 and REC2
(c) DED and CARD
(d) PDZ and SH3

17. Which molecule is used by the CRISPR-Cas system to recognize and bind to the specific DNA sequence to be targeted?
(a) Messenger RNA (mRNA)
(b) Guide RNA (gRNA)

(c) Transfer RNA (tRNA)
(d) Ribosomal RNA (rRNA)

18. What does PAM stand for in the CRISPR-Cas system?
(a) Protospacer Associated Motif
(b) Protospacer Adjacent Motif
(c) Prokaryotic Associated Motif
(d) Prokaryotic Adjacent Motif

19. PAM is a component of:
(a) invading virus
(b) bacterial CRISPR locus

20. Which Cas protein recognizes the PAM sequence in Type II CRISPR-Cas systems?
(a) Cas1
(b) Cas2
(c) Cas3
(d) Cas9

21. What is the PAM sequence recognized by *Streptococcus pyogenes* Cas9?
(a) NGG
(b) NNGRRT
(c) NNNVRYAC
(d) None of the above

22. What does sgRNA stand for, and why?
(a) Short genomic RNA, since it binds genomic DNA
(b) Simple guide RNA, because it makes gene editing simple
(c) Shorter guide RNA, because it is only 100 bp
(d) Single guide RNA, because it is a fusion of crRNA and tracrRNA

23. Which research teams discovered the potential of the type 2 CRISPR-Cas system for genome editing?
(a) Watson and Crick
(b) Doudna and Charpentier
(c) Franklin and Wilkins
(d) Mendel and Darwin

24. What is the purpose of the handle-shaped RNA scaffold in the gRNA?
(a) It cleaves the target DNA.
(b) It binds to the Cas9 protein.

(c) It ensures that the Cas9 enzyme cuts at the right point in the genome.
(d) It is not required for the function of the CRISPR system.

25. Which nuclease domain of Cas9 cleaves the DNA strand that is complementary and paired with the 20-letter nucleotide sequence in sgRNA?
(a) HNH nuclease domain
(b) RuvC nuclease domain
(c) Both HNH and RuvC nuclease domains
(d) None of the above

26. Which of the following is an example of a programmable endonuclease?
(a) Restriction enzymes
(b) Ligases
(c) Polymerases
(d) Cas9

27. Which repair pathway does not require template DNA for repairing double-stranded breaks?
(a) Mismatch repair pathway
(b) Nucleotide excision repair pathway
(c) Non-homologous end-joining repair pathway (NHEJ)
(d) Base excision repair pathway

28. What is the result of repairing double-stranded breaks through NHEJ?
(a) Introduction of insertion or deletion mutations in the targeted gene
(b) Correction of the targeted gene sequence
(c) Removal of the entire targeted gene
(d) None of the above

29. What is the term used to describe the process of making a gene inactive or inoperative?
(a) Knockdown
(b) Knockout
(c) Knockoff
(d) Knockover

30. What is the purpose of knocking out a gene?
(a) To increase the expression of the gene
(b) To study the function of the gene
(c) To introduce a new gene into the genome
(d) To decrease the transcription of the gene

31. How does the introduction of insertions or deletions in a gene affect the protein it encodes?
(a) It alters the reading frame and can create a defective protein
(b) It has no effect on the protein
(c) It increases the stability of the protein
(d) It results in the production of multiple proteins

32. What is the function of the MRN complex in the homology-directed repair pathway?
(a) To replace RPA protein
(b) To search for homologous template DNA
(c) To dissect 1000 bases from the 5' end
(d) To protect single-stranded DNA from nucleases

33. What is the function of RAD51 protein in the homology-directed repair pathway?
(a) To protect single-stranded DNA from nucleases
(b) To search for homologous template DNA
(c) To dissect 1000 bases from the 5' end
(d) To replace RPA protein

34. How can the efficiency of HDR be enhanced in cells?
(a) By suppressing the key molecules of NHEJ, KU70, KU80, or DNA ligase IV
(b) By enhancing the key molecules of NHEJ, KU70, KU80, or DNA ligase IV
(c) By suppressing the key molecules of MMR, MLH1, MSH2, or PMS2

35. What is the name of the human pathogen in which the CRISPR type II loci was first characterized?
(a) Streptococcus pyogenes
(b) Escherichia coli
(c) Bacillus subtilis
(d) Lactobacillus acidophilus

36. Which Cas9 protein was first used to cleave and edit a specific region of the genome of a human cell?
(a) SaCas9
(b) SpCas9
(c) Cpf1
(d) FnCpf1

37. In which region of the guide RNA: protospacer duplex can the SpCas9 enzyme tolerate certain mismatches?
(a) PAM-proximal region
(b) PAM-distal region

38. Which type of mismatches can reduce or eliminate the cleavage activity of Cas9?
(a) Mismatches in the PAM-proximal region
(b) Mismatches in the PAM-distal region

39. What is the advantage of using Cas9 nickase over wild-type SpCas9 for gene editing purposes?
(a) Reduces off-target effects
(b) Widens the range of potential target sequences
(c) Increases cleavage efficiency
(d) Increases binding affinity of gRNA

40. What is the difference between Cas9 enzyme and Cas9 nickase in terms of the type of double-stranded break they generate in the target DNA?
(a) Cas9 enzyme generates blunt-ended double-stranded breaks, while Cas9 nickase generates overhangs on each of the cleaved ends
(b) Cas9 enzyme and Cas9 nickase both generate overhangs on each of the cleaved ends
(c) Cas9 enzyme generates overhangs on each of the cleaved ends, while Cas9 nickase generates blunt-ended double-stranded breaks
(d) Cas9 enzyme and Cas9 nickase both generate blunt-ended double-stranded breaks

41. What is the difference between Cas9 enzyme and Cas9 nickase in terms of the number of guide RNAs required for target DNA recognition?
(a) Cas9 enzyme requires one guide RNA, while Cas9 nickase requires two guide RNAs
(b) Cas9 enzyme requires two guide RNAs, while Cas9 nickase requires one guide RNA
(c) Both Cas9 enzyme and Cas9 nickase require one guide RNA
(d) Both Cas9 enzyme and Cas9 nickase require two guide RNAs
42. How does the use of two guide RNAs with paired Cas9 nickases minimize off-target effects?
(a) By increasing the binding affinity of Cas9 nickase to target DNA
(b) By decreasing the sequence length required for target DNA recognition
(c) By increasing the cleavage activity of of Cas9 nickase

(d) By doubling the sequence length required for target DNA recognition

43. What is the difference between the original SpCas9 and the catalytically inactive dCas9?
(a) SpCas9 can cleave DNA, while dCas9 cannot
(b) dCas9 can cleave DNA, while SpCas9 cannot
(c) Both SpCas9 and dCas9 can cleave DNA
(d) Neither SpCas9 nor dCas9 can bind to DNA

44. How can Cas9 be engineered to recognize novel PAM sequences?
(a) By introducing mutations into the PAM-interacting domains of wild-type SpCas9
(b) By deleting the PAM-interacting domains of wild-type SpCas9
(c) By replacing the nuclease domains of wild-type SpCas9
(d) By using a different type of guide RNA

45. What is the limitation of combining multiple mutations into a single Cas9 variant?
(a) It reduces on-target catalytic activity
(b) It enhances on-target catalytic activity
(c) It reduces off-target cleavage activity
(d) It increases off-target effects

46. What is the advantage of using Cas9 orthologs with different PAM recognition sequences for gene editing?
(a) It reduces off-target effects
(b) It increases on-target specificity
(c) It widens the range of potential target sequences
(d) All of the above

47. Why are packaging sequences and transgene sequences expressed in separate plasmids for viral vector assembly?
(a) To increase the replication capacity of the viral particles
(b) To decrease the replication capacity of the viral particles
(c) To ensure that the viral particles formed only contain the transgene as their genome
(d) To increase the pathogenicity and immunogenicity of the viral particles

48. What is the potential risk of using integrating viral vectors for gene delivery?
(a) Uncontrolled cell division leading to cancer
(b) Decreased transgene expression in the host cell

(c) Decreased immunogenicity of the viral particles
(d) Reduced replication capacity of the viral particles

49. What is the primary advantage of using a gutless adenoviral vector compared to a conventional adenoviral vector?
(a) The gutless vector can carry larger genes
(b) The gutless vector is less likely to cause an immune response
(c) The gutless vector can integrate stably into the host genome
(d) The gutless vector can target specific cell types more efficiently

50. What is the purpose of the Herpes simplex virus thymidine kinase polyadenylation (TK pA) signal in the adenoviral vector?
(a) To facilitate transcriptional termination of the gRNA sequence
(b) To facilitate transcriptional termination of the Cas9 gene
(c) To facilitate transcriptional termination of the vector backbone
(d) To facilitate transcriptional termination of the packaging cells

51. What is the purpose of the PacI restriction sites in the adenoviral vector?
(a) To linearize the vector and remove its backbone from the viral sequence.
(b) To facilitate transcriptional termination of the gRNA sequence.
(c) To facilitate transcriptional termination of the Cas9 gene.
(d) To facilitate transcriptional termination of the vector backbone.

52. When is a dual gRNA vector required in CRISPR-mediated gene editing applications?
(a) When the goal is to generate a single-gene knockout.
(b) When the goal is to generate a deletion of a fragment between two DSBs.
(c) When the goal is to target two different genes simultaneously.
(d) Both b and c

53. Which of the following is a limitation of using AAV vectors for gene therapy?
(a) Limited packaging capacity
(b) High immunogenicity
(c) High cytotoxicity
(d) Low transduction efficiency

54. Which of the following cells can be transduced with lentiviral vectors?
(a) Non-dividing cells
(b) Dividing cells
(c) Both non-dividing and dividing cells
(d) Only bacterial cells

55. What is the main disadvantage of using lentiviral vectors for gene therapy?
(a) They can cause insertional mutagenesis
(b) They cannot deliver genes to non-dividing cells
(c) They have a small packaging capacity
(d) They are not able to efficiently transduce target cells

56. What sets prime editing apart from traditional CRISPR?
(a) It generates double-stranded DNA breaks.
(b) It does not require the presence of a Cas endonuclease.
(c) It enables targeted editing without generating double-stranded DNA breaks.
(d) It does not require the presence of a gRNA.

57. What is the full form of pegRNA in the prime editing system?
(a) PEGylated RNA
(b) Prime Editing Guide RNA
(c) Programmed Editing Genome RNA
(d) Polymerase Extension Guide RNA

58. What is the role of the prime editing repair template?
(a) To provide the template for reverse transcription
(b) To introduce nicks in the target DNA
(c) To bind to the Cas9 enzyme
(d) To edit the target DNA sequence

59. In prime editing, what is the function of the reverse transcriptase enzyme?
(a) To cleave the DNA at the target site
(b) To generate a DNA nick in the non-target strand
(c) To copy the RNA guide sequence into target DNA
(d) To repair the nick in the DNA strand after editing

60. Which of the following sequences in the guide RNA pairs up with the protospacer of the 3' DNA flap?
(a) The PAM sequence
(b) The primer binding sequence
(c) The Cas9 binding site
(d) The reverse transcriptase site

61. Which type of flap is relatively stable and does not get digested by host enzymes?
(a) 3' flaps
(b) 5' flaps
(c) Both 3' and 5' flaps
(d) None of the above

62. Which of the following is a potential limitation of prime editing?
(a) It is inefficient and requires high amounts of RNA and protein
(b) It can only be used to edit specific types of cells
(c) It can introduce off-target effects and unintended mutations
(d) It can only be used to introduce small deletions, not more complex changes.

63. What is the role of Cas9 nickase in base editing?
(a) It recognizes the target DNA sequence
(b) It removes the amino group from cytosine
(c) It protects the U base from being removed by uracil DNA glycosylase
(d) It nicks the strand opposite to the deaminated base

64. What is the advantage of using precision editing systems like base editing and prime editing?
(a) They have a higher chance of disrupting the function of the target gene
(b) They generate more off-target mutations
(c) They can precisely insert beneficial mutations
(d) They cannot be used in agriculture

65. How does the cytidine base editor (CBE) work?
(a) The CBE converts TA to CG base pair
(b) The CBE converts CG to TA base pair
(c) Both a and b
(d) None of the above

66. What is the role of the uracil DNA glycosylase inhibitor (UGI) in base editing?
(a) It removes the uracil base from the target DNA
(b) It protects the uracil base from being removed
(c) It promotes the conversion of the uracil base to a T base
(d) It enhances the efficiency of base editing

67. Which of the following is not a potential therapeutic use of base editing?
(a) Correcting point mutations associated with genetic diseases

(b) Inserting beneficial mutations
(c) Creating INDELs
(d) Targeting cancer mutations

Answers:

1. (a) 2. (b) 3. (a) 4. (d) 5. (a) 6. (a) 7. (b) 8. (a) 9. (d) 10. (c) 11. (a) 12. (b) 13. (c) 14. (c) 15. (a) 16. (a) 17. (b) 18. (b) 19. (a) 20. (d) 21. (a) 22. (d) 23. (b) 24. (b) 25. (a) 26. (d) 27. (c) 28. (a) 29. (b) 30. (b) 31. (a) 32. (c) 33. (b) 34. (a) 35. (a) 36. (b) 37. (b) 38. (a) 39. (a) 40. (a) 41. (a) 42. (d) 43. (a) 44. (a) 45. (a) 46. (d) 47. (c) 48. (a) 49. (a) 50. (b) 51. (a) 52. (d) 53. (a) 54. (c) 55. (a) 56. (c) 57. (b) 58. (a) 59. (c) 60. (b) 61. (a) 62. (d) 63. (d) 64. (c) 65. (b) 66. (b) 67. (c)

Chapter 15: CRISPR-mediated gene regulation

15.1. CRISPRa- Gene activation with dCas9

Although the CRISPR-Cas technique has revolutionized gene editing, scientists are always looking for new possibilities that the CRISPR-Cas system can do? CRISPR-Cas system has the potential to reversibly activate or silence genes with the help of the dead Cas9 enzyme, simply referred to as dCas9. We have already discussed in our previous chapters that dead Cas9 is a mutant in which both cleavage domains HNH and RuVC of Cas9 are inactivated. Although dCas9 can no longer cleave DNA, it can still bind target DNA with the same precision when guided by sgRNA.

When transcriptional activators are fused to dCas9, the resulting complex can activate the expression of the desired gene *(Fig 15-1)*. Gene activation by dCas9 is called **CRISPR activation** or simply written as **CRISPRa**.

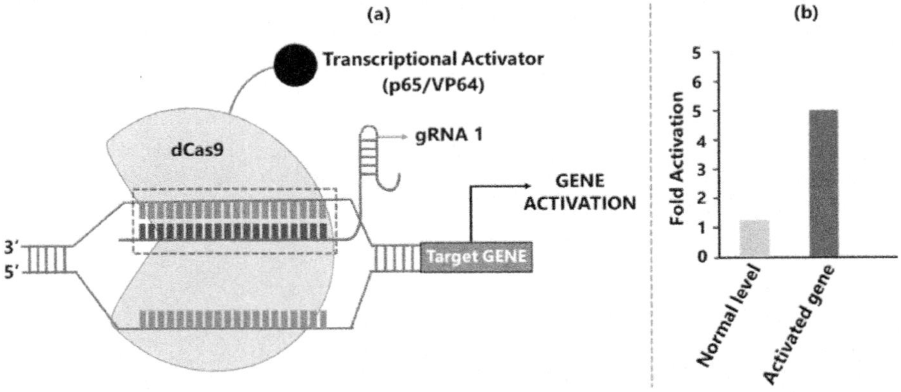

Fig 15-1: Gene activation when dCas9 is fused with transcriptional activators

Transcriptional Activators are the proteins that bind to the enhancers or promoter-proximal elements. After binding, they assist in the recruitment of transcription machinery, including transcriptional cofactors and RNA Polymerase, to the promoter for transcription of the desired gene, i.e., making RNA from the DNA template of the gene *(Fig 15-2)*.

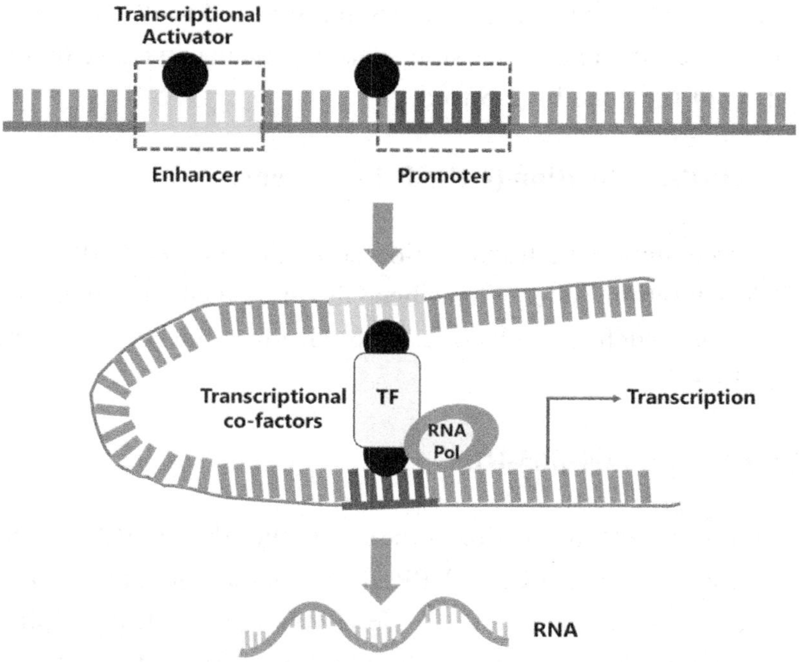

Fig 15-2: Gene transcription with transcriptional activators

Transcriptional activators consist of a DNA binding domain and a transcription activation domain. Through the activation domain, additional transcription factors or RNA polymerase can be recruited to the promoter site, thereby triggering the transcription of the targeted gene.

For activation of gene expression in eukaryotic cells, the transcriptional activators used are **VP64** or **p65**. VP64 is a synthetic tetramer of the activation domain of Herpes Simplex Viral Protein 16, which is a potent transcriptional activator that can activate the transcription of viral genes required for viral replication. On the other hand, the p65 is a subunit of the NF-kB transcription factor and is involved in many cellular processes, including inflammation and immune response.

When these transcriptional activators are attached to dCas9, they can activate the target gene expression. This approach can cause moderate activation of the gene by 2-5 fold.

15.2. CRISPR activation (CRISPRa) systems

In order to enhance the transcription activation power of dCas9, few CRISPR activation systems have been developed and are expressed in viral vectors such as adeno-associated virus (AAV) vectors or lentiviral vectors.

15.2.1. dCas9-VP64-p65-Rta

The first CRISPR activation system is the dCas9-VP64-p65-Rta system, abbreviated as dCas9-VPR. This system entails attaching the VP64 transcriptional activator to the C-terminus of dCas9, followed by the addition of other transcription factors, p65, and Rta, to the C-terminus of dCas9-VP64, as shown in *Fig 15-3* (Lin et al., 2015; Riedmayr et al., 2022).

For the expression of dCas9-VPR system, an all-in-one vector is used, which contains both gRNA cloning sites under the U6 promoter and a CMV promoter-driven dCas9-VP64-p65-Rta expression cassette. If multiple genes are required to be targeted, then multiple gRNAs are cloned in the vector.

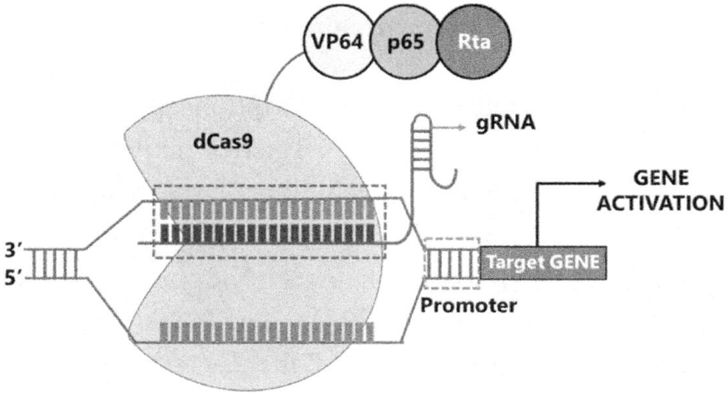

Fig 15-3: dCas9-VPR system

Once expressed, all three transcription factors - VP64, p65, and Rta are targeted to the same gene specified by the gRNA sequence (Moreno et al., 2018; McDougald et al., 2019). Using all three transcription factors, rather than only Vp64 or p65, leads to a higher expression level of the targeted genes. Studies have shown that when dCas9 was used to target various genes, they exhibited significantly higher levels of expression with dCas9-VPR than with dCas9-VP64 alone. To enhance the expression of multiple genes within a single cell utilizing the dCas9-VPR system, multiple sgRNAs are introduced into the cell.

The dCas9-VPR system has been used to activate the genes: neurogenin 2 and neurogenic differentiation 1, resulting in the differentiation of induced pluripotent stem cells into neurons that can be used for developing treatments for various neurodegenerative conditions, including Alzheimer's disease, stroke, and Parkinson's disease. The system has also been used to activate the *Myod1* gene in fibroblasts to successfully reprogram them into myocytes (Dominguez et al., 2016; Nihongaki, 2017; Chen and Qi, 2017; Omachi and Miner, 2021).

15.2.2. Synergistic Activation Mediator (SAM)

The second way to enhance the transcription activation power of dCas9 is by developing the synergistic activation mediator, abbreviated as SAM. This system is built upon the basic dCas9-VP64 structure but includes a modified sgRNA that incorporates two 138-nucleotide (ntd) RNA hairpin aptamers (Xiong et al., 2016; Vad-Nielsen et al., 2018; Hunt et al., 2021). The RNA hairpin aptamers form the binding sites for dimers of the bacteriophage MS2 coat proteins *(Fig 15-4)*. MS2 coat proteins can further recruit additional activators, such as p65 and the activation domain of human heat shock factor 1 (HSF1).

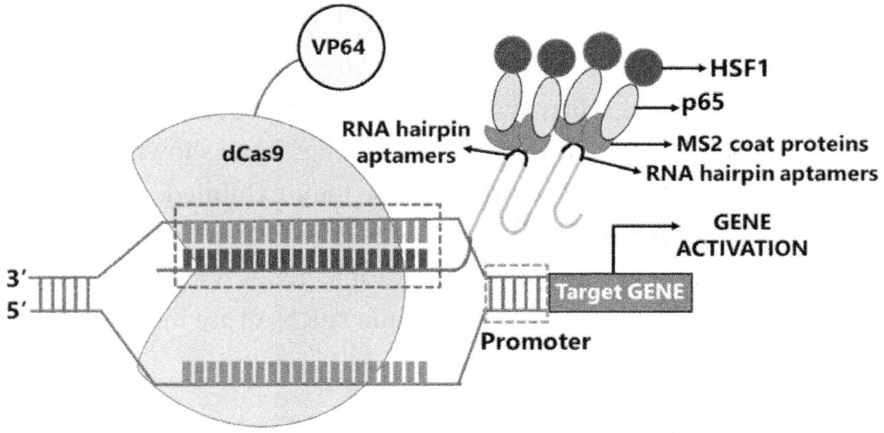

Fig 15-4: Synergistic Activation Mediator (SAM) system

For the expression of the SAM system, separate CRISPR vectors are required. One vector contains desired gRNA cloning sites under the U6 promoter and a CMV promoter-driven dCas9-VP64 expression cassette. The desired gRNA sequences are modified to include two 138-ntd hairpin RNA aptamers, which form binding sites for the bacteriophage MS2 coat proteins. If multiple genes are required to be targeted, then multiple modified gRNAs are cloned in the vector.

And the second vector drives the expression of MS2, p65, and HSF-1 proteins under CMV promoters.

When cells are co-transduced with these vectors, the dCas9 enzyme fused with VP64 forms a complex with the desired gRNA and binds to the upstream regions proximal to a promoter of the target gene. The RNA hairpin aptamers of gRNA recruit MS2 proteins, which further recruit p65 and HSF-1 proteins. These transcriptional activators then recruit other transcription factors and RNA polymerase to the promoter to initiate transcription of the gene.

Through synergistic interactions among the VP64, p65, and HSF1 activation domains, the dCas9-SAM system has the potential to amplify gene expression from 10 to multiple thousand-fold. The dCas9-SAM technique is being used by scientists to eradicate HIV-1 in infected organisms by reactivating the latent HIV gene by over-expressing viral proteins in the HIV host cells. Over-expression of viral proteins substantially triggered apoptosis of HIV-infected T cells for their destruction (Saayman et al., 2015; Zhang et al., 2015). Therefore, CRISPRa can provide an alternative therapeutic approach to excise HIV DNA from host DNA in HIV-infected T cells (discussed in *Chapter 8*).

15.2.3. Supernova Tagging System

The third way of enhancing the transcription activation of dCas9 is by using the Supernova Tagging system. In this system, the dCas9 protein is modified to be linked with the SunTag. The SunTag is a repeating polypeptide array with multiple copies of GCN4 peptide, which can recruit multiple copies of antibodies that are attached to transcriptional factors like VP64 or p65 (Tanenbaum et al., 2014; Konermann et al., 2015). By attaching transcriptional factors to the antibodies, the SunTag-dCas9 activating complex has the potential to amplify the gene expression by more than 50 fold.

Generally, the antibodies used are GCN4 antibodies which can bind to transcriptional factor VP64. For the expression of the Supernova Tagging system, three CRISPR vectors are required.

- One vector drives the expression of dCas9 and SunTag, i.e., multiple copies of GCN4 peptide.
- The second vector is responsible for inducing the expression of both GCN4 antibodies and the transcriptional activator VP64. Also, a Nuclear Localization Signal (NLS) tag is attached to transport the antibodies to the cell nuclei.
- And the third vector drives the expression of sgRNA.

When the cells are cotransduced with these vectors, the dCas9 enzyme fused with GCN4 peptides forms a complex with the desired gRNA and binds to the upstream regions proximal to a promoter of the target gene. Subsequently, the expressed GCN4 antibodies fused to the transcriptional factors VP64 bind to the GCN4 peptide fused to the dCas9 *(Fig 15-5)*. The attached multiple transcriptional activators then activate the target gene expression.

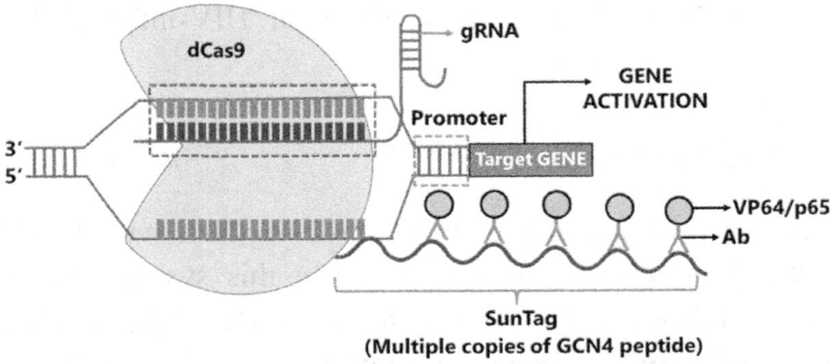

Fig 15-5: Supernova Tagging System

Studies have shown that the antibodies successfully bound to SunTag polypeptides, and transcriptional activator VP64 activated target *CXCR4* gene expression in K562 cell lines by 50-fold. Also, the dCas9-SunTag-VP64 system has been used to reactivate latent HIV-1 transcription in infected human T-cell lines (Ji et al., 2016; Kazi and Biswas, 2021).

A study has shown that when used with dCas9, the VPR, SAM, and SunTag activators are most effective at boosting gene expression across a range of fruit fly, mouse, and human cell lines.

15.3. Gene Therapy using CRISPRa

Gene therapy using the CRISPR activation system is under clinical trials. Healthy mice and humans have two copies of genes: one inherited from each parent. A defect in one copy can abolish gene function, leading to a reduced amount of protein and, consequently, human disease, a condition termed as haploinsufficiency. More than 660 genes are currently estimated to cause diseases due to haploinsufficiency, thus, leading to a wide range of phenotypes that include metabolic disorders, immunological diseases, neurological diseases, developmental disorders, kidney disease, cancer, infertility, and many others.

Let's take the example of obesity which is caused by haploinsufficiency of either the transcription factor Sim1 or the melanocortin 4 receptor (*Mc4r*) gene. Using a sgRNA, when dCas9 fused to an activator is targeted to either the *Sim1* promoter or its ~270 kb distant enhancer in the neuronal cells, it up-regulated Sim1 expression and rescued Sim1-mediated obesity in haploinsufficient mice (Matharu et al., 2019; Montefiori and Nobrega, 2019). On the other hand, targeting *Mc4r* promoter by means of a similar CRISPRa approach reduces weight gain in obese mice.

The other example of CRISPRa is eliciting antitumor immunity. The tumor antigens might not be expressed at levels sufficient to elicit an effective T cell-mediated response against cancer cells. In other cases, cancer cells downregulate antigen presentation to escape immune recognition. But delivering the CRISPRa system into tumor cells via AAV or lentiviral vectors activates the expression of endogenous genes, thereby enhancing the tumor antigens' presentation. This leads to increased antitumor immune responses like enhanced T cell recruitment to kill the tumor cells (Wang et al., 2019).

15.4. Gene silencing with dCas9 (CRISPRi)

In addition to being fused to transcriptional activators, dCas9 can also function as a repressor. Gene silencing by dCas9 is called **CRISPR interference** or written as **CRISPRi**. When the gRNA-dCas9 complex binds to the region downstream of the promoter or exon sequence of the target DNA, it interferes with transcription by preventing transcriptional factors and RNA polymerase from accessing the promoter of the target DNA, resulting in reversible gene silencing (Qi et al., 2013; Gilbert et al., 2014). However, the efficiency of gene repression by dCas9 alone is relatively low, possibly because it is insufficient to completely disrupt the action of eukaryotic RNA polymerases.

One strategy to improve the efficiency of gene repression is by fusing the dCas9 to the transcriptional repressor KRAB *(Fig 15-6)*. KRAB stands for Krüppel-associated box, which is the transcriptional repressor domain of the zinc finger proteins such as Kox1 or ZFN10 (Alerasool et al., 2020; Stoll et al., 2022). dCas9-KRAB-gRNA system can achieve the repression level of a targeted gene up to 70-90%.

An all-in-one vector is typically used for the expression of the CRISPRi system. This vector contains both gRNA cloning sites under the U6 promoter and a CMV promoter-driven dCas9-KRAB expression cassette. If multiple genes are required to be targeted, then multiple gRNAs are cloned in the vector. After the expression, dCas9 and KRAB are targeted to the same target gene defined by the gRNA sequence, resulting in its silencing.

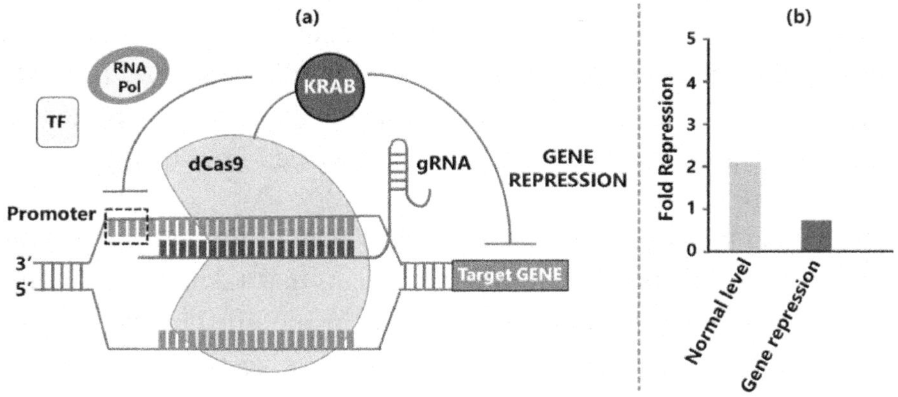

Fig 15-6: Gene silencing with dCas9 or CRISPRi

Other transcriptional repressors that can be fused to dCas9 are SALL1 and SDS3. CRISPRi has been used to treat familial hypercholesterolemia and obesity in mouse models in vivo by repressing the expression of genes that are involved in the disease pathology.

15.5. Epigenetic editing with dCas9

The second level of genomic regulation after transcription effectors is epigenetic gene regulation. This process involves a series of chemical modifications that results in selective silencing or activation of particular genes without direct altering the DNA primary sequence. Epigenetic modifications can occur within the DNA itself or on the histone proteins that assist in DNA packaging. These modifications play a crucial role in deciding whether genes are activated or suppressed, thereby influencing protein production within cells. This regulatory mechanism helps guarantee that each cell produces solely the proteins that are necessary for its proper function.

A common type of epigenetic modification is DNA methylation. DNA methylation involves the attachment of methyl groups to the C5 position of cytosine residues in CpG dinucleotide sequences of the DNA. The presence of methyl groups in the promoter blocks the binding of transcription factors, and RNA polymerase II; thus, the transcription of the gene is inhibited, causing the gene to be turned off or silenced, and therefore, no protein is produced from that gene.

Histone modification is another prevalent form of epigenetic modification. Histones are structural proteins that aid in the packaging of DNA within the cell nucleus. They can be modified by the addition or removal of chemical groups like methyl or acetyl groups. These chemical groups impact how tightly DNA is wrapped around the histones and whether a given genomic sequence is stretched out and accessible to the transcriptional machinery, affecting a gene to be turned on or off. For instance, the addition of acetyl groups to the lysine residues of histone proteins causes histones to drift away from DNA. This released structure facilitates the binding of transcription factors and RNA polymerase II to the target gene, thus activating its transcription. On the other hand, histone deacetylation, i.e., the removal of acetyl groups, blocks access to transcriptional machinery, thus, repressing the transcription of the target gene.

Also, the addition of methyl groups to the histone proteins can either activate or repress the transcription of the gene, thus turning it on or off. For example, trimethylation of histone H3 at lysine 4 (H3K4me3) in the promoter region activates the transcription of the gene. However, dimethylation of histone H3 at lysine 9 (H3K9me2), is a signal for transcriptional silencing.

Thus, the epigenome, which consists of chemical compounds that modify genomic DNA and histones, acts as the second level of gene expression regulation after the transcription effectors.

To achieve CRISPR-Cas9 mediated epigenetic regulation, dCas9 can be fused to various effectors. One such effector is histone acetyltransferase p300, which can be fused to dCas9, creating a dCas9-p300 fusion activator (Hilton et al., 2015; Chen et al., 2017; Kuscu et al., 2019). p300 adds acetyl groups to the histone proteins and facilitates the binding of transcription factors and RNA polymerase II to the target gene, thus significantly activating its transcription *(Fig 15-7)*. Furthermore, the dCas9-p300 fusion achieves higher activation than dCas9-VP64 alone.

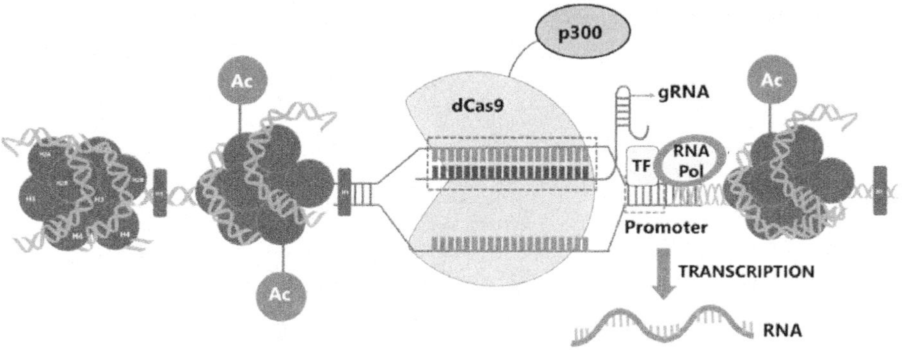

Fig 15-7: Gene activation with dCas9-p300

Another transcriptional activator is Tet1 demethylase, which, when fused to dCas9, decreases methylation and induces transcription at CpG regions of the promoter of the target gene (Gjaltema and Rots, 2020).

In contrast to transcriptional activators, transcriptional repressors DNA Methyltransferase 3 Alpha, abbreviated as DNMT3A, and Lysine-specific histone demethylase 1 abbreviated as LSD1, repress transcription of the target gene. DNA Methyltransferase 3 Alpha, when fused to dCas9, causes cytosine methylation of the promoter of the target gene, thus repressing its activation. On the other hand, LSD1 is a histone demethylase that, when fused to dCas9, targets histone

H3, methylated at lysine 4 residue, and demethylates it, leading to transcriptional repression of the target gene.

This is how dCas9 can regulate the target gene expression.

Chapter 16: CRISPR-mediated Chromatin Immunoprecipitation

16.1. Introduction to ChIP

Chromatin Immunoprecipitation, abbreviated as ChIP, is an antibody-based technology used to determine in vivo interactions occurring between DNA and proteins in the cell by immunoprecipitating chromatin with specific antibodies. The protein can be histones or regulatory proteins such as transcription factors, proteins binding to enhancers, etc. DNA-protein interactions play a key role in regulating several cellular functions, including DNA replication and recombination, DNA repair, gene transcription, chromosomal stability, cell cycle progression, and epigenetic silencing.

Crosslinking: The first step of ChIP assays begins with the covalent stabilization of protein-DNA complexes by crosslinking *(Fig 16-1a)*. Crosslinking is normally achieved by treating intact cells with formaldehyde.

Cell lysis: In the ChIP protocol, the second step involves extracting chromatin (*Fig 16-1b*). This can be achieved by either using a detergent-based solution to dissolve the cell membrane or by using glass beads to disrupt the membrane through bead beating. During the ChIP procedure, the stability of the protein-DNA complex remains unaffected by the presence of detergents or bead beating. This is because the covalent crosslinking maintains the stability of the complex throughout the process.

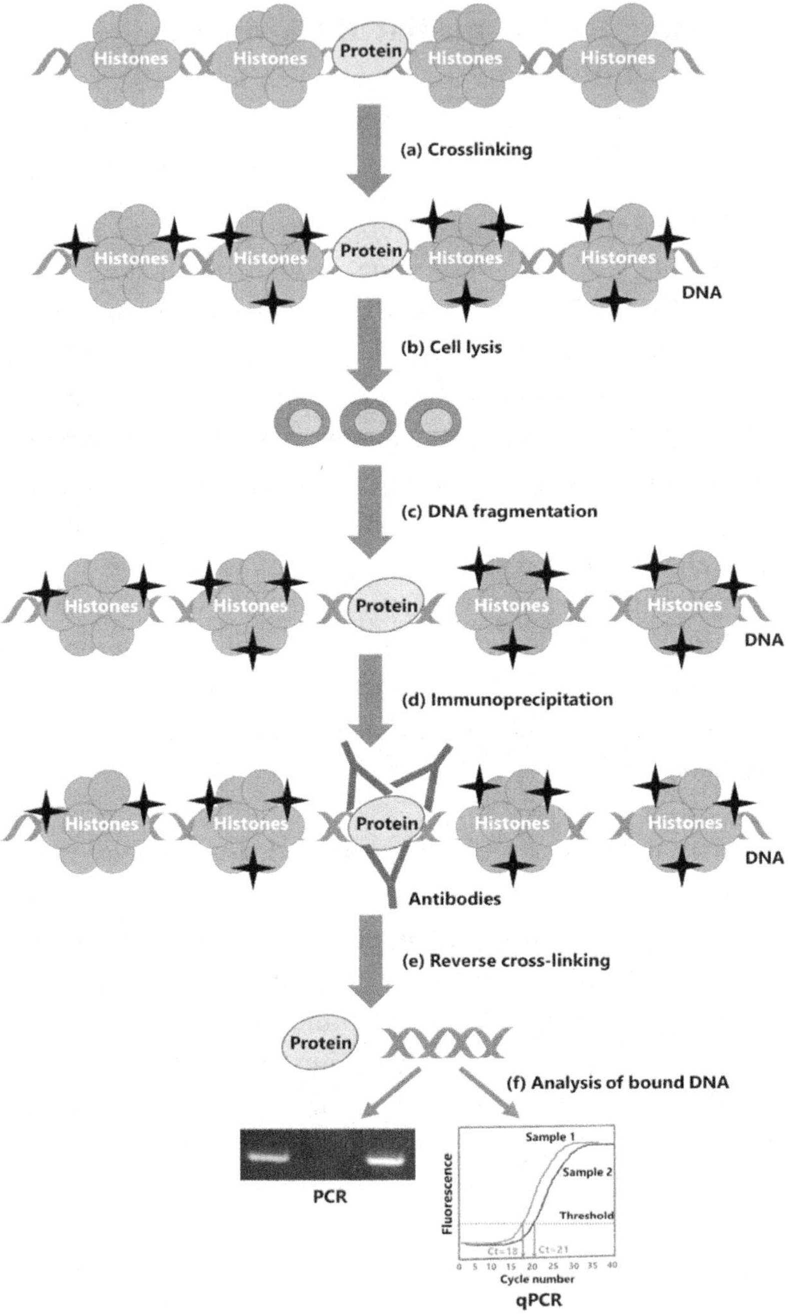

Fig 16-1: Chromatin Immunoprecipitation (ChIP) assay

DNA fragmentation: In order to analyze the sequences where proteins bind, it is necessary to fragment the extracted genomic DNA into smaller pieces *(Fig 16-1c)*. This fragmentation can be achieved by either mechanical methods, such as sonication, or enzymatic digestion using micrococcal nuclease.

Immunoprecipitation: The subsequent step in immunoprecipitation involves selectively enriching the desired protein-DNA complex and removing any other irrelevant cellular material. This is achieved by incubating the sonicated chromatin with antibodies that recognize the target protein and form an antibody-protein-DNA complex *(Fig 16-1d)*.

However, there are cases where specific antibodies against the target protein are not available. In such cases, fusion proteins tagged with affinity tags such as GST or FLAG can be used instead. To utilize this approach, the fusion protein is expressed in the cells of interest. After crosslinking and sonication, antibodies against the affinity tags are used to immunoprecipitate the target protein. Once the antibody-protein-DNA complex is formed, it is purified using an antibody-binding resin such as immobilized protein A, protein G, or protein A/G *(Fig 16-2)*. These resins selectively bind to the antibody portion of the complex and allow for the removal of any unbound material.

The choice of resin used depends on the antibody class or subclass used for immunoprecipitation. For example, protein A resin binds to the Fc region of the IgG antibodies, while protein G resin can bind to the Fc regions of IgG, IgM, and IgA antibodies. Protein A/G resin combines the binding properties of both protein A and protein G, allowing for a wider range of antibody classes and subclasses to be captured.

If the target protein is biotinylated, immobilized streptavidin can be used instead of antibody-binding resins.

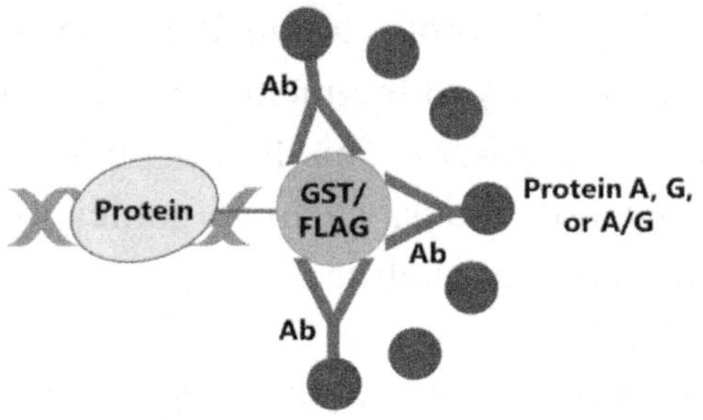

Fig 16-2: Affinity purification of DNA-bound tagged proteins-antibody complex using immobilized protein A, G or A/G

Reverse cross-linking: Subsequently, the crosslinks between DNA and protein are then reversed by either extensive heat incubations and/or by digesting the protein component using proteinase K *(Fig 16-1e)*. The heat treatment breaks the covalent bonds between the protein and DNA, while proteinase K cleaves peptide bonds in the protein components, both of which help to release the DNA from the protein-DNA complex. Next, the DNA is extracted from the complex with phenol-chloroform, followed by precipitation using ethanol. Phenol-chloroform is a mixture of two organic solvents that can separate DNA from other contaminants such as proteins, lipids, and carbohydrates. Ethanol precipitation is a process where the DNA is mixed with ethanol, which causes the DNA to clump together and form a visible pellet. The pellet is then washed with ethanol to remove any remaining contaminants, and the resulting purified DNA is resuspended in a buffer solution for further analysis.

After the DNA is purified from the complex, the next step in the ChIP process is to examine the **in vivo recruitment of the target protein**. This is done by using PCR amplification to detect specific DNA sequences that are bound to the protein of interest, such as promoter sequences *(Fig 16-1f)*. PCR, or polymerase chain reaction, is a powerful technique that can amplify small amounts of DNA into millions of copies. To perform the PCR, primers are designed to amplify the region of interest, which could be a promoter sequence or another region of DNA that is known to interact with the protein of interest. The purified DNA sample is then used as a template for the PCR reaction, along with the appropriate primers and other necessary components like buffer, polymerase enzyme, and dNTPs.

The enrichment of the PCR signal from the immunoprecipitated sample is compared to the signal from the mock sample with no antibody control. The mock sample is a negative control used to determine the background level of DNA amplification in the absence of the protein of interest. By comparing the signal from the immunoprecipitated sample to the mock sample, the region-specific recruitment of the protein can be determined.

If the immunoprecipitation was successful and specific to the protein of interest, there should be a higher level of amplification in the immunoprecipitated sample compared to the mock sample. This difference in amplification signals indicates that the protein is specifically bound to the region of interest in the genome.

By detecting specific DNA sequences bound to the protein of interest, the results obtained from PCR amplification can provide insights into the biological function of the protein and its role in various cellular processes. This information is crucial in understanding the complex mechanisms involved in gene regulation and other biological pathways.

16.2. enChIP: CRISPR-mediated ChIP

Using CRISPR, the researchers expanded chromatin immunoprecipitation (ChIP) to identify proteins that are associated with a genomic sequence specified by a particular gRNA (Fujita and Fujii, 2013). This technique is also called engineered DNA-binding molecule-mediated ChIP, abbreviated as enChip. In the enChIP system, catalytically inactive dCas9 is used, which is fused to tags like FLAG (Fujita et al., 2018; Hamidian et al., 2018). A guide RNA (gRNA) is designed to specifically recognize and bind to the target genomic sequence, such as a promoter or other regions of interest. The gRNA then recruits FLAG-tagged dCas9 to a locus of interest, where it forms a complex with the DNA *(Fig 16-3 a and b)*.

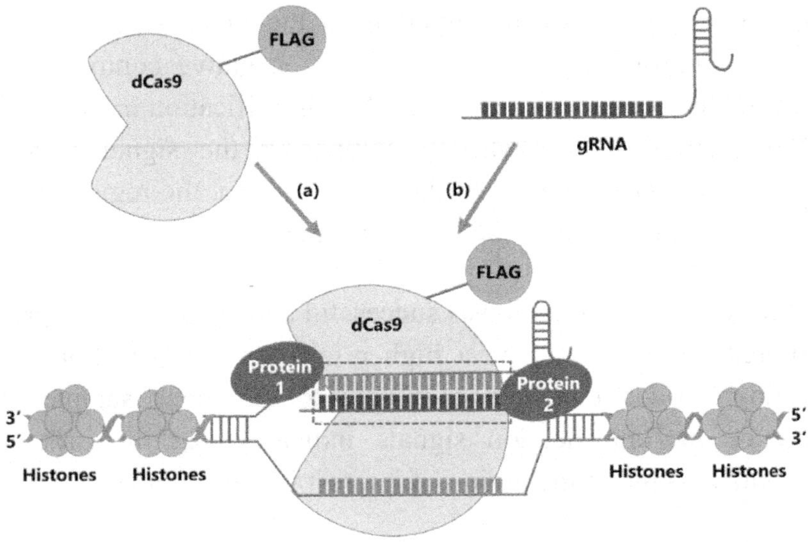

Fig 16-3: Recruitment of FLAG-tagged dCas9 to target DNA using gRNA

After which, the chromatin is crosslinked to fix the DNA-protein complexes in place *(Fig 16-4a)*, fragmented *(Fig 16-4b)*, and dCas9-bound DNA-protein complexes are immunoprecipitated via FLAG-tag specific antibodies, which bind to the FLAG tag on the dCas9 protein *(Fig 16-4c)*.

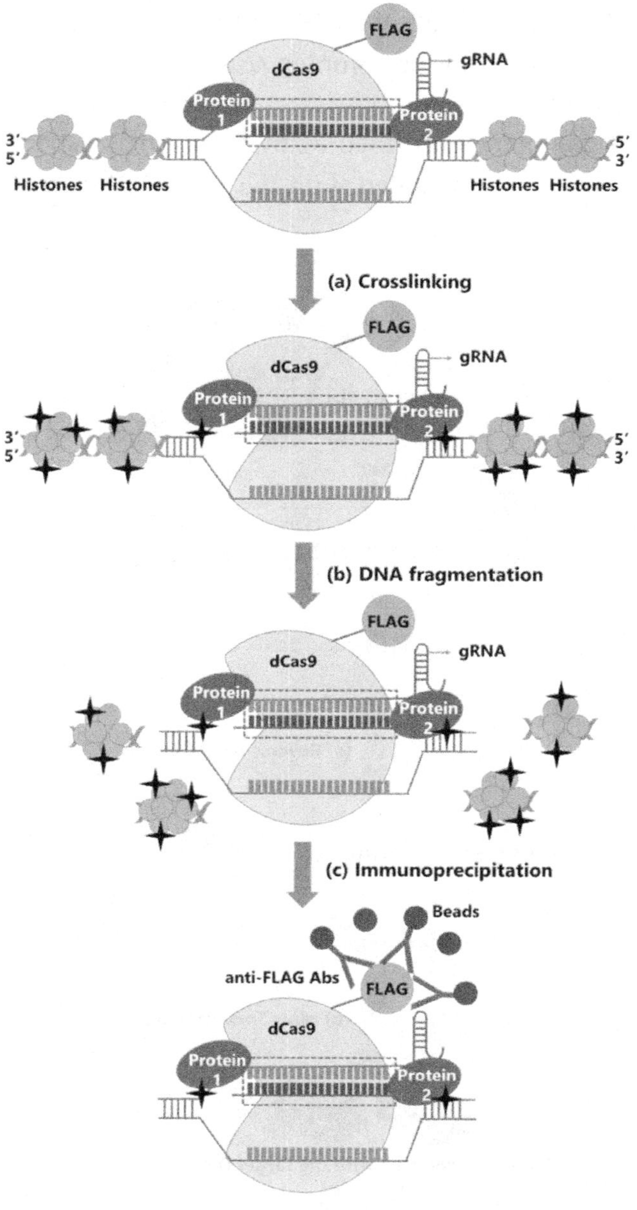

Fig 16-4: Immunoprecipitation of dCas9-bound DNA-protein complexes

Subsequently, the FLAG antibody-dCas9-DNA complex is purified using an antibody-binding resin/beads such as immobilized proteins

A, G, or A/G. The crosslinks are then reversed to release the DNA and protein components from the complex *(Fig 16-5a)*.

The isolated complexes are then analyzed by mass spectrometry to identify the proteins associated with the target DNA sequence *(Fig 16-5b)*.

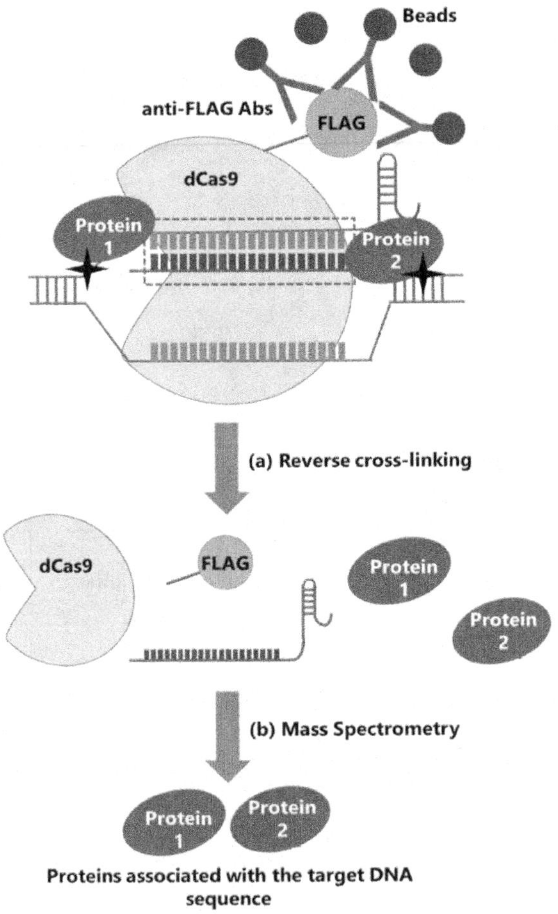

Fig 16-5: Analysis of proteins associated with the target DNA sequence

Mass spectrometry is a powerful analytical technique used to identify and quantify molecules based on their mass-to-charge ratio (m/z). In the context of enChIP, mass spectrometry is used to identify the proteins that are bound to the target DNA sequence. After the immunoprecipitation step, the protein-DNA complexes are isolated and then subjected to protease digestion, which cleaves the proteins into smaller peptides. The resulting peptides are then separated by liquid chromatography based on their chemical and physical properties, such as size, charge, and affinity. The separated peptides are then introduced into the mass spectrometer, where they are ionized, i.e., given a positive or negative charge by applying a high voltage or using a laser. Using an electric or magnetic field, the resulting ions are separated based on their mass-to-charge ratio (m/z). The lighter ions will be deflected more than the heavier ones, allowing for separation based on their m/z ratio. The detector then records the intensities of the ions, producing a mass spectrum that represents the peptides present in the sample.

The mass spectra are then analyzed using database searching software that matches the observed mass spectra to theoretical spectra of known proteins. The software generates a list of proteins that are present in the sample based on the peptide masses and their fragmentation patterns.

The identified proteins are then further validated using additional experimental techniques such as Western blotting or co-immunoprecipitation to confirm their association with the target DNA sequence. Thus, mass spectrometry is a highly sensitive and powerful tool for identifying the proteins associated with a particular genomic region in the enChIP technique.

Tagging dCas9 with BirA: Another approach to tagging dCas9 is by fusing dCAs9 to the biotin ligase BirA (Schmidtmann et al., 2016; Liu et al., 2017; Ugur et al., 2020). After supplementing the cells with exogenous biotin, biotin ligase results in the addition of biotin to lysine residues of proteins located in the proximity of the targeted locus *(Fig 16-6)*. The cells are then lysed, and the chromatin-bound by the dCas9-BirA protein complex is fragmented.

The biotinylated proteins are then captured using streptavidin-coated beads, which have a high affinity for biotin. After capturing the biotinylated proteins, they are eluted and can be analyzed using mass spectrometry to identify the proteins that are associated with the targeted DNA sequence.

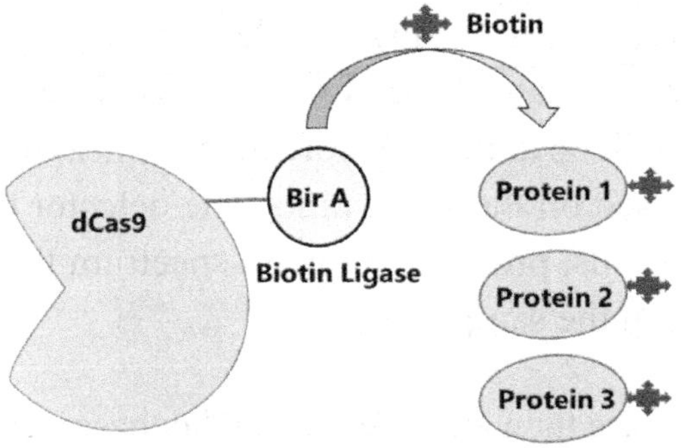

Fig 16-6: Addition of biotin to lysine residues of proteins located in the proximity of the targeted locus

16.3. How is enChIP different from traditional ChIP?

EnChIP and traditional ChIP both aim to identify proteins associated with a specific region of DNA, but the approaches differ in several key ways.

(i) Target specificity: In traditional ChIP, a specific antibody is used to capture the protein of interest, which is bound to the target DNA. The DNA is then fragmented, and the protein-DNA complexes are immunoprecipitated using the antibody against the protein of interest.

On the other hand, enChIP involves recruiting a catalytically inactive dCas9 protein to a specific region of DNA using a gRNA. The dCas9 is fused to tags such as FLAG, and the protein-DNA complex is crosslinked, fragmented, and immunoprecipitated using antibodies specific to the tag.

Because enChIP uses a gRNA to target a specific genomic region, it is generally more specific than traditional ChIP.

(ii) Detection and analysis: Traditional ChIP often uses PCR, microarrays, or next-generation sequencing to identify the regions of DNA that are bound by the protein of interest. Whereas in enChIP, the isolated protein-DNA complex are analyzed using mass spectrometry to identify the proteins associated with the target DNA sequence.

(iii) Sensitivity: enChIP is generally more sensitive than traditional ChIP because it can detect weaker protein-DNA interactions that may be missed by traditional ChIP.

Overall, enChIP offers several advantages over traditional ChIP in terms of target specificity and sensitivity, thus making it a valuable tool for studying protein-DNA interactions and identifying the proteins associated with specific genomic regions.

Chapter 17: CRISPR-mediated live cell imaging

17.1. Fluorescence In Situ Hybridization (FISH)

Cytogenetic techniques like Fluorescence In Situ Hybridization (FISH) can detect and locate the precise location of particular DNA sequences on chromosomes. In this process, detecting the target DNA sequence can be compared to searching for a needle in a haystack, with the DNA sequence being the needle and the set of chromosomes being the haystack. The process is significantly facilitated by using a fluorescent probe that is complementary to the DNA sequence of interest. Therefore, the initial step in this process is to generate fluorescent probes through the nick translation technique. The probes used are generally short double-stranded DNA.

Nick Translation method

(i) The nick translation method uses the enzymes DNase I and DNA polymerase I. The enzyme DNAse creates random cuts or single-stranded nicks in one or both strands of the DNA to leave free 3' hydroxyl groups and 5' phosphate termini. Now labeled nucleotides that have a fluorophore attached are provided. The free 3' hydroxyl group at the nicking site is used as a primer by the polymerase I enzyme. DNA polymerase I enzyme performs 2 activities:

- One is 5'-3' exonuclease activity that degrades DNA and removes the native nucleotides from free 5' phosphate termini in the 5′→3′ direction *(Fig 17-1a and b)*
- and the other activity of DNA polymerase I enzyme is 5'-3' polymerase activity that uses free 3' hydroxyl group as a primer and the opposite strand as the template to synthesize a new strand complementary to the template *(Fig 17-1c)*. The ligase enzyme then seals the gaps *(Fig 17-1d)*.

Since the polymerase enzyme incorporates fluorescent nucleotides, the result is the replacement of unlabeled nucleotides with the labeled nucleotides in the ds DNA. This labeled DNA is called a probe.

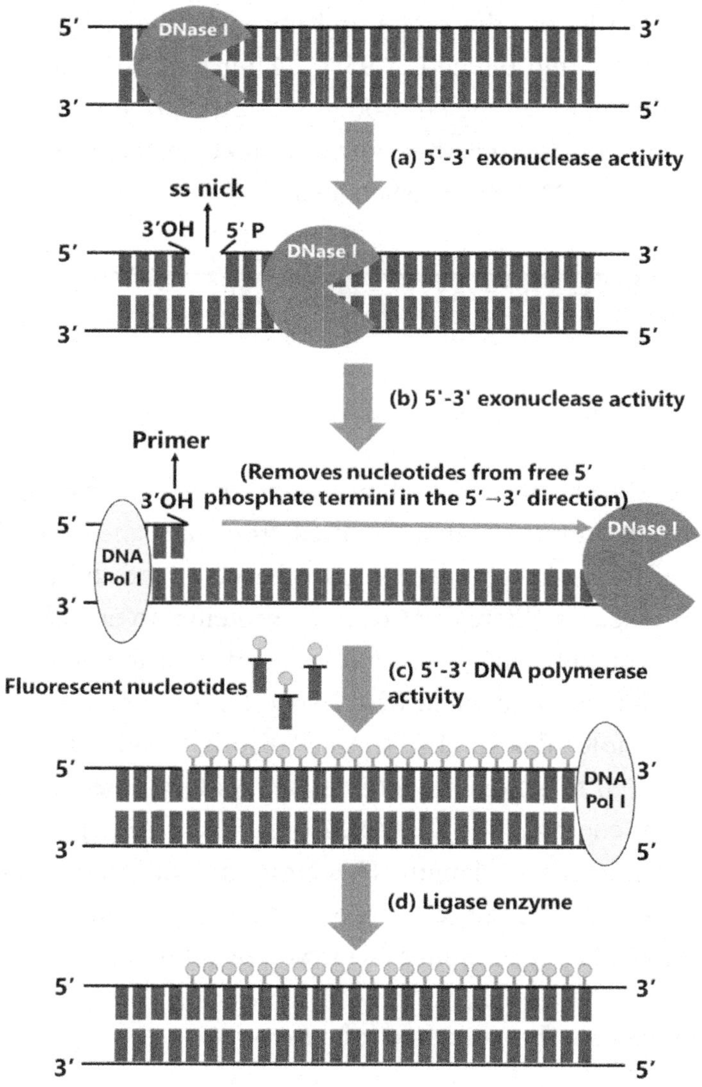

Fig 17-1: Nick translation to generate fluorescent probes

(ii) The second step is the fixation of the cells with the formalin-based fixative that causes protein-protein and protein-DNA crosslinks. Then the fluorescent probes are added.

(iii) In the third step, the probe and chromosomal target DNA are denatured by heating the cell at 95°C. After that, the temperature is reduced, allowing the labeled probe to specifically bind to the target sequence *(Fig 17-2)*. Once hybridization takes place, probes that are not bound to any DNA are washed away.

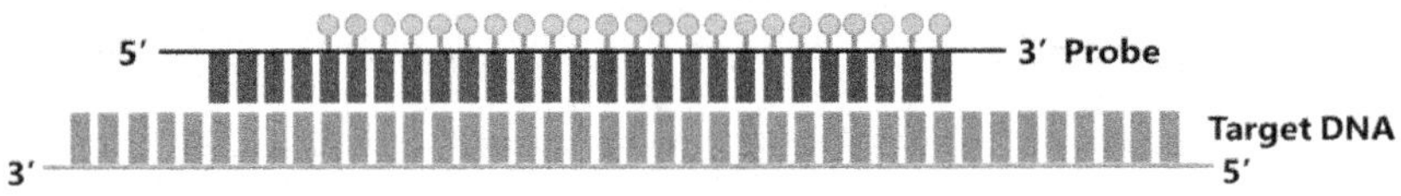

Fig 17-2: Hybridization of fluorescent probes to the target DNA

(iv) The results can be analyzed on a fluorescence microscope, which detects the fluorescent signals emitted from the labeled probe. The principle of fluorescence microscopy is based on the ability of certain molecules, such as fluorescent dyes or proteins, to absorb light of a specific wavelength and then emit light of a different wavelength. When a sample is illuminated with a specific wavelength of light, the fluorescent molecules in the sample absorb this light energy and become excited to a higher energy state. As the molecules return to their original energy state, they release the excess energy in the form of light of a longer wavelength. This emitted light is then captured by a camera or other detector, which converts the light into an image that can be analyzed by a computer or viewed by a human observer.

Components of a fluorescence microscope:

(i) Excitation Light Source: A light source (usually a high-intensity mercury lamp or a laser) emits light of higher energy and a shorter wavelength to excite the fluorescent molecules in the sample *(Fig 17-3)*.

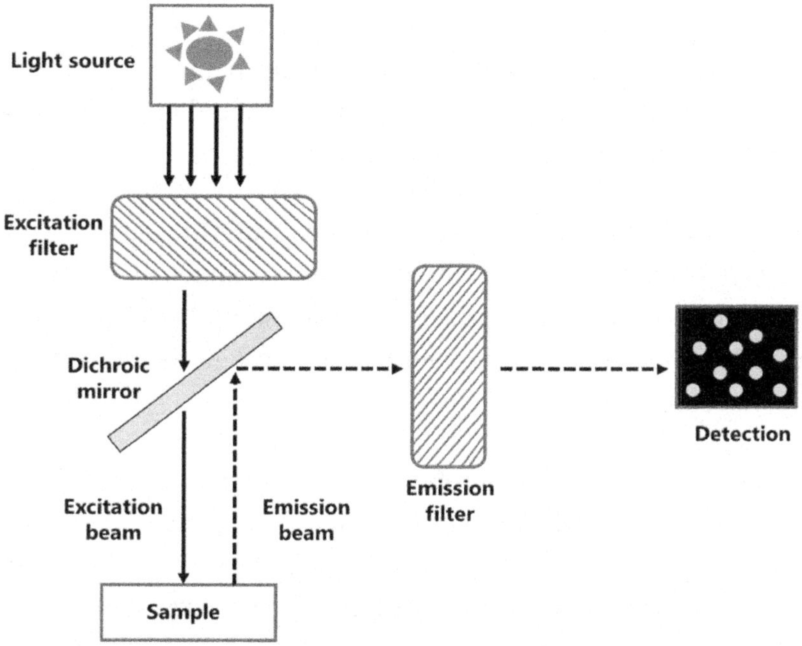

Fig 17-3: Principle of fluorescence microscope

(ii) Excitation Filter: The excitation filter allows only the short wavelength of light to pass through it and removes the non-specific wavelengths of light.

(iii) Dichroic Mirror: The filtered light is then reflected by the dichroic filter and falls on the sample, which is fluorophore-labeled. The fluorophore molecules absorb shorter wavelength rays and get excited to a higher energy state. As the molecules return to their original energy state, they release the excess energy in the form of light of a longer wavelength.

(iv) Emission Filter: The emission filter blocks any residual excitation light and passes the desired longer emission wavelengths to the detector.

(v) Detector: A camera or other detector captures the emitted fluorescent light and forms glowing images of the fluorophore-labeled molecules against a dark background.

In the case of FISH, a fluorescent probe is used to label a specific DNA sequence of interest. The fluorophore attached to the probe emits a fluorescence signal when it is excited by the excitation light source. The emission light is then passed through the emission filter to the detector.

The specificity of the FISH probe is crucial for accurate results. If the probe is designed to bind only if it is 100% complementary to the gene and that sequence is present in the sample. In that case, the resulting image should show a bright spot of fluorescence corresponding to the location of the target DNA sequence. But if any sort of mutation occurred in this gene, no hybridization is possible, and no fluorescence signal will be detected at that location.

FISH technique can be used in the clinical diagnosis of various chromosomal abnormalities, including deletions, duplications, and translocations.

17.2. CRISPR-mediated live-cell imaging of genomic loci

Despite the widespread application, there are several drawbacks associated with FISH. First, the need for cell fixation makes the technique cumbersome for studying chromatin dynamics. Additionally, whether the state of chromatin architecture is properly preserved during FISH processing has always been questionable since the DNA duplex must be denatured to allow probes to hybridize with the target sequence.

The CRISPR-based imaging system is a recently-developed technology that has the potential to overcome the limitations of FISH and enable live-cell imaging of the genome. For this, dead Cas9 or dCas9 is used, which retains the ability to interact with sgRNA and to bind to target DNA. Either the dCas9 protein or sgRNA is modified to develop DNA imaging probes that integrate fluorescent proteins in a manner that doesn't interfere with either dCas9 binding to sgRNA or sgRNA binding to the genomic sequence.

17.3. Labeling dCas9 and sgRNA with fluorescent proteins

Let's now understand how dCas9 and sgRNA can be labeled. The dCas9 can be labeled by fusing it with eGFP fluorescent protein. dCas9-eGFP, when introduced into the cell with sgRNA, can image the highly repetitive elements of the telomere region in the chromosome (Chen et al., 2013; Wu et al., 2019). The repetitive sequences can be imaged using a single gRNA, but imaging of nonrepetitive sequences requires simultaneous delivery of different gRNA per target *(Fig 17-4)*.

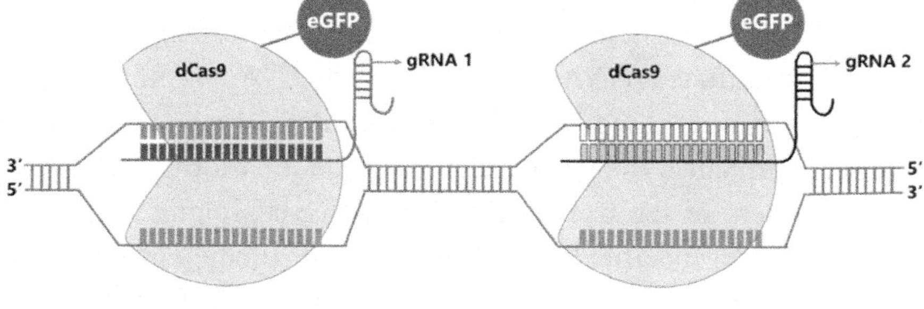

Fig 17-4: Different gRNAs are required to be delivered with eGFP tagged dCas9 for imaging nonrepetitive sequences

For instance, the dCas9-eGFP protein has been used by scientists to image nonrepetitive regions of the *MUC4* gene through the use of an array of at least 26 different sgRNAs.

To further improve genome imaging or gene detection, dCas9 can be fused with multiple fluorescent proteins through the use of the **supernova tagging system**. In this system, the dCas9 protein is modified to be linked with the SunTag, which is a repeating polypeptide array with multiple copies of GCN4 peptide *(Fig 17-5)*. GCN4 peptide can recruit multiple copies of antibodies that are attached to a number of fluorescent proteins.

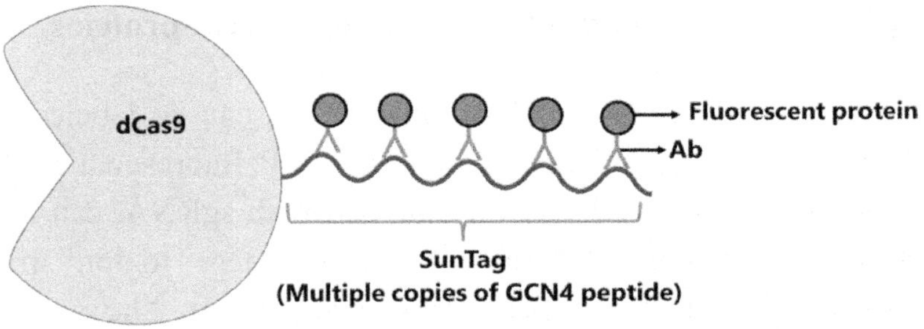

Fig 17-5: dCas9 protein linked with SunTag can recruit multiple copies of antibodies attached to fluorescent proteins

Alternatively to fused dCas9-fluorescent protein, a number of research groups have demonstrated the feasibility of imaging genome loci using **modified sgRNAs** that incorporate RNA hairpin aptamers. The aptamers can then recruit RNA-binding proteins fused to fluorescent proteins. The most widely used aptamer forms the binding sites for dimers of the bacteriophage MS2 coat proteins. MS2 coat proteins then can specifically bind to its cognate binding protein MCP fused to a fluorescent protein *(Fig 17-6)*. Thus, each dCas9-sgRNA complex can then be tagged by multiple fluorescent proteins through MS2–MCP interactions (Tanenbaum et al., 2014; Ye et al., 2017).

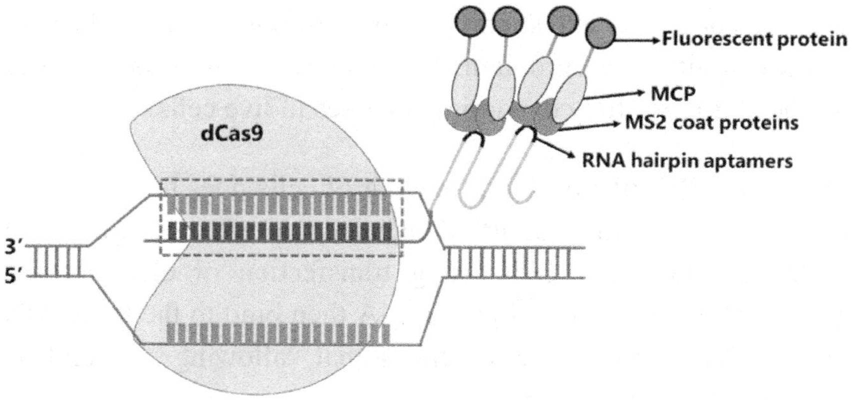

Fig 17-6: Tagging dCas9-sgRNA complex by multiple fluorescent proteins through MS2–MCP interactions

A second approach, termed **Casilio**, employs the modification of sgRNA to harbor a unique 8-mer RNA sequence called PUF binding sequence or PBS (Cheng et al., 2016; Clow et al., 2020). Like MCP, PUF domain can also be fused to a fluorescent protein while retaining its capacity to bind to PBS *(Fig 17-7)*. Engineering sgRNA with tandem repeats of PBS makes it possible to label the dCas9-sgRNA complex with multiple PUF fusion fluorescent proteins.

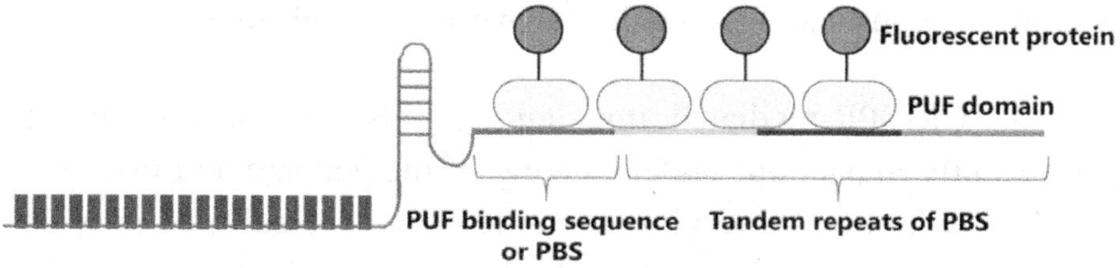

Fig 17-7: Casilio approach to label the dCas9-sgRNA complex with multiple fluorescent protein-PUF fusion proteins

17.4. Advantages of CRISPR-mediated imaging

CRISPR-mediated imaging of cells offers several advantages over traditional imaging methods. Traditional imaging methods involve fixing cells or tissues and staining them with fluorescent probes. This process can alter the chromatin architecture during processing and limit the ability to study dynamic processes in live cells.

In contrast, CRISPR-mediated imaging of cells does not require cell fixation. Instead, the modified labeled dCas9 and gRNA are introduced into living cells using transfection or electroporation methods. The labeled dCas9 and gRNA then bind to the target DNA sequence and emit a fluorescent signal, allowing for real-time visualization of the genomic locus.

One significant advantage of this technique is the ability to study the dynamics of chromatin structure, gene expression, and other genomic processes in real-time. By observing changes in chromatin structure, gene expression, and other genomic events as they occur, researchers can better understand the dynamic mechanisms that regulate gene expression and genome function.

Another advantage of this technique is the ability to label multiple loci simultaneously using different fluorescent proteins. This approach enables researchers to study the interactions between different genomic regions and observe their dynamics in real-time.

Overall, CRISPR-mediated imaging of cells has the potential to significantly impact our understanding of the genome and its role in health and disease. Observing genomic events in real-time provides researchers with a powerful tool for understanding the complex processes that underlie gene expression and genomic function.

Chapter 18: CRISPR-based diagnosis of infectious diseases

The CRISPR-Cas system has recently emerged as a significant tool in the field of infectious disease diagnosis. With the emergence and reemergence of new viral diseases compromising healthcare worldwide, there is a critical need for reliable and rapid point-of-care (POC) testing kits to diagnose infected patients. Unlike RT-PCR, which necessitates a specialized central laboratory infrastructure, POC tests can be performed at or near the point of care, such as clinics and hospitals. This eliminates the need for sample transportation and allows communities without specialized laboratory infrastructure to detect infected patients quickly. This can help prevent delays in diagnosis and enable timely intervention, particularly in regions where laboratory facilities or highly trained personnel are not readily available. Several respiratory viruses, including SARS CoV-1, MERS, and SARS CoV-2 (COVID-19), as well as other viruses like Influenza A H1N1 (swine flu), Ebola, and Zika, have caused major viral outbreaks globally, emphasizing the need for reliable and rapid POC testing kits. The CRISPR-Cas system has shown promise in developing such POC kits, making it a valuable tool in the fight against infectious diseases and a crucial asset in the field of healthcare.

18.1. Cas enzymes for diagnosing infectious diseases

A team of scientists found that two cousins of Cas9, called, Cas12 and Cas13, can be harnessed to detect human disease. While Cas12 targets DNA, Cas13 targets RNA. Part of the mechanism of their action is similar to that of Cas9. A guide RNA that is complementary to the target sequence is required for specific binding of Cas enzymes, and then the Cas12 or Cas13 proteins cleave at the target site. An additional interesting feature of Cas12 and Cas13 enzymes is that they show trans or collateral cutting activity *(Fig 18-1)*. It means that on finding the target, the cleavage activity of Cas12 and Cas13 enzymes is not just restricted to the target DNA or RNA; they can also cut any single-stranded non-targeted nucleic acid molecules in the vicinity. Although this collateral DNase and RNase activity might appear to be a disadvantage in terms of specific gene editing, it has made these enzymes a powerful tool for developing CRISPR-based diagnostics. The collateral cutting activity of these enzymes has attracted scientists to build reporter systems in CRISPR diagnostics.

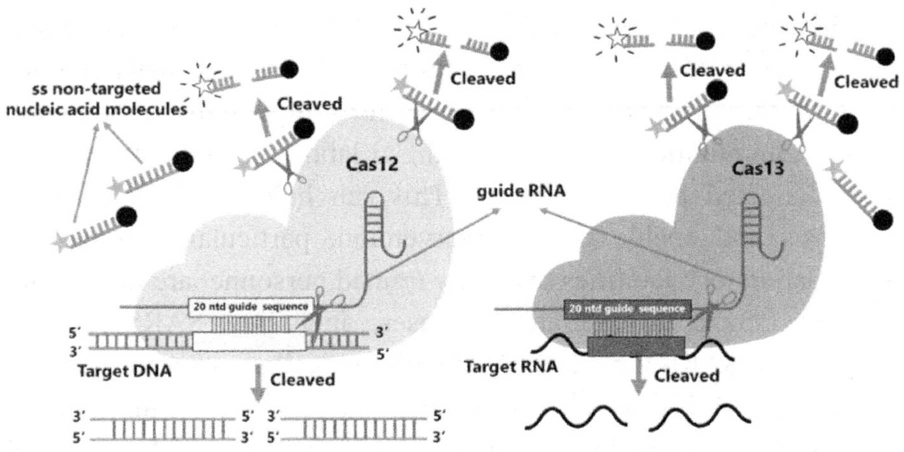

Fig 18-1: Cas12 and Cas13 enzymes show trans or collateral cutting activity

Before the COVID-19 pandemic, the field of CRISPR diagnostics was in its infancy. However, with the outbreak of the pandemic and the widespread impact of the virus, there has been a significant surge in this area of research to develop rapid and accurate diagnostic tools for detecting SARS-CoV-2, the virus that causes COVID-19.

18.2. Sherlock (CRISPR-Cas13 based) diagnostic test

On 6th May 2020, Massachusetts-based biotech firm Sherlock Biosciences with its "Sherlock CRISPR diagnostic kit," became the first company that received FDA approval for the CRISPR-based COVID-19 diagnostic test. SHERLOCK is an acronym for "Specific

High-sensitivity Enzymatic Reporter unLOCKing. This kit uses the Cas13 enzyme, which targets the RNA. For the diagnosis, a guide RNA (gRNA) is designed that recognizes a specific RNA sequence found in the SARS-CoV-2 genome (Li et al., 2018). Cas13 and the gRNA then form a complex and search for the sequence match in the RNA of the SARS-CoV-2. When the gRNA binds to the programmed sequence of the target RNA, the Cas13 enzyme cuts the target SARS-CoV-2 RNA. After cleaving the target RNA, Cas13 doesn't get inactivated; instead, HEPN motifs of Cas13 get further activated to cut the surrounding unrelated single-stranded RNA reporter molecules that may be nearby in the reaction solution. In the test, the cleavage of these single-stranded RNA reporter molecules resulting from the collateral activity of the Cas13 enzyme is determined (Kellner et al., 2019).

Mechanism of action

Diagnosing COVID-19 requires taking a sample from the patient. The sample taken is usually a nasopharyngeal or oral pharyngeal swab. To execute CRISPR-based diagnosis, the viral RNA is extracted from the patient's sample. The extracted RNA may contain other viral, bacterial, or patient's own RNA as well. Thus, to increase the test's sensitivity, the SARS-CoV-2 RNA is amplified with the Reverse Transcription- Recombinase Polymerase Amplification process, abbreviated as RT-RPA (Zhang et al., 2020; Joung et al., 2020).

In the first step, the reverse transcriptase enzyme converts the SARS-CoV-2 RNA into complementary DNA, referred to as cDNA. The cDNA is then amplified by the RPA process using primers specific to the SARS-CoV-2 genes, for example, S gene, Orf1ab gene, N gene, etc *(Fig 18-2)*.

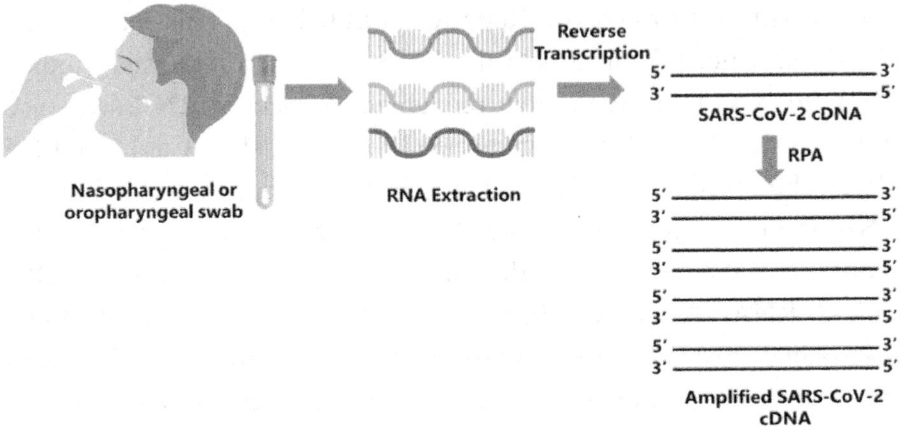

Fig 18-2: Sample Isolation and Reverse Transcription of SARS-CoV-2 RNA

Recombinase Polymerase Amplification (RPA) process: The Recombinase Polymerase Amplification is an isothermal nucleic acid amplification. Unlike polymerase chain reaction PCR, the RPA reaction occurs at a single temperature, so there's no need for a thermocycler, making the RPA an excellent candidate for developing low-cost, rapid, point-of-care diagnosis (POC). And is ideally suited to fields and other settings with minimal resources for diagnosing infectious diseases like COVID-19, food contaminations, etc. RPA is as specific as PCR amplification but is much, much faster. Results are typically generated within 3-10 minutes.

The RPA process occurs by three enzymes: recombinase, single-stranded DNA binding proteins (SSB), and strand displacing polymerase *(Fig 18-3)*.

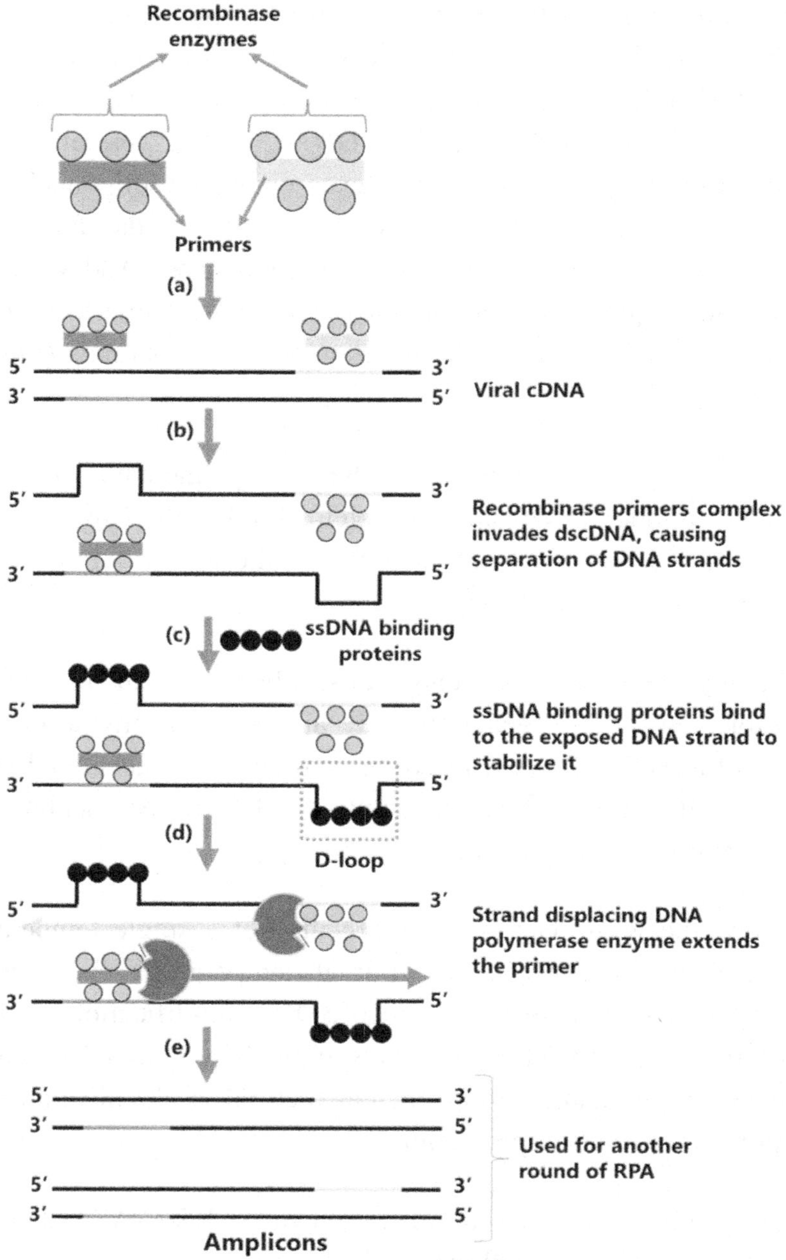

Fig 18-3: Recombinase Polymerase Amplification (RPA) process

(i) In traditional PCR, the denaturation step in which double-stranded template DNA is separated into two single strands is performed at 94°C. But in RPA, this step is substituted by the two enzymes: recombinase and single-stranded DNA binding proteins. The recombinase enzymes form complexes with the oligonucleotide primers *(Fig 18-3a)*, then these complexes scan the ds viral cDNA target, searching for the homologous sequences. And once the homologous sequences are found, the recombinase primer complexes invade the dsDNA, causing the separation of DNA strands *(Fig 18-3b)*.

(ii) When the primers are paired to their complementary sequences, the single-stranded DNA binding proteins bind to the exposed DNA strand to stabilize it. The local separation of these DNA strands forms a D-loop structure *(Fig 18-3c)*.

(iii) Finally, the strand displacing DNA polymerase enzyme extends the primer, eventually generating the amplicons from the original strands of template DNA *(Fig 18-3d)*. These newly generated DNA strands are then used for another round of RPA for exponential viral cDNA amplification *(Fig 18-3e)*.

Typically, the RPA reactions are executed at a single temperature ranging from 37°C-42°C. At optimal temperature, the reaction progresses rapidly and results in specific DNA amplification from just a few target copies to detectable levels, typically within 3-10 minutes. No other sample manipulation, such as thermal or chemical melting, is required to initiate amplification.

Cas13 activity: After the amplification of SARS-CoV-2 cDNA is done, it is then **transcribed into RNA** because the kit uses detection by Cas-13 enzyme, which targets the RNA. This amplified RNA is then mixed with the Cas13 enzyme, guide RNA, and single-stranded

RNA reporter molecules in a reaction tube. The guide RNA matches a specific RNA sequence found in the SARS-CoV-2 genome.

Each of the single-stranded RNA reporter molecules is coupled with the fluorescent FAM molecule at one end and a biotin molecule at the other end. When the reporter molecules are intact, biotin acts as a quencher, which suppresses the fluorescence emitted by the FAM molecule *(Fig 18-4)*.

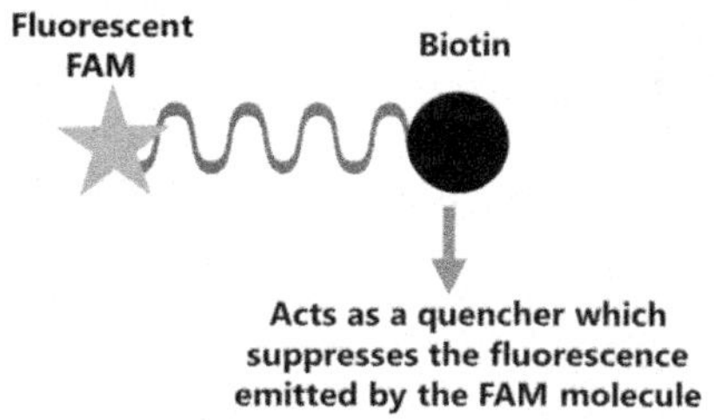

Fig 13-4: Single-stranded RNA reporter molecules

Imagine, if the patient sample contains the SARS-CoV-2 virus, then the Cas13 cleaves the viral RNA and shows collateral activity. Because of collateral activity, Cas13 cleaves the surrounding FAM-biotinylated ssRNA reporter molecules, leading to the separation of the FAM molecules from the biotin molecules *(Fig 18-5a)*. When excited with light, the freed fluorescent FAM molecules produce a quantifiable fluorescence that can be detected by a detector, indicating the presence of SARS-CoV-2 RNA in the sample *(Fig 18-5b)*.

Lateral Flow Assay: Visualization of Cas13 activity can also be achieved using lateral flow strips that are designed to capture labeled nucleic acids. The interpretation of lateral flow strips is very intuitive and easy.

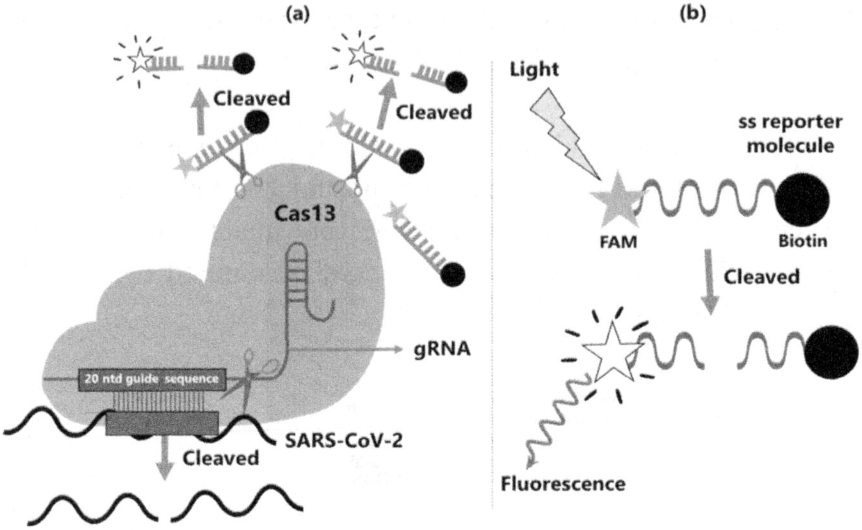

Fig 18-5: (a) Collateral activity of Cas13 enzyme and (b) cleaved ssRNA reporter molecule produces fluorescence when excited with light

Similar to a pregnancy test, the strip changes color if the virus is detected *(Fig 18-6)*.

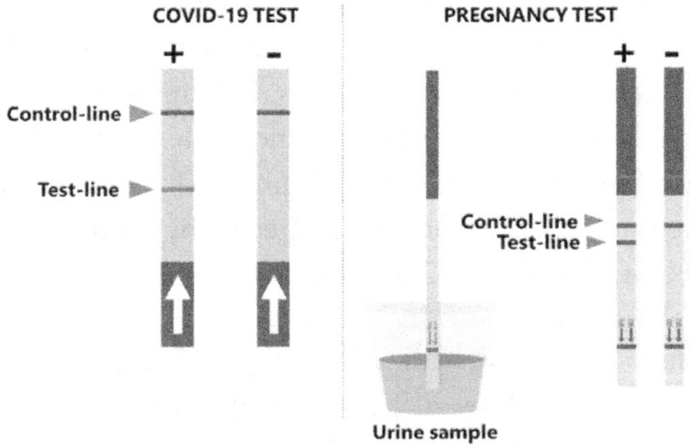

Fig 18-6: Visualization of Cas13 activity is similar to a pregnancy test

Typically, lateral flow test strips are composed of a sample pad, a conjugate pad, a testing area, and an absorbent pad *(Fig 18-7)*.

- The sample pad is where the test strip receives samples in the form of liquid drops.
- The absorbent pad generates a suction force, pulling the sample from the sample pad towards the conjugate pad and then to the test area.
- The conjugate pad is covered with a large number of colloidal gold particles, conjugated with anti-FAM antibodies.
- The testing area is marked with the letters T and C. T stands for the test area, and C stands for the control area. The generation of colored lines on T or C areas indicates whether the result is positive or negative. In the control area, the streptavidin is immobilized, and in the test area, the antibodies specific for the anti-FAM antibodies are immobilized.

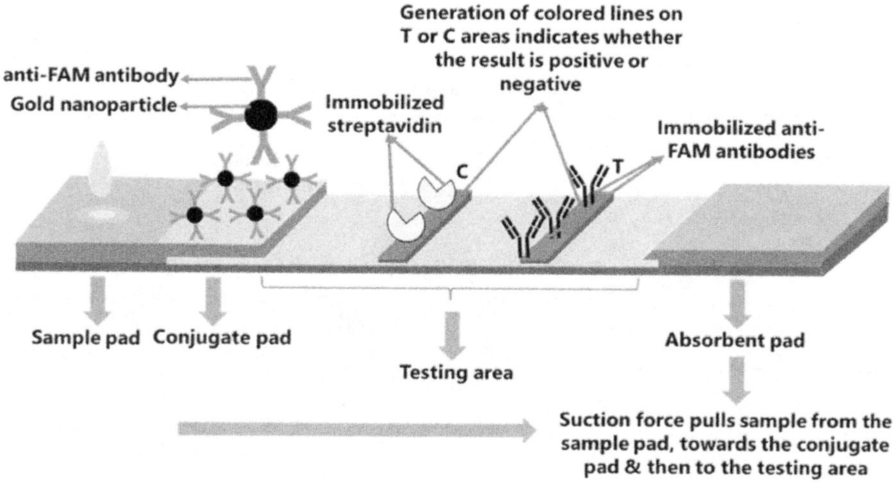

Fig 18-7: Components of lateral flow strip

Negative Test: Take the scenario where the sample is negative, which means SARS-CoV-2 is not present in the patient's sample. In this case, the Cas13 enzyme remains inactive, and thus no collateral cleavage of the FAM-biotinylated ssRNA reporter molecules occurs. When the sample containing the intact reporter molecule is loaded on the sample pad at one end of the strip, it flows to the conjugate pad. The conjugate pad is covered with a large number of colloidal gold particles, conjugated with anti-FAM antibodies. Thus, on reaching the conjugate pad, the intact reporter molecules by their FAM labels get attached to the anti-FAM antibodies conjugated to the gold particles.

The colloidal gold particles are then pulled into the control (C) area by the suction force generated by the absorbent pad. When the gold particles reach the C area, the attached intact reporter molecules accumulate at the C area by the interaction of the biotin labels with immobilized streptavidin. The interaction leads to the appearance of a visible colored line in the C area *(Fig 18-8)*. The bound gold particles are unable to move further into the test area. Thus, a single-colored line appears in the C area on the strip, indicating that the **test is negative**.

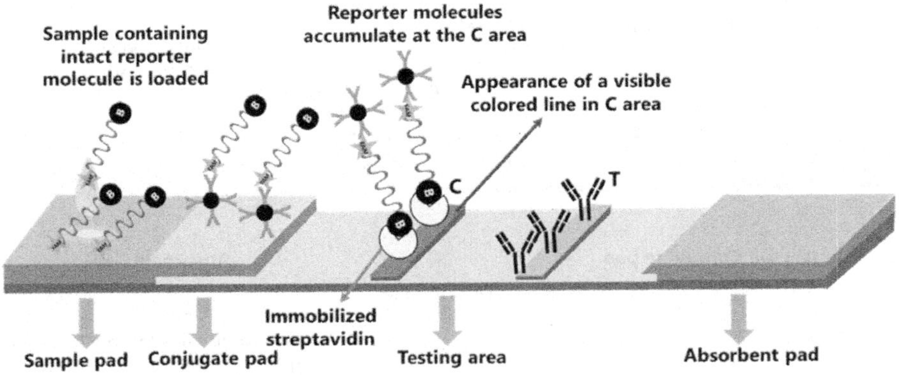

Fig 18-8: Appearance of a single visible colored line in the control (C) area indicates the negative test

Note: This process does not involve the use of dyes or enzymes to generate the visible signal. Instead, the gold particles themselves create a colored line on the strip because they absorb and scatter light. When many of these tiny gold particles gather in one spot, their colors combine, making a visible line that's easy to see.

Positive Test: On the other hand, if the patient sample contains SARS-CoV-2, then the Cas13 cleaves the viral RNA and shows collateral activity. Because of the collateral activity, Cas13 cleaves the FAM-biotinylated ssRNA reporter molecules, leading to the separation of biotin molecules and FAM molecules. Therefore, when the sample containing the cleaved reporter molecules is loaded on the sample pad at one end of the strip, it flows to the conjugate pad, where the separated FAM labels get attached to the anti-FAM antibodies conjugated to the gold particles.

The bound colloidal gold particles and unbound biotin label molecules are then pulled into the control area by the suction force generated by the absorbent pad. When they reach the C area, the biotin labels bind to the C area by interacting with the immobilized streptavidin *(Fig 18-9)*. On the other hand, the FAM label bound gold particles are not captured in the C area, but their presence creates a visible colored line in the C area, indicating that the test has worked correctly.

Then, they continue to move to the next location, the test (T) area of the lateral flow strip containing the anti-rabbit antibody, which binds anti-FAM antibodies. This results in the accumulation of gold particles and the appearance of a second visible colored line in the T area, confirming the positive test result.

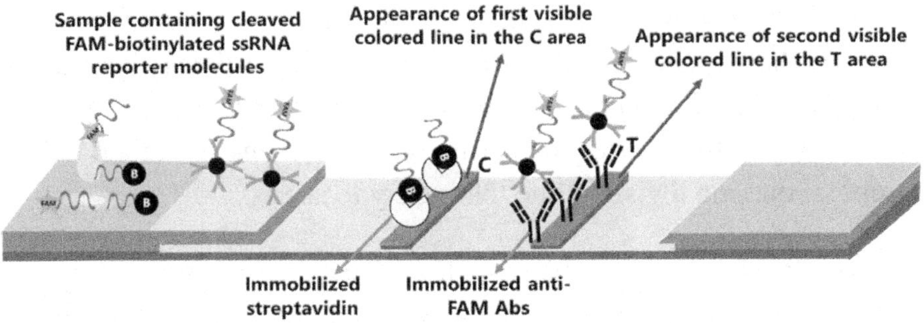

Fig 18-9: Appearance of visible colored lines in the control (C) and test (T) areas indicate positive test

Thus, the appearance of 2 lines in the Test (T) and Control (C) area indicates the positive test; in other words, the patient is infected with the SARS-CoV-2. On the other hand, the appearance of a single colored line in the C area of the lateral flow strip indicates the negative test; in other words, the patient is not infected with SARS-CoV-2.

The Sherlock CRISPR-Cas13 based diagnostic test produces results within an hour.

18.3. DETECTR (CRISPR-Cas12 based) diagnostic test

Another CRISPR-based diagnostic kit is "DETECTR," developed by **Mammoth Biosciences**. DETECTR is an acronym for DNA Endonuclease Targeted CRISPR Trans Reporter. This kit works similarly to SHERLOCK. The difference is in the types of Cas enzymes used. While SHERLOCK uses the Cas13 enzyme, which targets RNA, DETECTR uses the Cas12a enzyme (Chen et al., 2018), which targets DNA. For the diagnosis, a guide RNA (gRNA) is designed that recognizes a specific DNA sequence, for example, one of the genes found in the SARS-CoV-2 genome.

Like the Cas13 enzyme, Cas12a also possesses collateral activity, which means the Cas12a enzyme first cuts the specific target DNA.

The enzyme is then further activated to chop the surrounding unrelated single-stranded DNA molecules regardless of their complementarity with these ssDNA molecules. In other words, Cas12a acts like a paper shredder that shreds the surrounding DNA irrespective of the complementarity (Broughton et al., 2020a; Broughton et al., 2020b; Mustafa and Makhawi, 2020).

Loop Mediated Isothermal Amplification (LAMP): Viral RNA is first extracted from the patient's sample to perform the CRISPR-based diagnosis. This extracted RNA might include other viral, bacterial, or the patient's own RNA. In order to enhance the test's sensitivity, the SARS-CoV-2 RNA is amplified with the **Reverse Transcription Loop-Mediated Amplification process, abbreviated as RT-LAMP**.

In the first step, the reverse transcriptase enzyme converts the SARS-CoV-2 RNA into complementary DNA (cDNA). The LAMP process then amplifies the cDNA. The reaction is performed at 62 °C for 20–30 min. Like RPA, LAMP is also an **isothermal DNA amplification method**. It also occurs at a single temperature, so there's no need for a thermocycler.

For performing the LAMP process, four different primers are used that can specifically recognize six distinct regions of the target DNA i.e., SARS-CoV-2 cDNA. In the *Fig 18-10*, suppose two lines represent the target DNA. The six regions on template strand 3'-5' that are required to be recognized by primers are represented as F3c, F2c, F1c, B1, B2, and B3. Therefore, on the complementary coding strand 5'-3', these regions are represented as F3, F2, F1, B1c, B2c, and B3c.

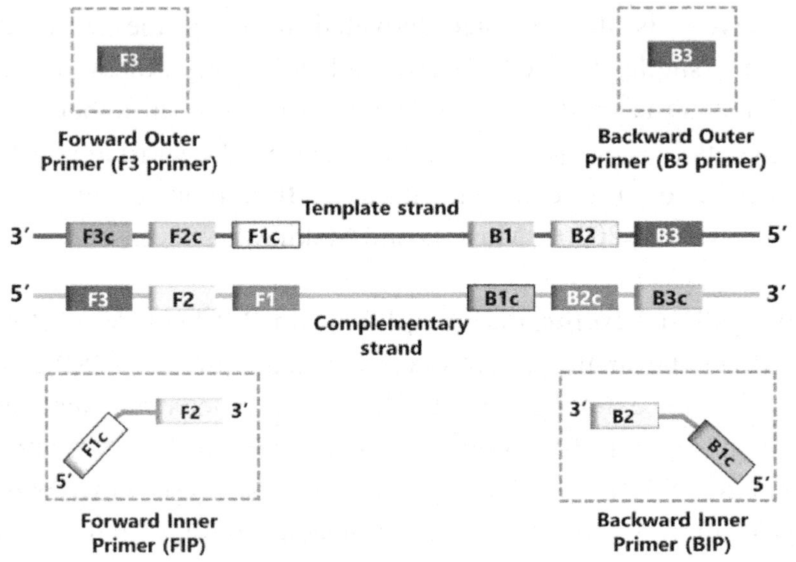

Fig 18-10: Primers for LAMP process

Primers required for the amplification process:

1. The first primer is the forward inner primer designated as FIP. It consists of an F2 region at the 3' end and an F1c region at the 5'end. The F2 region of the primer is complementary to the template sequence's F2c region. And the F1c region is identical to the F1c region of the template sequence *(Fig 18-11)*.

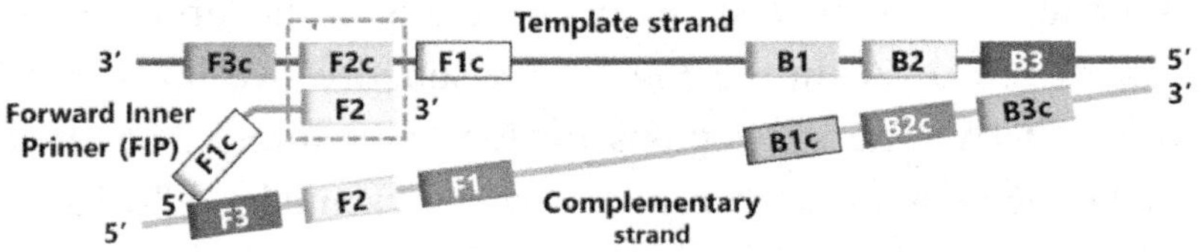

Fig 18-11: Forward Inner Primer (FIP)

2. The second primer is the forward outer primer, also called the F3 primer, because it consists of only one region, i.e., an F3 region

complementary to the F3c region of the template sequence *(Fig 18-12)*.

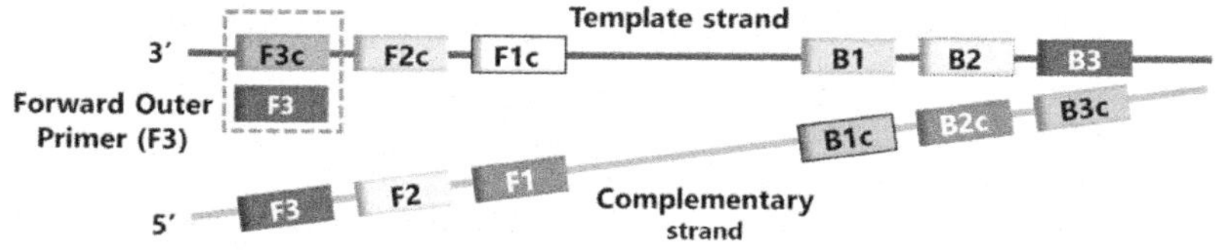

Fig 18-12: Forward Outer Primer (F3)

3. The third primer is the backward inner primer designated as BIP. It consists of a B2 region at the 3' end and a B1c region at the 5' end. The B2 region is complementary to the template sequence's B2c region. And the B1c region is identical to the B1c region of the template sequence *(Fig 18-13)*.

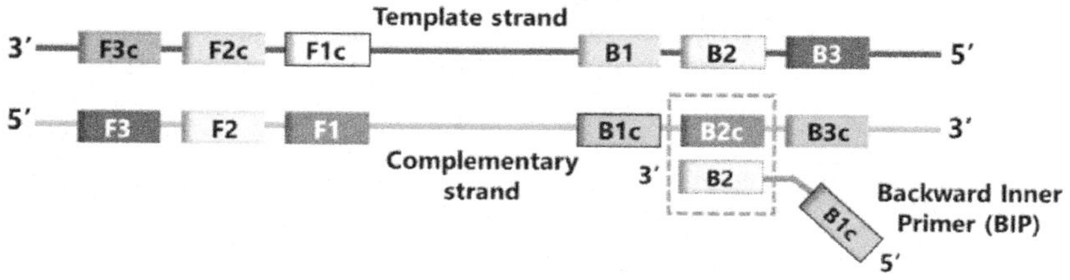

Fig 18-13: Backward Inner Primer (BIP)

4. The fourth primer is the backward outer primer, also called as B3 primer. It consists of a B3 region, which is complementary to the B3c region of the template sequence *(Fig 18-14)*.

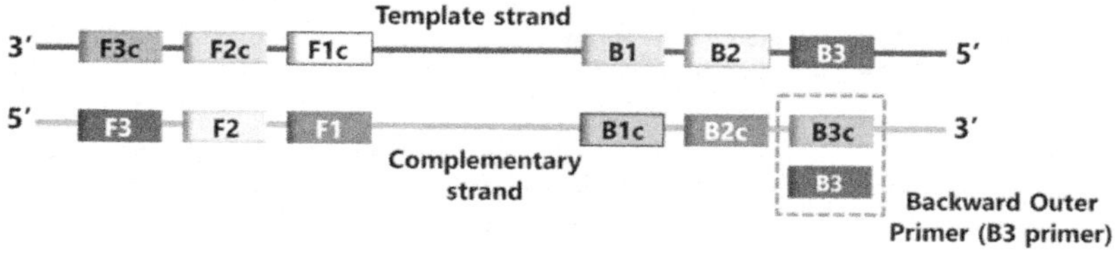

Fig 18-14: Backward Outer Primer (B3)

DNA Amplification

The amplification reaction is initiated from the strand invasion by the forward inner primer (FIP).

1. F2 region of FIP hybridizes to the F2c region of the target DNA, and then the strand displacing DNA polymerase extends the primer leading to the synthesis of the complementary strand. This leads to the separation of ds target DNA *(Fig 18-15)*.

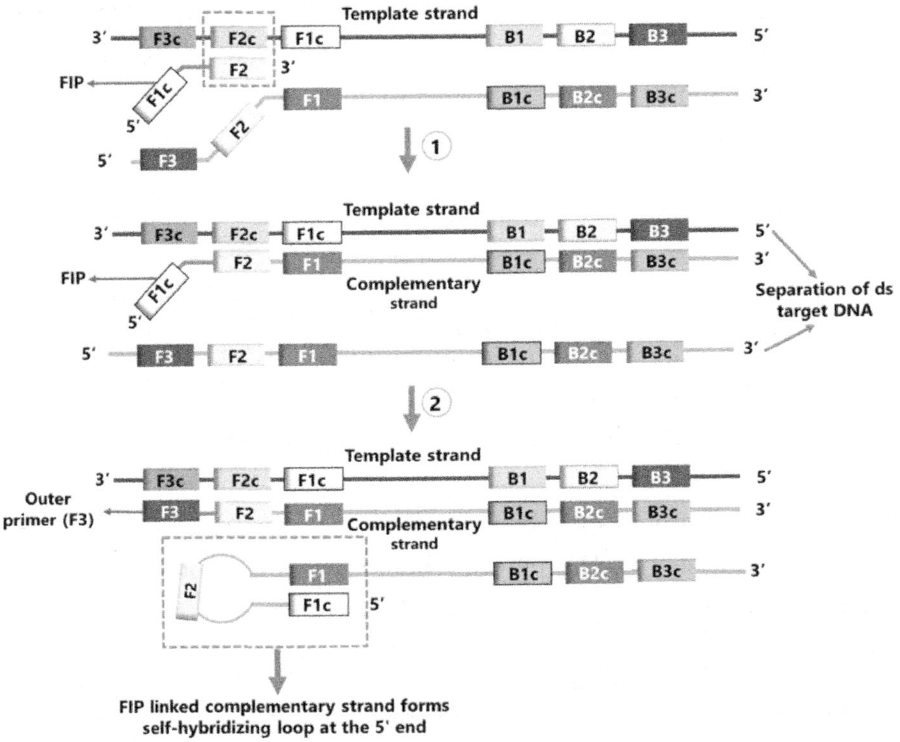

Fig 18-15: Steps 1 and 2 of the LAMP process

2. To this first product, outer primer F3 binds to the F3c region of the target DNA. Then strand displacing DNA polymerase extends the F3 primer, which leads to displacing the FIP linked complementary

strand. As the strand is displaced, it forms a self-hybridizing loop at the 5' end due to the inclusion of a reverse complementary sequence F1c in the primer sequence *(Fig 18-15)*.

3. This displaced single-stranded DNA with a loop at the 5' end now serves as a template for backward inner primer (BIP) that consists of a B2 region at the 3' end and a B1c region at the 5' end. B2 region of the primer hybridizes to the B2c region of the template DNA. DNA synthesis is then initiated, leading to the creation of a complementary strand and opening of the 5' end loop *(Fig 18-16)*.

4. Now, the outer backward primer B3 hybridizes to the B3c region of the target DNA and is extended by the DNA polymerase, displacing the backward inner primer-linked complementary strand *(Fig 18-16)*.

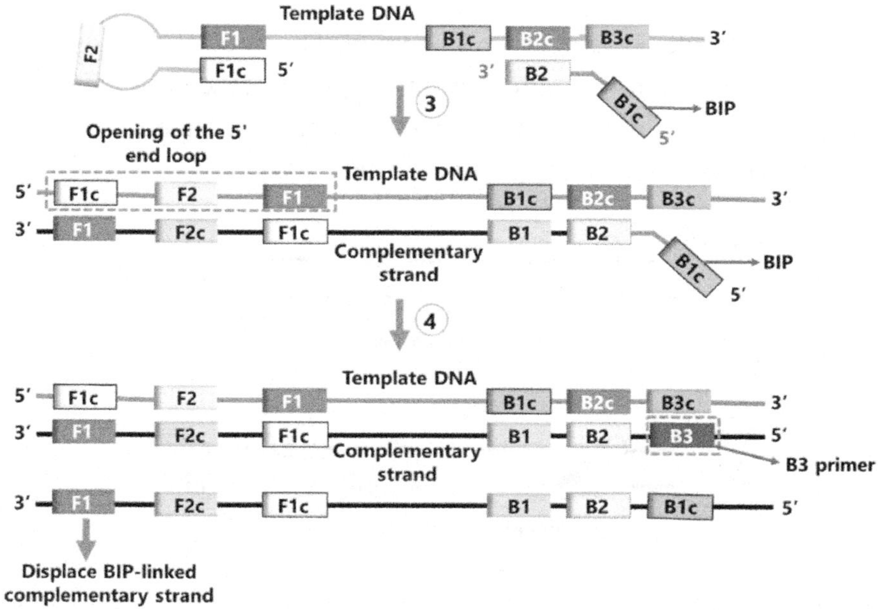

Fig 18-16: Steps 3 and 4 of the LAMP process

As the backward inner primer (BIP)-linked strand is displaced, it results in the formation of a dumbbell-shaped DNA because of the presence of reverse complementary sequences B1c and F1. The dumbbell structure DNA then forms a seed for exponential LAMP amplification, which is the second stage of the LAMP reaction. This dumbbell structure contains multiple sites for the initiation of DNA synthesis *(Fig 18-17).*

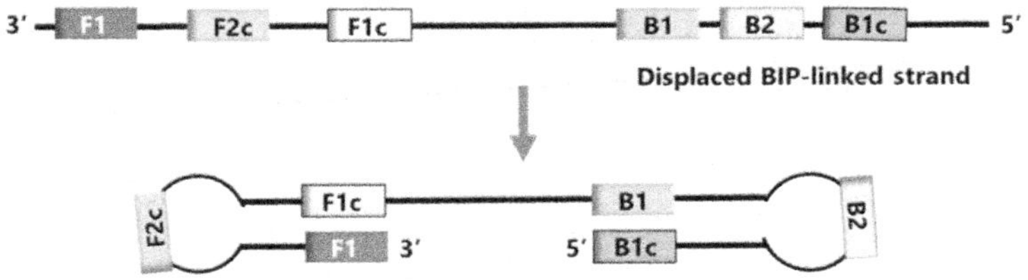

Fig 18-17: Displaced BIP-linked strand forms dumbbell-shaped DNA

(i) One is from the 3' end of the F1 region. The nucleotides are added to the 3' end of the F1 region by DNA polymerase, extending and opening up the loop at the 5' end. The dumbbell-shaped DNA now gets converted to a stem-loop structure *(Fig 18-18).*

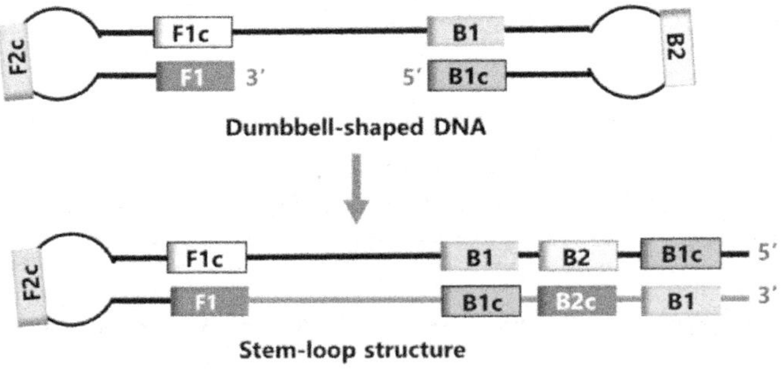

Fig 18-18: Dumbbell-shaped DNA gets converted into a stem-loop structure

(ii) To initiate LAMP cycling, the forward inner primer (FIP) hybridizes to the loop of the stem-loop DNA structure. Strand synthesis is initiated here. As the strand synthesis occurs, the F1 strand is displaced and forms a new loop at the 3' end *(Fig 18-19)*.

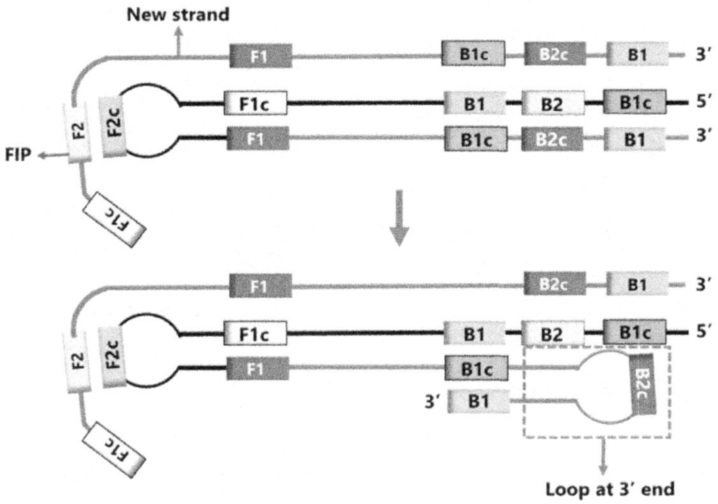

Fig 18-19: Formation of a new loop at the 3' end

Now, nucleotides are added to the 3' end of B1, leading to a long concatemer formation. The extension leads to the displacement of the FIP linked strand *(Fig 18-20)*.

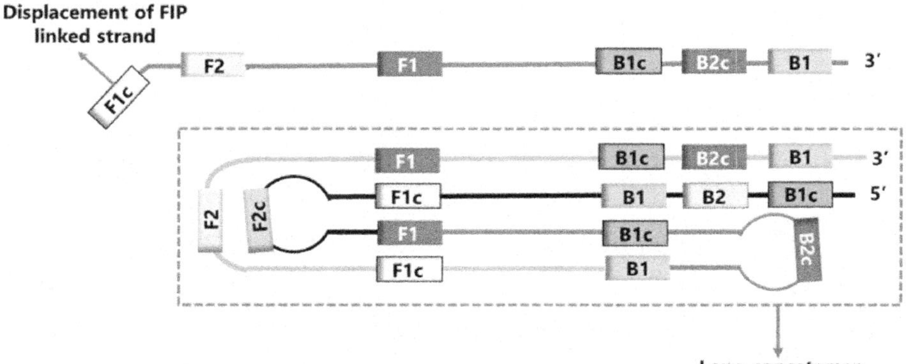

Fig 18-20: Displacement of FIP-linked strand and formation of long concatemer

This displaced strand again forms a dumbbell-shaped DNA. Now the dumbbell-shaped DNA and the concatemers formed have multiple sites for annealing of primers, leading to DNA synthesis initiation *(Fig 18-21)*.

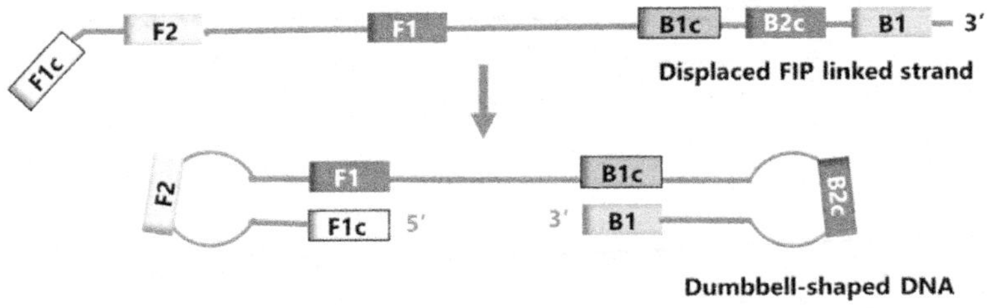

Fig 18-21: Displaced FIP-linked strand forms dumbbell-shaped DNA

The final products obtained due to amplification are a mixture of stem-loop DNA with various stem lengths and different dumbbell structures with loops *(Fig 18-22)*.

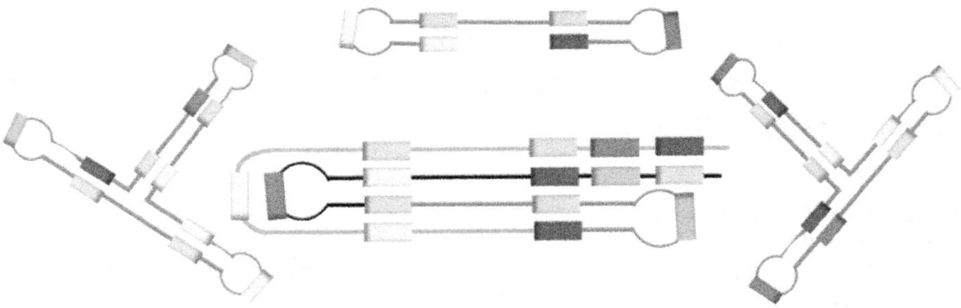

Fig 18-22: Mixture of different stem-loop DNA and dumbbell structures

The LAMP reaction occurs in a single tube containing target DNA, buffer, primers and DNA polymerase enzyme. The tube is incubated at 64°C in a regular laboratory water bath or heat block that maintains a constant temperature.

Cas12 activity: For the detection of SARS-CoV-2 in the sample, the amplified cDNA is then mixed with the Cas12a enzyme, guide RNA, and single-stranded DNA reporter molecules in a reaction tube. The guide RNA matches a specific DNA sequence found in the SARS-CoV-2 genome, for example, one of the genes found in the SARS-CoV-2 genome. Each of the single-stranded DNA reporter molecules are coupled with the fluorescent FAM molecule at one end and a biotin molecule at the other end. When the reporter molecules are intact, biotin acts as a quencher, which suppresses the fluorescence emitted by the FAM molecule. If the patient sample contains SARS-CoV-2, then the Cas12a cleaves the amplified viral cDNA and shows collateral activity. Because of the collateral activity, Cas12a cleaves the surrounding FAM-biotinylated ssDNA reporter molecules, leading to the separation of the FAM molecules from the biotin molecules. The freed fluorescent FAM molecules, when excited with light, produce a quantifiable fluorescence which can be detected by a detector, indicating the presence of SARS-CoV-2 cDNA in the sample.

Detection by Lateral Flow Strip: Detection of Cas12a activity and the viral cDNA can also be done via a lateral flow strip, similar to the Sherlock test (Broughton et al., 2020c). Lateral flow test strips are composed of a sample pad, a conjugate pad, a testing area, and an absorbent pad. The testing area contains two types of lines or areas: The test area designated as T and the control area designated as C. In the control area, the streptavidin is immobilized, and in the test area, the antibodies specific for the anti-FAM antibodies are immobilized.

Negative Test: If the sample is negative, i.e., SARS-CoV-2 is not present in the patient's sample, the Cas12a enzyme remains inactive, and thus, no collateral cleavage of the FAM-biotinylated ssDNA reporter molecules occurs. When the sample containing the intact reporter molecule is loaded on the sample pad of the strip, on reaching the conjugate pad, the intact reporter molecules get attached to the anti-FAM antibodies conjugated to the gold particles by their FAM labels. When the gold particles reach the C area, the attached intact reporter molecules accumulate at the C area by the interaction of biotin labels with immobilized streptavidin. The bound gold particles are unable to move further into the T area. Thus, a single-colored line appears in the C area on the strip, indicating that the **test is negative** *(Fig 18-23)*.

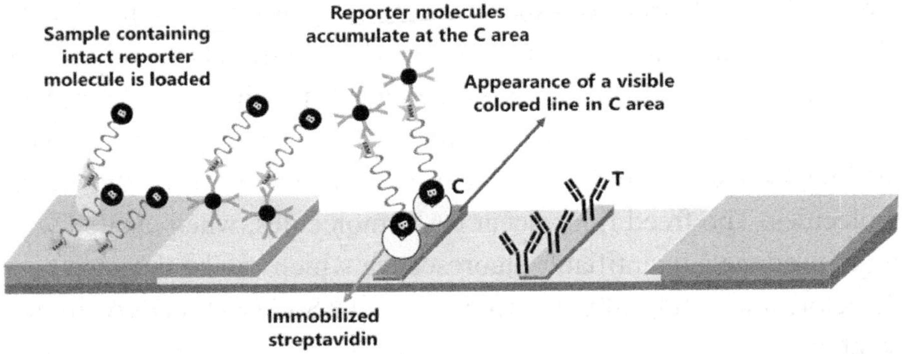

Fig 18-23: Appearance of a single visible colored line in the C area indicates the negative test

Positive Test: On the other hand, if the patient sample contains SARS-CoV-2, then the Cas12a cleaves the viral cDNA and shows collateral activity; thus cleaves the FAM-biotinylated ssDNA reporter molecules, leading to the separation of biotin molecules and FAM molecules. Therefore, when the sample containing the cleaved reporter molecules is loaded on the sample pad, it flows to the conjugate pad, where the separated FAM labels get attached to the anti-FAM antibodies conjugated to the gold particles.

The bound colloidal gold molecules and unbound biotin label molecules are then pulled into the C area. After reaching the C area, the biotin labels bind to the C area by interacting with the immobilized streptavidin. Whereas the FAM label bound gold particles are not captured in the C area, but their presence creates a visible colored line in the C area, indicating that the test has worked correctly.

Then, they continue to move to the T area of the lateral flow strip, which contains anti-rabbit antibodies that bind to anti-FAM antibodies. This interaction causes gold particles to accumulate, forming a second visible colored line in the T area *(Fig 18-24)*.

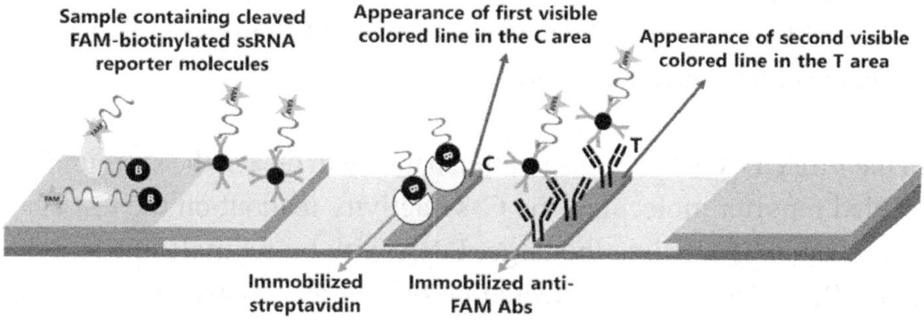

Fig 18-24: Appearance of visible colored lines in the control and test area indicates the positive test

Thus the appearance of 2 lines in the T and C area indicates a positive test, i.e., the patient is infected with SARS-CoV-2. On the other hand, the appearance of a single colored line in the C area of the lateral flow strip indicates a negative test, indicating the patient is not infected with SARS-CoV-2.

The DETECTR CRISPR-Cas12a-based diagnostic test produces results within 30-40 minutes. The LAMP process takes 20-30 minutes, and the Cas detection reaction takes another 10 minutes; thus, one can get the result within 30-40 minutes.

18.4. FELUDA (CRISPR-Cas9 based) diagnostic test

The other type of CRISPR-based diagnostic test is FELUDA, made by Indian researchers at CSIR Institute of Genomics and Integrative

Biology (CSIR-IGIB). FELUDA is an acronym for FnCas9 Editor Linked Uniform Detection Assay. FELUDA test is different from other CRISPR based tests SHERLOCK and DETECTR in the types of Cas enzymes used. SHERLOCK and DETECTR use the Cas13 and Cas12a enzymes, which show collateral activity. On the other hand, the FELUDA test uses specially adapted Cas9 protein, derived from *Francisella novicida* bacteria, and it does not show collateral activity like Cas12 and Cas13 enzymes (Acharya et al., 2019). It means Cas9 is incapable of cleaving the surrounding ssFAM-biotinylated reporter molecules like Cas12 and Cas13.

In the FELUDA test, because of the inability to cleave the surrounding labeled reporter molecules, the Cas9 activity readout on the test strip can be done by labeling the target DNA with biotin molecule and the guide RNA with FAM molecule (Azhar et al., 2020a; Azhar et al., 2020b; Suvvari et al., 2020).

Cas9 activity: First, the patient sample is collected, and then the RNA is extracted from the sample. This is followed by reverse transcription wherein SARS-CoV-2 RNA is converted into cDNA, which is further amplified by the PCR or RPA. The amplification is performed using a normal forward primer and a reverse biotinylated primer. The biotinylated reverse primers label the amplified SARS-CoV-2 cDNA at the 5' end *(Fig 18-25)*.

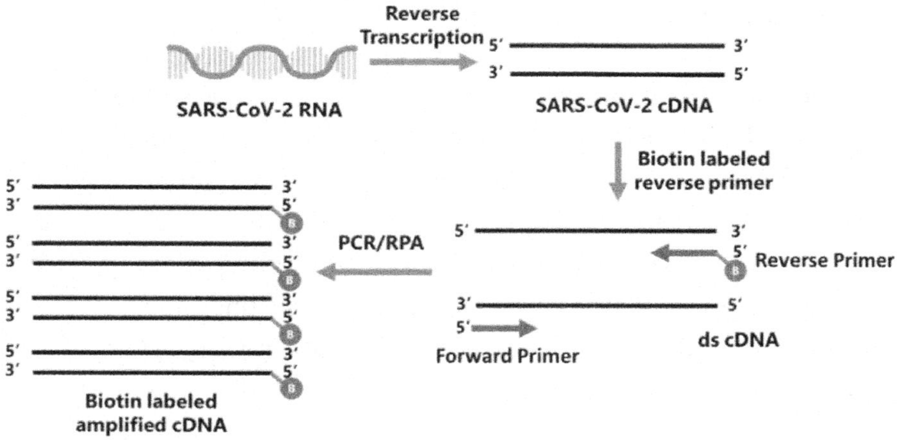

Fig 18-25: Biotinylated reverse primers label the amplified SARS-CoV-2 cDNA at the 5' end

On the other hand, the guide RNA is labeled with FAM molecules *(Fig 18-26)*. In the next step, the biotin-labeled amplified cDNA molecules are then mixed with Cas9 protein and FAM-labeled guide RNA. The Cas9 and FAM-labeled gRNA forms a search complex, which searches for the sequence match in the amplified cDNA of SARS-CoV-2. After the binding of gRNA, Cas9 protein introduces double-stranded breaks in the SARS-CoV-2 cDNA.

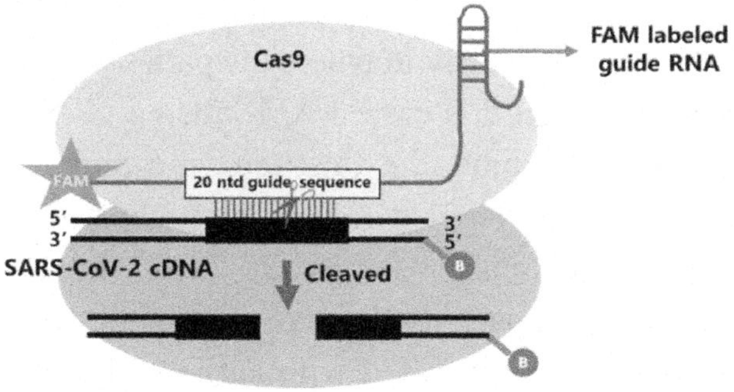

Fig 18-26: Cas9 and FAM-labeled gRNA forms a search complex

Detection by Lateral Flow Strip:

Detection of Cas9 activity is also done via a lateral flow paper strip, similar to SHERLOCK and DETECTR assays. In the paper strip used for the FELUDA test, positions of the test and control lines are swapped compared to paper strips used for DETECTR and SHERLOCK *(Fig 18-27a)*. It leads to a completely different assessment of the resulting signals. On the test line, streptavidin is immobilized, and on the control line, antibodies specific for anti-FAM antibodies are immobilized *(Fig 18-27b)*.

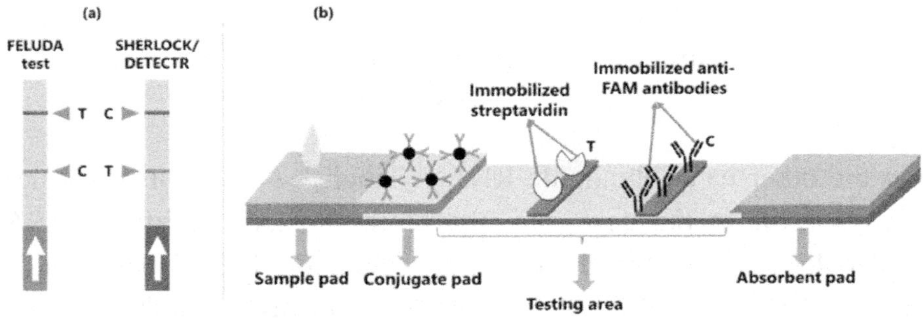

Fig 18-27: (a) Positions of test and control lines are swapped in FELUDA test and (b) streptavidin is immobilized on the test line, and antibodies specific for anti-FAM antibodies are immobilized on the control line

Negative Test: Take the scenario where the patient's sample does not contain SARS-CoV-2. In this case, FAM-labeled gRNA cannot find the matching sequence; thus, the Cas9 enzyme remains inactive. It also means that the biotin label is not present in the sample. When the sample that doesn't contain SARS-CoV-2 is loaded on the sample pad, it flows to the conjugate pad which is covered with a large number of colloidal gold particles conjugated with anti-FAM antibodies. Thus, on reaching the conjugate pad, the search complex by the FAM labels (gRNA is FAM-labeled) gets attached to the anti-FAM antibodies conjugated to the gold particles.

The FAM-labeled search complex bound colloidal gold molecules are then pulled into the test area by the suction force generated by the absorbent pad. When the gold particles reach the test area, they are unable to bind to immobilized streptavidin. Thus, they are pulled further into the control area, where anti-FAM antibodies are immobilized. On reaching the C area, the FAM-labeled search complex bound gold molecules accumulate, thereby interacting with anti-FAM antibodies, leading to the appearance of a visible colored line in the control area. It indicates that the patient sample does not contain SARS-CoV-2; in other words, the **test is negative** *(Fig 18-28)*.

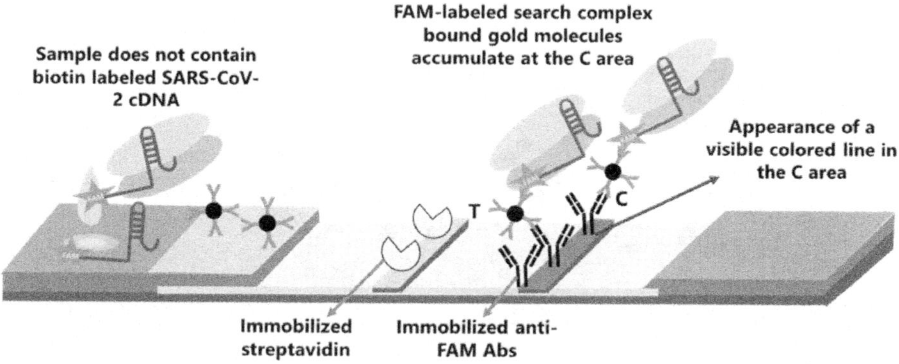

Fig 18-28: Appearance of a single visible colored line in the control area indicates negative test

Positive Test: On the other hand, if the patient's sample contains SARS-CoV-2, then FAM-labeled gRNA can find the matching sequence; thus, the Cas9 enzyme gets activated and cleaves the biotinylated SARS-CoV-2 cDNA. It also means that both the FAM and biotin labels are present in the sample. When the sample containing cleaved SARS-CoV-2 cDNA and FAM-labeled search complex is loaded on the sample pad at one end of the strip, it flows to the conjugate pad containing colloidal gold particles, conjugated with anti-FAM antibodies. Thus, on reaching the conjugate pad, the

FAM-labeled search complex gets attached to the anti-FAM antibodies conjugated to the gold particles.

The FAM-labeled search complex bound gold molecules, and cleaved biotinylated cDNA are then pulled into the test area by the suction force generated by the absorbent pad. When they reach the test area, the cleaved biotinylated cDNA accumulates at the test area because of the interaction of the biotin labels with immobilized streptavidin. Some of the FAM-labeled search complexes recognize the cleaved cDNA and bind to it through complementary sequence interactions. This binding leads to the accumulation of gold nanoparticles at the T area, resulting in the appearance of a colored line *(Fig 18-29)*.

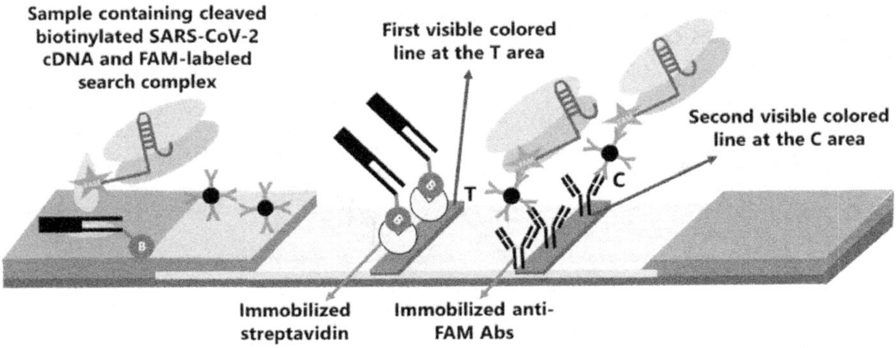

Fig 18-29: Appearance of visible colored lines in the control and test area indicates the positive test

On the other hand, the FAM-labeled search complex gold particles that were not captured at the T area move to the control area of the lateral flow strip. On reaching the C area, these complexes encounter immobilized antibodies that specifically recognize and bind to anti-FAM antibodies. The interaction leads to the accumulation of gold nanoparticles at the C area, producing a second visible colored line. It indicates that the patient sample contains SARS-CoV-2; in other words, the **test is positive**.

Thus, the appearance of 2 lines in the T and C area indicate the positive test; in other words, the patient is infected with the SARS-CoV-2. On the other hand, the appearance of a single-colored line in the C area of the lateral flow strip indicates the negative test; in other words, the patient is not infected with SARS-CoV-2.

This diagnostic test kit can deliver test results in 40-45 minutes.

CRISPR-based diagnostic tests meet high-quality benchmarks with 96% sensitivity and 98% specificity for detecting the SARS-CoV-2. Sensitivity can be defined as the ability of a test to correctly identify individuals with the disease, while specificity is the assay's ability to accurately identify those without the disease.

Additionally, the CRISPR-based diagnostic tests do not require RT-PCR-based confirmation for samples that are confirmed as positives or negatives by the CRISPR SARS-CoV-2 test.

18.5. Pros and Cons of CRISPR technology in diagnostics

The CRISPR-based diagnostic assay is as accurate, specific, and sensitive as real-time RT–PCR but is a more rapid approach for Point of Care diagnosis. Like RT-PCR, the requirement of personal protective equipment, extraction kits, reagents, sample collection, and RNA extraction also exists for the CRISPR-based diagnosis. However, the CRISPR-based approach has some more advantages over real-time RT–PCR.

Pros of CRISPR-based diagnosis

(i) Point of care applications: Besides their accuracy is on par with RT-PCR, CRISPR-based diagnostic tests can perform the test at one temperature using a type of isothermal nucleic acid amplification process called LAMP (loop-mediated isothermal amplification) or RPA (Recombinase Polymerase Amplification) methods. For performing amplification, CRISPR-based diagnostic tests do not require a thermocycler (PCR machine). The reaction can occur in a single tube containing buffer, target DNA, DNA polymerase, and primers. The tube is then incubated at the required temperature in a regular laboratory water bath or heat block to maintain a constant temperature. Because of this, the CRISPR based tests can be used for point of care (POC).

On the contrary, RT-PCR uses thermocyclers, expensive laboratory equipment, and well-trained personnel, which are hard to adapt for point of care (POC).

Point of care diagnostic tests are extremely important for diagnosing diseases in countries or settings where laboratory facilities or extremely trained personnel are unavailable. CRISPR-based tests can be just performed in a test tube, and then the results can be interpreted on the paper strip, similar to pregnancy tests. Because of the ease of performing the test and simple interpretation of results, CRISPR-based assay can be used for diagnosis at airports, ports, clinics, schools, etc., for better disease diagnosis, monitoring, management, and containment of diseases like COVID-19.

(ii) Speed: Another critical benefit of CRISPR-based diagnostics is the speed at which it generates test results compared to RT-PCR. The CRISPR-based diagnostic test provides results within 45 minutes compared to RT-PCR, which provides results within 3-4 hours. Laboratory-based RT-PCR testing takes around 3-4 hours, but when transportation of clinical samples is required, the diagnosis time often exceeds 24-48 hours, resulting in delays in diagnosis. But the CRISPR based diagnostic kits can diagnose the samples even in minimal settings; the test can be performed at the hospital or clinic, where the sample is collected. Thus transportation of samples to specific laboratories is not required, because of which delay in diagnosis can be avoided. Therefore, it can help healthcare professionals to decide treatment or course of action for a patient as soon as the patient is diagnosed.

Additionally, with an efficiency equivalent to RT-PCR, CRISPR-based diagnostics can run more tests in roughly the same amount of time.

(iii) Cost-effective: The third benefit of CRISPR-based diagnostic tests is that they provide a more cost-effective approach to diagnosis than the RT-PCR technique.

Cons of CRISPR-based diagnosis

(i) Off-target effects: However, one of the main disadvantages of the CRISPR-Cas system is the non-specific binding of the guide RNA to the target genome, which in turn, can lead to poor signaling and misinterpretation of the results.

Therefore, to reduce the off-target effects of CRISPR-Cas complex

- Guide RNAs must be carefully designed using specialized bioinformatics tools to select the best ones.
- Or two guide RNAs can be used instead of 1 guide RNA to detect the viral genome.

(ii) Separate nucleic acid-preamplification steps: SHERLOCK, DETECTR, and FELUDA tests require separate nucleic acid pre-amplification steps by RPA or LAMP processes and several manual handling steps, which undoubtedly complicates the testing procedures and potentially increases the risk of carry-over contaminations due to amplification products transferring.

Therefore, to minimize these limitations, the researchers have developed more efficient POC CRISPR-based diagnostic tests like:

18.6. All-in-one dual CRISPR-Cas12a assay

CRISPR-based nucleic acid detection methods like SHERLOCK and DETECTR typically require separate nucleic acid pre-amplification by RPA or LAMP processes and multiple manual operations. To overcome these limitations, a team of researchers led by associate professor Changchun Liu, University of Connecticut, developed an all-in-one dual CRISPR-Cas12a (AIOD-CRISPR) assay, in which two crRNAs are used instead of 1 crRNA for the detection of SARS-CoV-2. The two individual crRNAs are without PAM sequence limitation and are designed to bind two different sites that are close to the recognition sites of primers in the target sequence (Ding et al., 2020a; Ding et al., 2020b; Ding et al., 2020c).

Suppose two lines represent target DNA, and the region that is to be recognized by the first Cas12a-crRNA complex is represented as T1 on the 5'-3' strand. Thus on the complementary strand 3'-5', this region is represented as T1c. On the other hand, the region that is to be recognized by the second Cas12a-crRNA is represented as T2 on the 3'-5' strand. Thus, on the complementary strand 5'-3', this region is represented as T2c *(Fig 18-30)*.

As already discussed, in the case of RPA reactions, denaturation of the target DNA is performed by the two enzymes: recombinase and single-stranded DNA binding protein. The recombinase enzymes form complexes with the forward (F) and reverse (R) oligonucleotide primers, then these complexes scan the ds viral cDNA target, searching for the homologous sequences. The region to which the forward primer-recombinase enzyme complex binds is represented as Fc on the 3'-5' strand. Therefore, on the complementary strand 5'-3', this region is labeled as F. Likewise, the region to which the reverse primer-recombinase enzyme complex pairs is represented as Rc on the

5'-3' strand. Therefore on the complementary strand 3'-5', this region is labeled as R.

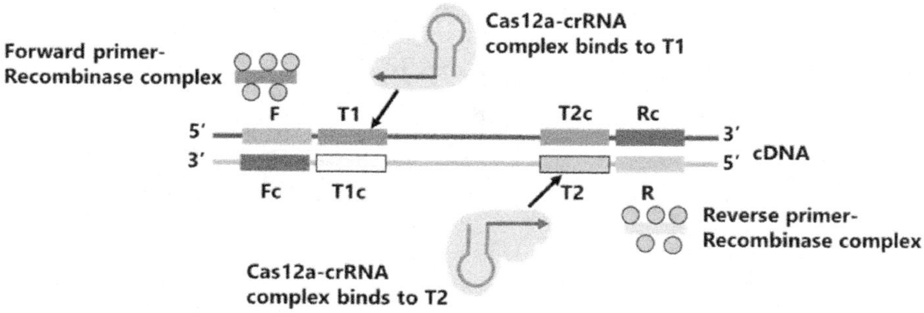

Fig 18-30: Primers-recombinase complexes and Cas12a-crRNA complexes for RPA process

To initiate the RPA reaction, the two Cas12a-crRNA complexes are then mixed with RPA primers, FAM-biotinylated ssDNA reporter molecules, RPA enzymes (recombinase, single-stranded DNA binding proteins, strand-displacement DNA polymerase), and the viral cDNA in a single reaction tube (Ding et al., 2020c).

(i) When the reaction tube is incubated at 37 °C, the RPA amplification is initiated. The recombinase enzymes form complexes with the forward (F) and reverse (R) primers; then, these complexes scan the ds viral cDNA target. Once the homologous sequences are found, the recombinase primer complex invades the dsDNA, causing the separation of DNA strands *(Fig 18-31a)*.

(ii) When the primers are paired to their complementary sequences, the single-stranded DNA binding proteins bind to the exposed DNA strand to stabilize it *(Fig 18-31b)*. The local separation of these DNA strands forms a D-loop like structure, which exposes the binding sites for the attachment of Cas12a-crRNA complexes *(Fig 18-31c)*.

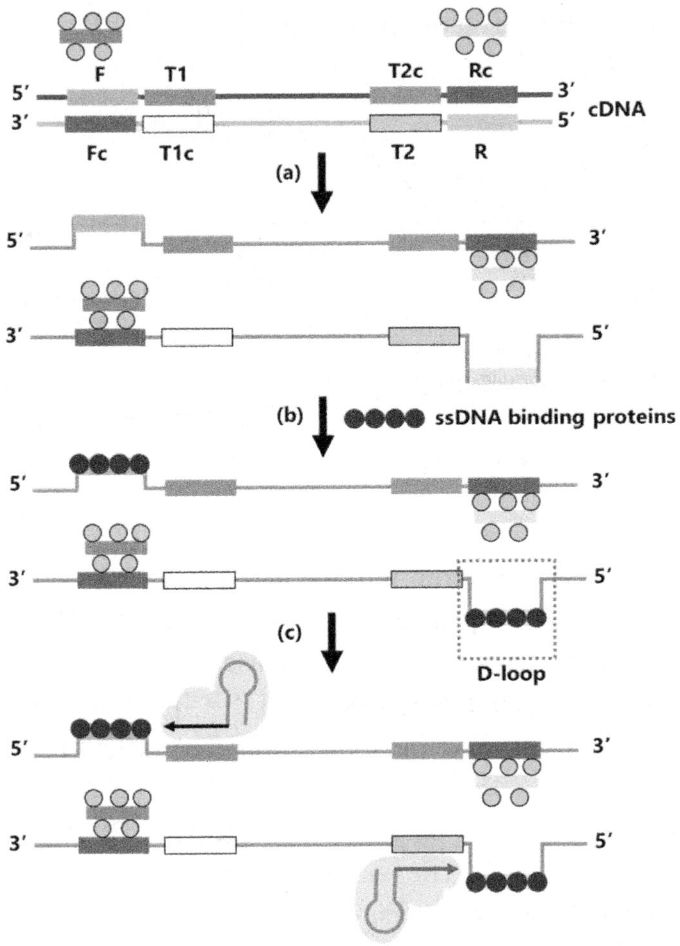

Fig 18-31: D-loop like structure exposes the binding sites for the attachment of Cas12a-crRNA complexes

(iii) When the Cas12a-crRNA complexes bind the target sites, the Cas12a endonuclease is activated and cleaves the FAM-biotinylated ssDNA reporter molecules, generating strong fluorescence signals *(Fig 18-32)*.

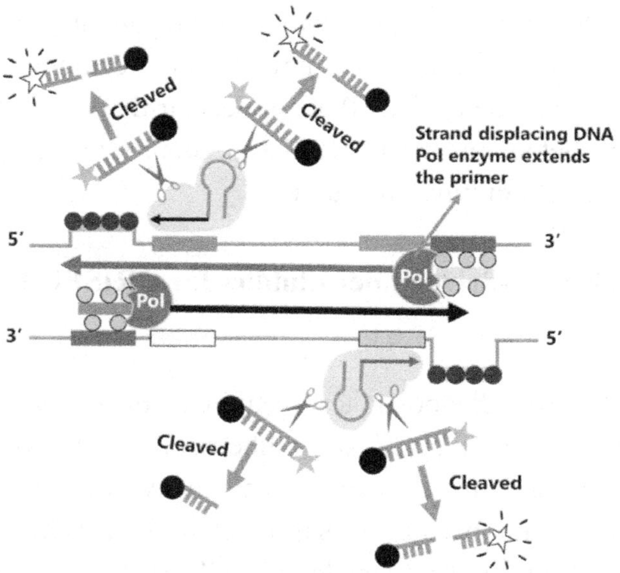

Fig 18-32: Fluorescence produced due to collateral cleavage of FAM biotinylated ssDNA reporter molecules by activated Cas12a enzyme

(iv) When strand displacing DNA polymerase enzyme extends the primer, it eventually generates the amplicons from the original strands of template DNA.

(v) The newly generated DNA strands are then used for another round of RPA and the binding of Cas12a-guide RNA complexes leading to the activation of Cas12a endonuclease. Activated Cas12a enzyme then again shows collateral activity, causing the cleavage of the surrounding FAM-biotinylated ssDNA reporter molecules, leading to the separation of the fluorescent FAM molecules from the biotin molecules. This same process is repeated continuously on amplified products of RPA. This gives leverage to amplify the fluorescence signal. This process is possible in a single tube that eliminates the need for separate preamplification and amplified product transferring steps.

(vi) After 40 minutes, the fluorescence signal can be detected by keeping the tube under a blue LED or UV light illuminator. The presence of fluorescence signals indicates the patient is infected with SARS-CoV-2. Because of the simplicity and accuracy, the AIOD-CRISPR assay can be used as POC molecular diagnostic technology for the rapid detection of COVID-19 in small clinics, schools, and other settings with minimal resources.

18.7. Electric field-driven microfluidics for CRISPR-based diagnostics

The CRISPR-based diagnostic tests require a considerable volume of reagents and certain manual steps like viral RNA extraction and nucleic acid amplification. Such reagents many times become scarce during the pandemic because of supply chain issues. Keeping this in mind, scientists at Stanford developed a CRISPR- Cas12 assay using electric field and microfluidics technologies to extract nucleic acids, thus, reducing time, manual steps, and reagents for the assay (Ramachandran et al., 2020).

The assay involves three main steps.

(i) First, the human sample is placed on the microfluidic chip, and SARS-CoV-2 RNA is purified from the patient's sample with the help of a technique called isotachophoresis (ITP). Isotachophoresis is a form of electrophoresis that separates and concentrates charged molecules in solution, solely based on their electrophoretic mobility or the speed at which the molecules move within the applied electric field. Electric fields are applied across the sample to selectively extract and then concentrate viral RNA *(Fig 18-33)*.
(ii) Next, the extracted RNA is manually pipetted off the chip and put into a standard test tube. The test tube contains the reverse transcriptase enzyme that converts RNA to cDNA, primers, and the DNA polymerase enzyme that amplifies it through the isothermal amplification process (LAMP or RPA).

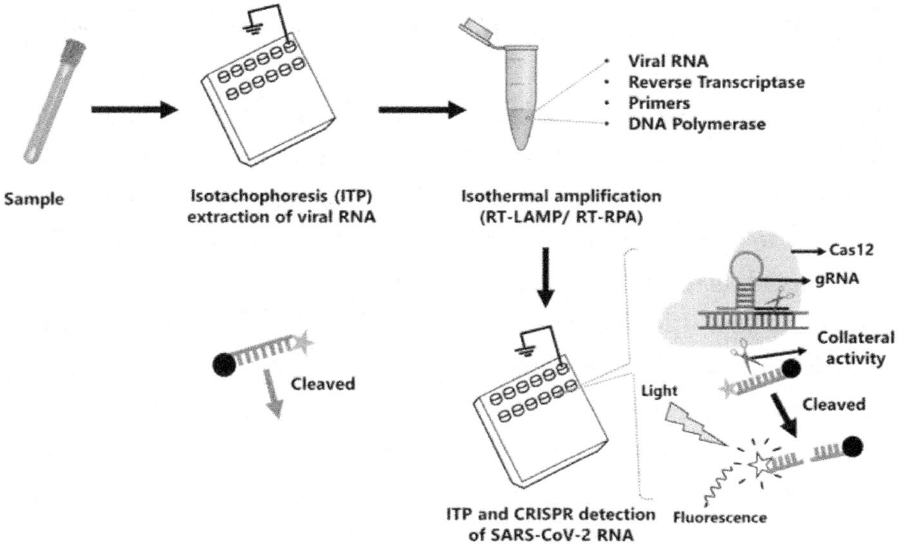

Fig 18-33: Electric field-driven microfluidics for CRISPR-based detection of SARS-CoV-2 RNA in the sample

(iii) Finally, the amplified cDNA is pipetted back onto the microfluidic chip, where the CRISPR enzymatic assay is performed. In this step, electric fields are again applied to concentrate critical components, such as the viral cDNA, guide RNA-Cas12 complex, and ssDNA reporter molecules into a tiny well. This increases the chances that the guide RNA and Cas12 enzyme will interact with the target cDNA, speeding up the reaction. Activated Cas12 enzyme shows collateral activity, and because of collateral activity, it cleaves the FAM-biotinylated ssDNA reporter molecules, leading to the separation of fluorescent FAM molecules. The freed fluorescent FAM molecules, when excited with light, produce a quantifiable fluorescence which can be detected by a detector, indicating the presence of SARS-CoV-2 RNA in the sample.

Additionally, this method consumes a minimal volume of reagents (around 100-fold lower than other CRISPR based methods).

18.8. CRISPR-based diagnostic tests use smartphone cameras

Scientists at Gladstone Institutes and UC Berkeley have developed a CRISPR-based COVID-19 diagnostic test that can provide the test results with a smartphone's help in 15-30 minutes. The test eliminates the need to convert RNA to DNA steps and then amplify the DNA segments, thus reducing the time needed to complete the analysis. The diagnostic test utilizes **three guide RNAs CRISPR-Cas13 protein complexes**, which bind and cleave viral RNA. CRISPR Cas13 proteins are “programmed” to recognize three different regions of SARS-CoV-2 viral RNA and then combined with reporter molecules that become fluorescent when cleaved. Cas13 proteins get activated upon binding of guide RNAs to SARS-CoV-2 RNA, and the activated Cas13 proteins then start to cleave the reporter molecules, leading to the separation of the fluorescent label from the quencher *(Fig 18-34)*.

The mobile phone camera converted into a fluorescent microscope can detect the fluorescence emitted by the cleaved reporter molecules. The higher the fluorescence detected, the higher the number of virus particles in the sample, and vice versa.

Besides rapid testing, the high sensitivity of mobile phone cameras, along with the connectivity, GPS, and data-processing capabilities of the smartphone, has the potential to make the CRISPR-based test portable and an attractive tool for point-of-care disease (POC) diagnosis in low-resource regions.

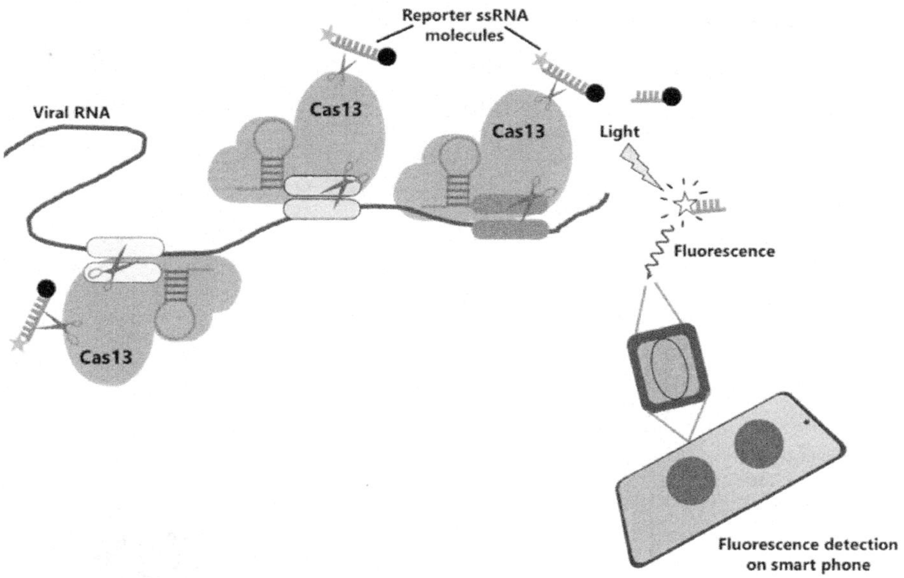

Fig 18-34: Detection of SARS-CoV-2 with CRISPR-based smartphone diagnostic test

Additionally, the test can detect tiny amounts of the virus, in under 30 minutes and much larger concentrations of the virus, such as those in highly contagious people, in under 5 minutes.

How is a smartphone converted into a fluorescent microscope?

Fluorescence microscope is a type of microscope that works on the principle of fluorescence. A substance is said to be fluorescent when it absorbs the energy of invisible shorter wavelength radiation (such as UV light) and emits longer wavelength radiation of visible light (such as green or red light). High-energy and short-wavelength light (such as UV or blue light) is generated from a light source and passed through an excitation filter *(Fig 18-35)*. This filter allows only specific short-wavelength light to pass through and removes non-specific wavelengths. The filtered light is reflected by a dichroic filter and illuminates the fluorophore-labeled sample. The fluorophore absorbs the short-wavelength light and emits longer-wavelength light, which passes through an emission filter. The emission filter blocks any residual excitation light and allows the desired longer-wavelength emission to be detected, resulting in glowing images of the fluorophore-labeled molecules against a dark background.

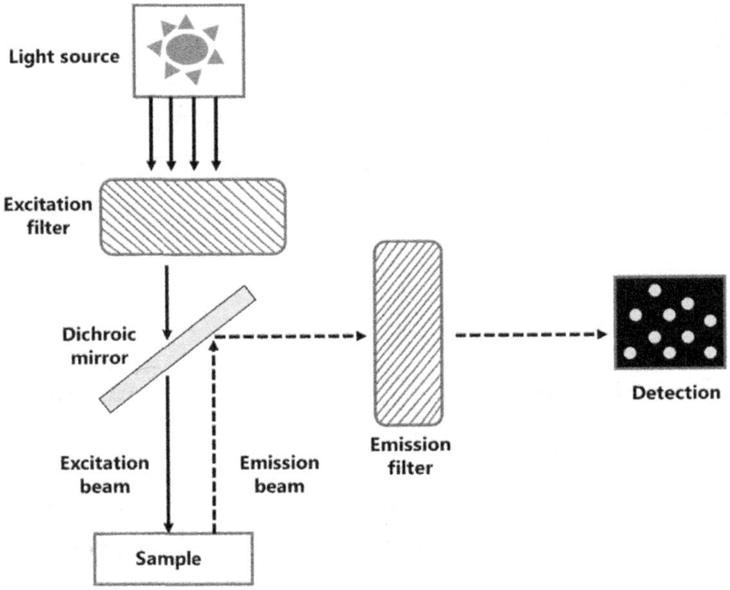

Fig 18-35: Principle of fluorescence microscopy

The smartphone is converted into a fluorescent microscope by clipping the lens onto the phone or tablet camera to enable magnification *(Fig 18-36)*. The lens collects and focuses the fluorescent light emitted by the sample, which is then captured by the camera. The emission filter is placed over the lens to filter out unwanted wavelengths of light, allowing only the fluorescent light emitted by the specimen to pass through. This helps to enhance the contrast and clarity of the fluorescent image, making it easier to visualize and analyze.

Simple LEDs can be used to excite the fluorescent reporter molecules. The excitation light can scatter or reflect off the sample and create background noise, which can interfere with the accuracy and quality of the images obtained. Thus a simple plastic filter is used to reject the scattered excitation light, creating a dark field background that is necessary for viewing fluorescent images.

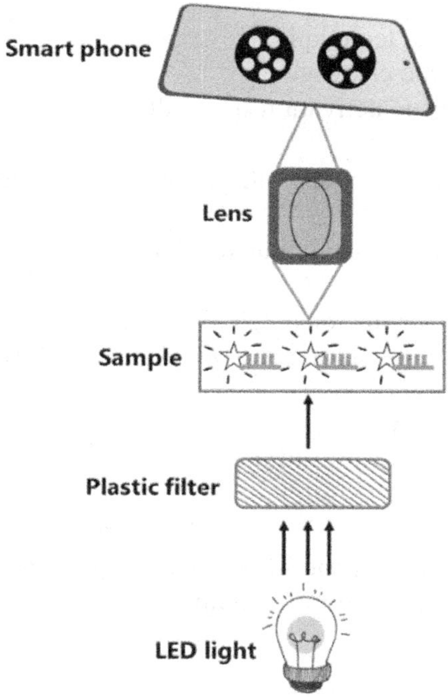

Fig 18-36: Smartphone converted into the fluorescent microscope

Moreover, the CRISPR-based diagnostic kits that we have discussed are not just for diagnosing SARS-CoV-2 but **can also be used to detect other infectious diseases**. All needed is to design guide RNA specific to the genome of the organism that is required to be detected.

Practice Test 3

1. Which transcriptional activators are commonly used for gene activation in eukaryotic cells?
(a) VP64 and p65
(b) VP16 and p68
(c) NF-kB and NF-AT
(d) NF-AT and p68

2. What is the CRISPRa technique?
(a) A technique for gene cleavage
(b) A technique for gene activation using dCas9 fused with transcriptional activators
(c) A technique for gene activation using Cas9 nickase fused with transcriptional activators
(d) A technique for gene silencing using RNAi

3. Which of the following is a CRISPR activation system (CRISPRa) designed to enhance transcription activation power of dCas9?
(a) CRISPR-Cas9
(b) dCas9-VP64-p65-Rta
(c) dCas9-KRAB
(d) RNAi

4. Which system utilizes a modified sgRNA that includes RNA hairpin aptamers to recruit transcriptional activators?
(a) dCas9-SPRY
(b) dCas9-VPR
(c) dCas9-SAM
(d) dCas9-KRAB

5. What is the purpose of the RNA hairpin aptamers in the SAM system?
(a) To recruit the dCas9 protein to the target gene
(b) To amplify gene expression
(c) To form binding sites for MS2 coat proteins
(d) To prevent off-target effects

6. What is the role of the SunTag in the Supernova Tagging system?
(a) It helps to transport CRISPR components to the target cell
(b) It increases the specificity of dCas9

(c) It activates the target gene expression by recruiting transcriptional activators
(d) It cuts the target DNA sequence

7. Which antibodies are used in the Supernova Tagging system?
(a) VP64 antibodies
(b) NLS antibodies
(c) GCN4 antibodies
(d) CRISPR antibodies

8. What is haploinsufficiency?
(a) A condition in which one copy of a gene is not functional
(b) A condition in which both copies of a gene are not functional
(c) A condition in which a gene is overexpressed

9. How does CRISPRa help in eliciting antitumor immunity?
(a) It activates the expression of tumor antigens
(b) It destroys cancer cells
(c) It delivers immune cells to the tumor site
(d) It inhibits the growth of tumor cells

10. Which transcriptional repressor can be fused to dCas9 to improve the efficiency of repression?
(a) KRAB
(b) SALL1
(c) SDS3
(d) All of the above

11. What is epigenetic gene regulation?
(a) Chemical modification of the DNA sequence
(b) Direct alteration of the DNA primary sequence
(c) Chemical modification that results in selective silencing or activation of particular genes without direct altering the DNA primary sequence
(d) Direct activation of all genes without chemical modification

12. Which type of histone modification activates the transcription of a gene?
(a) Histone methylation at lysine 9
(b) Histone acetylation
(c) Histone deacetylation
(d) All of the above

13. Which effector can be fused to dCas9 to induce transcriptional activation of a gene?
(a) DNMT3A
(b) LSD1
(c) p300 histone acetyltransferase
(d) None of the above

14. What is the purpose of ChIP assays?
(a) To sequence the entire genome of a cell
(b) To determine in vivo interactions between DNA and proteins
(c) To measure protein expression levels in cells
(d) To determine interactions between lipids and proteins

15. What is the purpose of affinity purification in a ChIP assay?
(a) To remove irrelevant cellular material
(b) To selectively enrich the protein-DNA complex of interest
(c) To reverse the crosslinks between DNA and protein
(d) To extract DNA from the complex

16. What is the purpose of fusing dCas9 to the biotin ligase BirA in enChIP?
(a) To fragment chromatin
(b) To add biotin to lysine residues of proteins in the proximity of the targeted locus
(c) To reverse crosslinks
(d) To analyze proteins associated with the target DNA sequence

17. What is the purpose of enChIP?
(a) To identify proteins associated with a genomic sequence specified by a particular gRNA
(b) To identify proteins associated with RNA molecules
(c) To identify proteins associated with lipids
(d) To identify proteins associated with the cell membrane

18. What is the purpose of mass spectrometry in enChIP?
(a) To fragment the DNA for analysis
(b) To analyze the biotinylated proteins
(c) To detect promoter or other DNA sequences bound to a protein
(d) To identify the proteins associated with the target DNA sequence

19. What can FISH and other in situ techniques be used for in clinical diagnosis?
(a) Detecting chromosomal abnormalities

(b) Treating chromosomal abnormalities
(c) Preventing chromosomal abnormalities
(d) Causing chromosomal abnormalities

20. What is the purpose of using a fluorescent probe in the FISH technique?
(a) To denature the target DNA
(b) To create single-stranded nicks in the DNA
(c) To locate the precise location of particular DNA sequences on chromosomes
(d) To fix the cells with formalin-based fixative

21. What is the result of the incorporation of fluorescent nucleotides in the nick translation method?
(a) The replacement of labeled nucleotides with unlabeled nucleotides in the ds DNA
(b) The replacement of unlabeled nucleotides with labeled nucleotides in the ds DNA
(c) The creation of single-stranded nicks in the DNA
(d) The denaturation of the target DNA

22. How is dCas9 labeled in the CRISPR-based imaging system?
(a) By fusing it with eGFP fluorescent protein.
(b) By fusing it with MS2 coat protein.
(c) By modifying it to harbor a unique 8-mer RNA sequence.
(d) By fusing it with a PUF domain.

23. How is dCas9-sgRNA complex labeled with multiple fluorescent proteins using the Casilio approach?
(a) GCN4 antibodies
(b) SunTag
(c) PUF binding sequence
(d) None of the above

24. How can non-repetitive sequences be imaged using dCas9-eGFP?
(a) By using a single gRNA
(b) By introducing a combination of gRNAs per target

25. What is the function of the SunTag in the CRISPR-Cas system?
(a) To bind to sgRNA and target DNA
(b) To recruit multiple copies of antibodies attached to fluorescent proteins
(c) To modify sgRNA with a unique RNA sequence called PUF binding sequence

(d) To cleave reporter molecules

26. Sherlock CRISPR-Cas based diagnostic kit uses which enzyme?
(a) Cas13, which targets RNA
(b) Cas12, which targets DNA
(c) Cas9, which targets DNA

27. Which of the following statements is incorrect?
(a) SHERLOCK works by amplifying RNA using recombinase polymerase amplification (RPA)
(b) Recombinase polymerase amplification (RPA) is an isothermal nucleic acid amplification
(c) RPA reaction occurs at a single temperature
(d) Thermocycler is required for RPA reaction.

28. Which enzymes are required for the Recombinase polymerase amplification (RPA) process?
(a) recombinase, single stranded DNA binding proteins (SSB) and strand displacing DNA polymerase
(b) recombinase, primase and strand displacing DNA polymerase
(c) topoisomerase, recombinase. DNA polymerase
(d) Helicase, recombinase, DNA polymerase

29. Which of the enzymes perform denaturation in the RPA process?
(a) recombinase and DNA polymerase
(b) recombinase and single stranded DNA binding proteins
(c) DNA polymerase and single stranded DNA binding proteins
(d) single stranded DNA binding proteins and topoisomerase

30. When does Cas13 enzyme show collateral activity?
(a) If the patient sample contains SARS-CoV-2
(b) If the patient sample does not contain SARS-CoV-2

31. When do the ssRNA reporter molecules emit fluorescence?
(a) If Cas13 cleaves the FAM-biotinylated ssRNA reporter molecules, the freed FAM molecules when excited with light, produce a quantifiable fluorescence
(b) If Cas13 cleaves the FAM-biotinylated ssRNA reporter molecules, the freed biotin molecules when excited with light, produce a quantifiable fluorescence

(c) If Cas13 cleaves the FAM-biotinylated ssRNA reporter molecules, both the freed biotin and FAM molecules when excited with light, produce a quantifiable fluorescence
(d) None of the above

32. How do the lateral flow strips indicate if the patient is infected with SARS-CoV-2?
(a) 2 colored lines appear in the T and C area
(b) Single colored line appears in the C area
(c) No colored line is observed on the strip

33. Which of the following statements are correct?
(a) If the patient's sample doesn't contain SARS-CoV-2, then no collateral cleavage of the FAM-biotinylated ssRNA reporter molecules occurs by Cas13.
(b) The intact reporter molecules accumulate at the control area by the interaction of the biotin labels with immobilized streptavidin, leading to the appearance of a visible colored line in the control area.
(c) The bound gold particles are unable to move further into the test area, thus a single-colored line appears in the control area on the strip, indicating that the test is negative.
(d) All of the above

34. Which of the following events happens if the test is positive?
(a) Because of collateral activity, Cas13 cleaves the FAM-biotinylated ssRNA reporter molecules, leading to separation of biotin molecules and FAM molecules.
(b) Biotin labels bind to the C area by the interaction with the immobilized streptavidin, leading to colored line formation in the C area.
(c) FAM label bound gold particles move to the test area of the lateral flow strip containing the anti-rabbit antibody, which binds anti-FAM antibodies. This generates a second visible colored line in the test area.
(d) All of the above

35. DETECTR CRISPR-Cas based diagnostic kit uses which enzyme?
(a) Cas13, which targets RNA
(b) Cas12, which targets DNA
(c) Cas9, which targets DNA

36. Which of the following statements is correct?
(a) DETECTR works by amplifying SARS-CoV-2 cDNA using loop-mediated amplification (LAMP) process

(b) LAMP is an isothermal nucleic acid amplification
(c) LAMP reaction occurs at a single temperature (62 °C), so there's no need for a thermocycler
(d) All of the above

37. Which structure of DNA forms a seed for exponential LAMP amplification?
(a) Dumbbell structure DNA
(b) Stem loop structure DNA
(c) Linear DNA
(d) DNA Hairpin structure

38. In how much time DETECTR provides test results?
(a) 5-10 minutes
(b) 30-40 minutes
(c) 4 hours
(d) 1 day

39. FELUDA is:
(a) CRISPR-Cas12 based diagnostic test
(b) CRISPR-Cas13 based diagnostic test
(c) CRISPR-Cas9 based diagnostic test
(d) CRISPR-Cas1 based diagnostic test

40. Which of the Cas enzymes don't show collateral activity?
(a) Cas12
(b) Cas9
(c) Cas13
(d) None of the above

41. Which of the following statements is/are correct about the FELUDA test?
(a) Target DNA is labeled with biotin molecule and the guideRNA with FAM molecule
(b) Target DNA is labeled with FAM molecule and the guideRNA with biotin molecule
(c) ssFAM biotinylated reporter molecules are used
(d) All of the above
42. How are the amplified cDNA products labeled for the FELUDA test?
(a) By using the biotinylated forward primers
(b) By using the biotinylated reverse primers
(c) By using forward and reverse primers which are both biotinylated

(d) By using the biotinylated nucleotides

43. How is the FELUDA test different from SHERLOCK and DETECTOR tests?
(a) Cas9 enzyme used in FELUDA test doesn't show collateral activity
(b) Positions of T and C lines are swapped
(c) The SARS-CoV-2 amplified genome and gRNA both are required to be labeled
(d) All of the above are correct

Answers:

1. (a) 2. (b) 3. (b) 4. (c) 5. (c) 6. (c) 7. (c) 8. (a) 9. (a) 10. (d) 11. (c) 12. (b) 13. (c) 14. (b) 15. (b) 16. (b) 17. (a) 18. (d) 19. (a) 20. (c) 21. (b) 22. (a) 23. (c) 24. (b) 25. (b) 26. (a) 27. (d) 28. (a) 29. (b) 30. (a) 31. (a) 32. (a) 33. (d) 34. (d) 35. (b) 36. (d) 37. (a) 38. (b) 39. (c) 40. (b) 41. (a) 42. (b) 43. (d)

References

Acharya, S., Mishra, A., Paul, D., Ansari, A. H., Azhar, M., Kumar, M., ... & Chakraborty, D. (2019). Francisella novicida Cas9 interrogates genomic DNA with very high specificity and can be used for mammalian genome editing. *Proceedings of the National Academy of Sciences*, *116*(42), 20959-20968.

Ahmad, P., Ashraf, M., Younis, M., Hu, X., Kumar, A., Akram, N. A., & Al-Qurainy, F. (2012). Role of transgenic plants in agriculture and biopharming. *Biotechnology advances*, *30*(3), 524-540.

Ahmad, S., Wei, X., Sheng, Z., Hu, P., & Tang, S. (2020). CRISPR/Cas9 for development of disease resistance in plants: recent progress, limitations and future prospects. *Briefings in Functional Genomics*, *19*(1), 26-39.

Aida, T., & Feng, G. (2020). The dawn of non-human primate models for neurodevelopmental disorders. *Current opinion in genetics & development*, *65*, 160-168.

Al Saber, M., Biswas, P., Dey, D., Kaium, M. A., Islam, M. A., Tripty, M. I. A., ... & Kim, B. (2021). A comprehensive review of recent advancements in cancer immunotherapy and generation of CAR T cell by CRISPR-Cas9. *Processes*, *10*(1), 16.

Al Shakaki, A. (2022). Ethical Deliberations on the Gene-Editing CRISPR-Cas9 Designer Babies: Islamic Perspective. *QScience Connect*, *2022*(3), 4.

Alba, R., Bosch, A., & Chillon, M. (2005). Gutless adenovirus: last-generation adenovirus for gene therapy. *Gene therapy*, *12*(1), S18-S27.

Alerasool, N., Segal, D., Lee, H., & Taipale, M. (2020). An efficient KRAB domain for CRISPRi applications in human cells. *Nature Methods*, *17*(11), 1093-1096.

Amitai, G., & Sorek, R. (2016). CRISPR–Cas adaptation: insights into the mechanism of action. *Nature Reviews Microbiology*, *14*(2), 67.

Annoni, M., & Pievani, T. (2021). What is wrong in extinguishing a species? Charting the Ethical Challenges of using Gene-Drive Technologies to eradicate A. gambiae vector populations. *BioLaw Journal-Rivista di BioDiritto*, (1S), 429-444.

Anzalone, A. V., Koblan, L. W., & Liu, D. R. (2020). Genome editing with CRISPR–Cas nucleases, base editors, transposases and prime editors. *Nature biotechnology*, *38*(7), 824-844.

Anzalone, A. V., Randolph, P. B., Davis, J. R., Sousa, A. A., Koblan, L. W., Levy, J. M., ... & Liu, D. R. (2019). Search-and-replace genome editing without double-strand breaks or donor DNA. *Nature*, *576*(7785), 149-157.

Azhar, M., Phutela, R., Ansari, A. H., Sinha, D., Sharma, N., Kumar, M., ... & Maiti, S. (2020a). Rapid, field-deployable nucleobase detection and identification using FnCas9. *BioRxiv*.

Azhar, M., Phutela, R., Kumar, M., Ansari, A. H., Rauthan, R., Gulati, S., ... & Maiti, S. (2020b). Rapid, accurate, nucleobase detection using FnCas9. *medRxiv*.

Barnett, B. G., Crews, C. J., & Douglas, J. T. (2002). Targeted adenoviral vectors. *Biochimica et Biophysica Acta (BBA)-Gene Structure and Expression*, *1575*(1-3), 1-14.

Baylis, F. (2019). *Altered inheritance: CRISPR and the ethics of human genome editing*. Harvard University Press.

Behr, M., Zhou, J., Xu, B., & Zhang, H. (2021). In vivo delivery of CRISPR-Cas9 therapeutics: progress and challenges. *Acta Pharmaceutica Sinica B*, *11*(8), 2150-2171.

Belling, J. N., Heidenreich, L. K., Tian, Z., Mendoza, A. M., Chiou, T. T., Gong, Y., ... & Jonas, S. J. (2020). Acoustofluidic sonoporation for gene delivery to human hematopoietic stem and progenitor cells. *Proceedings of the National Academy of Sciences*, *117*(20), 10976-10982.

Bhaya, D., Davison, M., & Barrangou, R. (2011). CRISPR-Cas systems in bacteria and archaea: versatile small RNAs for adaptive defense and regulation. *Annual review of genetics*, *45*, 273-297.

Birling, M. C., Herault, Y., & Pavlovic, G. (2017). Modeling human disease in rodents by CRISPR/Cas9 genome editing. *Mammalian genome*, *28*, 291-301.

Brocken, D. J., Tark-Dame, M., & Dame, R. T. (2018). dCas9: a versatile tool for epigenome editing. *Current issues in molecular biology*, *26*(1), 15-32.

Brokowski, C., & Adli, M. (2019). CRISPR ethics: moral considerations for applications of a powerful tool. *Journal of molecular biology*, *431*(1), 88-101.

Broughton, J. P., Deng, W., Fasching, C. L., Singh, J., Chiu, C. Y., & Chen, J. S. (2020a). A protocol for rapid detection of the 2019 novel coronavirus SARS-CoV-2 using CRISPR diagnostics: SARS-CoV-2 DETECTR. *Medrxiv: the Preprint Server for Health Sciences*.

Broughton, J. P., Deng, X., Yu, G., Fasching, C. L., Servellita, V., Singh, J., ... & Chiu, C. Y. (2020b). CRISPR–Cas12-based detection of SARS-CoV-2. *Nature biotechnology*, *38*(7), 870-874.

Broughton, J. P., Deng, X., Yu, G., Fasching, C. L., Singh, J., Streithorst, J., ... & Chiu, C. Y. (2020c). Rapid detection of 2019 novel coronavirus SARS-CoV-2 using a CRISPR-based DETECTR lateral flow assay. *MedRxiv*.

Burt, A., Coulibaly, M., Crisanti, A., Diabate, A., & Kayondo, J. K. (2018). Gene drive to reduce malaria transmission in sub-Saharan Africa. *Journal of Responsible Innovation*, *5*(sup1), S66-S80.

Cai, L., Bai, H., Mahairaki, V., Gao, Y., He, C., Wen, Y., ... & Cheng, L. (2018). A universal approach to correct various HBB gene mutations in human stem cells for gene therapy of beta-thalassemia and sickle cell disease. *Stem cells translational medicine*, *7*(1), 87-97.

Camerlengo, F., Frittelli, A., & Pagliarello, R. (2022). CRISPR towards a Sustainable Agriculture. *Encyclopedia*, *2*(1), 538-558.

Carroll, D. (2011). Genome engineering with zinc-finger nucleases. *Genetics*, *188*(4), 773-782.

Cebrian-Serrano, A., & Davies, B. (2017). CRISPR-Cas orthologues and variants: optimizing the repertoire, specificity and delivery of genome engineering tools. *Mammalian Genome*, *28*, 247-261.

Cencic, R., Miura, H., Malina, A., Robert, F., Ethier, S., Schmeing, T. M., ... & Pelletier, J. (2014). Protospacer adjacent motif (PAM)-distal sequences engage CRISPR Cas9 DNA target cleavage. *PloS one*, *9*(10), e109213.

Cereseto, A., Cradick, T. J., & Davies, K. (2021). Base Editors Flex Sights on Sickle-Cell Disease. *The CRISPR Journal*, *4*(2), 166-168.

Chakravorty, M., Nanda, M., Arora, N., Singh, S., Kumar, V., & Deshwal, S. (2022). CRISPR-Cas9 mediated genome tailoring to improve nutritional quality and shelf life in crops: A review. *Plant Gene*, 100369.

Champer, J., Buchman, A., & Akbari, O. S. (2016). Cheating evolution: engineering gene drives to manipulate the fate of wild populations. *Nature Reviews Genetics*, *17*(3), 146-159.

Chandrasekaran, J., Brumin, M., Wolf, D., Leibman, D., Klap, C., Pearlsman, M., ... & Gal-On, A. (2016). Development of broad virus resistance in non-transgenic cucumber using CRISPR/Cas9 technology. *Molecular plant pathology*, *17*(7), 1140-1153.

Che, L. R., He, Z. B., Liu, Y., Yan, Z. T., Han, B. Z., Chen, X. J., ... & Qiao, L. (2021). Electroporation-mediated nucleic acid delivery during non-embryonic stages for gene-function analysis in Anopheles sinensis. *Insect Biochemistry and Molecular Biology*, *128*, 103500.

Chen, B., Gilbert, L. A., Cimini, B. A., Schnitzbauer, J., Zhang, W., Li, G. W., ... & Huang, B. (2013). Dynamic imaging of genomic loci in living human cells by an optimized CRISPR/Cas system. *Cell*, *155*(7), 1479-1491.

Chen, F., Pruett-Miller, S. M., Huang, Y., Gjoka, M., Duda, K., Taunton, J., ... & Davis, G. D. (2011). High-frequency genome editing using ssDNA oligonucleotides with zinc-finger nucleases. *Nature methods*, *8*(9), 753-755.

Chen, J. S., Ma, E., Harrington, L. B., Da Costa, M., Tian, X., Palefsky, J. M., & Doudna, J. A. (2018). CRISPR-Cas12a target binding unleashes indiscriminate single-stranded DNase activity. *Science*, *360*(6387), 436-439.

Chen, L., Zhu, B., Ru, G., Meng, H., Yan, Y., Hong, M., ... & Li, D. (2022). Re-engineering the adenine deaminase TadA-8e for efficient and specific CRISPR-based cytosine base editing. *Nature Biotechnology*, 1-10.

Chen, M., & Qi, L. S. (2017). Repurposing CRISPR system for transcriptional activation. *RNA Activation*, 147-157.

Chen, P. J., & Liu, D. R. (2022). Prime editing for precise and highly versatile genome manipulation. *Nature Reviews Genetics*, 1-17.

Chen, S., Jiao, Y., Pan, F., Guan, Z., Cheng, S. H., & Sun, D. (2022). Knock-in of a Large Reporter Gene via the High-Throughput Microinjection of the CRISPR/Cas9 System. *IEEE Transactions on Biomedical Engineering*, *69*(8), 2524-2532.

Chen, T., Gao, D., Zhang, R., Zeng, G., Yan, H., Lim, E., & Liang, F. S. (2017). Chemically controlled epigenome editing through an inducible dCas9 system. *Journal of the American Chemical Society*, *139*(33), 11337-11340.

Chen, W., Li, H., Liu, Z., & Yuan, W. (2016). Lipopolyplex for therapeutic gene delivery and its application for the treatment of Parkinson's disease. *Frontiers in Aging Neuroscience*, *8*, 68.

Chen, Y. C., Sheng, J., Trang, P., & Liu, F. (2018). Potential application of the CRISPR/Cas9 system against herpesvirus infections. *Viruses*, *10*(6), 291.

Cheng, A. W., Jillette, N., Lee, P., Plaskon, D., Fujiwara, Y., Wang, W., ... & Wang, H. (2016). Casilio: a versatile CRISPR-Cas9-Pumilio hybrid for gene regulation and genomic labeling. *Cell research*, *26*(2), 254-257.

Chylinski, K., Le Rhun, A., & Charpentier, E. (2013). The tracrRNA and Cas9 families of type II CRISPR-Cas immunity systems. *RNA biology*, *10*(5), 726-737.

Chylinski, K., Makarova, K. S., Charpentier, E., & Koonin, E. V. (2014). Classification and evolution of type II CRISPR-Cas systems. *Nucleic acids research*, *42*(10), 6091-6105.

Clow, P. A., Jillette, N., Zhu, J. J., & Cheng, A. W. (2020). CRISPR-mediated multiplexed live cell imaging of nonrepetitive genomic loci. *bioRxiv*, 2020-03.

Cowan, P. J., Hawthorne, W. J., & Nottle, M. B. (2019). Xenogeneic transplantation and tolerance in the era of CRISPR-Cas9. *Current Opinion in Organ Transplantation*, *24*(1), 5-11.

Cribbs, A. P., & Perera, S. M. (2017). Focus: Genome editing: Science and bioethics of CRISPR-Cas9 gene editing: An analysis towards separating facts and fiction. *The Yale journal of biology and medicine*, *90*(4), 625.

Cyranoski, D. (2016). CRISPR gene-editing tested in a person for the first time. *Nature*, *539*(7630).

Davis, A. J., & Chen, D. J. (2013). DNA double strand break repair via non-homologous end-joining. *Transl cancer res*, *2*(3), 130-143.

de Araujo, M. (2017). Editing the Genome of Human Being: CRISPR-Cas9 and the Ethics of Genetic Enhancement. *Journal of Ethics and Emerging Technologies*, *27*(1), 24-42.

Deltcheva, E., Chylinski, K., Sharma, C. M., Gonzales, K., Chao, Y., Pirzada, Z. A., ... & Charpentier, E. (2011). CRISPR RNA maturation by trans-encoded small RNA and host factor RNase III. *Nature*, *471*(7340), 602-607.

Demirci, Y., Zhang, B., & Unver, T. (2018). CRISPR/Cas9: an RNA-guided highly precise synthetic tool for plant genome editing. *Journal of cellular physiology*, *233*(3), 1844-1859.

Dever, D. P., Bak, R. O., Reinisch, A., Camarena, J., Washington, G., Nicolas, C. E., ... & Porteus, M. H. (2016). CRISPR/Cas9 β-globin gene targeting in human haematopoietic stem cells. *Nature*, *539*(7629), 384-389.

Dimitri, A., Herbst, F., & Fraietta, J. A. (2022). Engineering the next-generation of CAR T-cells with CRISPR-Cas9 gene editing. *Molecular Cancer*, *21*(1), 78.

Ding, X., Yin, K., Li, Z., & Liu, C. (2020a). All-in-One dual CRISPR-cas12a (AIOD-CRISPR) assay: a case for rapid, ultrasensitive and visual detection of novel coronavirus SARS-CoV-2 and HIV virus. *BioRxiv*.

Ding, X., Yin, K., Li, Z., Lalla, R. V., Ballesteros, E., Sfeir, M. M., & Liu, C. (2020b). Ultrasensitive and visual detection of SARS-CoV-2 using all-in-one dual CRISPR-Cas12a assay. *Nature communications*, *11*(1), 1-10.

Ding, X., Yin, K., Li, Z., Lalla, R. V., Sfeir, M. M., & Liu, C. (2020c). All-in-one dual CRISPR-Cas12a (AIOD-CRISPR) assay protocol for SARS-CoV-2 detection.

Dominguez, A. A., Lim, W. A., & Qi, L. S. (2016). Beyond editing: repurposing CRISPR–Cas9 for precision genome regulation and interrogation. *Nature reviews Molecular cell biology*, *17*(1), 5-15.

Dong, W., & Kantor, B. (2021). Lentiviral vectors for delivery of gene-editing systems based on CRISPR/Cas: current state and perspectives. *Viruses*, *13*(7), 1288.

Doudna, J. A., & Charpentier, E. (2014). Genome editing. The new frontier of genome engineering with CRISPR-Cas9. Sci. 346, 1258096.

Doudna, J. A., & Charpentier, E. (2014). The new frontier of genome engineering with CRISPR-Cas9. *Science*, *346*(6213), 1258096.

Douglas, J. T. (2007). Adenoviral vectors for gene therapy. *Molecular biotechnology*, *36*, 71-80.

Drury, D. W., Dapper, A. L., Siniard, D. J., Zentner, G. E., & Wade, M. J. (2017). CRISPR/Cas9 gene drives in genetically variable and nonrandomly mating wild populations. *Science advances*, *3*(5), e1601910.

Dupuis, M. È., Villion, M., Magadán, A. H., & Moineau, S. (2013). CRISPR-Cas and restriction–modification systems are compatible and increase phage resistance. *Nature communications*, *4*(1), 1-7.

Durai, S., Mani, M., Kandavelou, K., Wu, J., Porteus, M. H., & Chandrasegaran, S. (2005). Zinc finger nucleases: custom-designed molecular scissors for genome engineering of plant and mammalian cells. *Nucleic acids research*, *33*(18), 5978-5990.

Enriquez, P. (2018). Editing humanity: On the precise manipulation of DNA in human embryos. *NCL Rev.*, *97*, 1147.

Esvelt, K. M., Smidler, A. L., Catteruccia, F., & Church, G. M. Emerging Technology: Concerning RNA-guided gene drives for the alteration of wild populations. eLife [Internet]. 2014 [cited 2021 Sep 25]. Doi: 10.7554. *eLife*, *3401*.

Fajrial, A. K., He, Q. Q., Wirusanti, N. I., Slansky, J. E., & Ding, X. (2020). A review of emerging physical transfection methods for CRISPR/Cas9-mediated gene editing. *Theranostics*, *10*(12), 5532.

Frangoul, H., Altshuler, D., Cappellini, M. D., Chen, Y. S., Domm, J., Eustace, B. K., ... & Corbacioglu, S. (2021). CRISPR-Cas9 gene editing for sickle cell disease and β-thalassemia. *New England Journal of Medicine*, *384*(3), 252-260.

Fujita, T., & Fujii, H. (2013). Efficient isolation of specific genomic regions and identification of associated proteins by engineered DNA-binding molecule-mediated chromatin immunoprecipitation (enChIP) using CRISPR. *Biochemical and biophysical research communications*, *439*(1), 132-136.

Fujita, T., Kitaura, F., Oji, A., Tanigawa, N., Yuno, M., Ikawa, M., ... & Fujii, H. (2018). Transgenic mouse lines expressing the 3x FLAG-dC as9 protein for enCh IP analysis. *Genes to cells*, *23*(4), 318-325.

Gaj, T., Gersbach, C. A., & Barbas III, C. F. (2013). ZFN, TALEN, and CRISPR/Cas-based methods for genome engineering. *Trends in biotechnology*, *31*(7), 397-405.

Gao, Y., Wu, H., Wang, Y., Liu, X., Chen, L., Li, Q., ... & Zhang, Y. (2017). Single Cas9 nickase induced generation of NRAMP1 knockin cattle with reduced off-target effects. *Genome biology*, *18*(1), 1-15.

Garland-Thomson, R. (2020). How we got to CRISPR: The dilemma of being human. *Perspectives in biology and medicine*, *63*(1), 28-43.

Gasiunas, G., Young, J. K., Karvelis, T., Kazlauskas, D., Urbaitis, T., Jasnauskaite, M., ... & Siksnys, V. (2020). A catalogue of biochemically diverse CRISPR-Cas9 orthologs. *Nature communications*, *11*(1), 5512.

Gaudelli, N. M., Komor, A. C., Rees, H. A., Packer, M. S., Badran, A. H., Bryson, D. I., & Liu, D. R. (2017). Programmable base editing of A• T to G• C in genomic DNA without DNA cleavage. *Nature*, *551*(7681), 464-471.

Ghaffari, S., Khalili, N., & Rezaei, N. (2021). CRISPR/Cas9 revitalizes adoptive T-cell therapy for cancer immunotherapy. *Journal of Experimental & Clinical Cancer Research*, *40*(1), 269.

Ghanbarlou, M. M., Abdoli, S., Bamehr, H., Qazizadeh, L., Omid, H., Iranpour, H., ... & Shokrgozar, M. A. (2021). Delivery of dCas9 CRISPR System Into the Hard Transfection Cells by Magnetofection Approach.

Ghorbal, M., Gorman, M., Macpherson, C. R., Martins, R. M., Scherf, A., & Lopez-Rubio, J. J. (2014). Genome editing in the human malaria parasite Plasmodium falciparum using the CRISPR-Cas9 system. *Nature biotechnology*, *32*(8), 819-821.

Ghorbani, A., Hadifar, S., Salari, R., Izadpanah, K., Burmistrz, M., Afsharifar, A., ... & Neely, G. G. (2021). A short overview of CRISPR-Cas technology and its application in viral disease control. *Transgenic Research*, *30*, 221-238.

Gilbert, L. A., Horlbeck, M. A., Adamson, B., Villalta, J. E., Chen, Y., Whitehead, E. H., ... & Weissman, J. S. (2014). Genome-scale CRISPR-mediated control of gene repression and activation. *Cell*, *159*(3), 647-661.

Gjaltema, R. A., & Rots, M. G. (2020). Advances of epigenetic editing. *Current opinion in chemical biology*, *57*, 75-81.

Gleditzsch, D., Pausch, P., Müller-Esparza, H., Özcan, A., Guo, X., Bange, G., & Randau, L. (2019). PAM identification by CRISPR-Cas effector complexes: diversified mechanisms and structures. *RNA biology*, *16*(4), 504-517.

Gong, H., Liu, M., Klomp, J., Merrill, B. J., Rehman, J., & Malik, A. B. (2017). Method for dual viral vector mediated CRISPR-Cas9 gene disruption in primary human endothelial cells. *Scientific Reports*, *7*(1), 42127.

Groot, A. T., & Dicke, M. (2002). Insect-resistant transgenic plants in a multi-trophic context. *The Plant Journal*, *31*(4), 387-406.

Gu, X., Liu, L., & Zhang, H. (2021). Transgene-free genome editing in plants. *Frontiers in Genome Editing*, *3*, 805317.

Guilinger, J. P., Thompson, D. B., & Liu, D. R. (2014). Fusion of catalytically inactive Cas9 to FokI nuclease improves the specificity of genome modification. *Nature biotechnology*, *32*(6), 577-582.

Guo, M., Ren, K., Zhu, Y., Tang, Z., Wang, Y., Zhang, B., & Huang, Z. (2019). Structural insights into a high fidelity variant of SpCas9. *Cell Research*, *29*(3), 183-192.

Hamidian, A., Vaapil, M., von Stedingk, K., Fujita, T., Persson, C. U., Eriksson, P., ... & Mohlin, S. (2018). Promoter-associated proteins of EPAS1 identified by enChIP-MS–a putative role of HDX as a negative regulator. *Biochemical and biophysical research communications*, *499*(2), 291-298.

Hammond, A. M., & Galizi, R. (2017). Gene drives to fight malaria: current state and future directions. *Pathogens and global health*, *111*(8), 412-423.

Hefferin, M. L., & Tomkinson, A. E. (2005). Mechanism of DNA double-strand break repair by non-homologous end joining. *DNA repair*, *4*(6), 639-648.

Hillary, V. E., & Ceasar, S. A. (2022). Prime editing in plants and mammalian cells: Mechanism, achievements, limitations, and future prospects. *BioEssays*, *44*(9), 2200032.

Hille, F., Richter, H., Wong, S. P., Bratovič, M., Ressel, S., & Charpentier, E. (2018). The biology of CRISPR-Cas: backward and forward. *Cell*, *172*(6), 1239-1259.

Hilton, I. B., D'ippolito, A. M., Vockley, C. M., Thakore, P. I., Crawford, G. E., Reddy, T. E., & Gersbach, C. A. (2015). Epigenome editing by a CRISPR-Cas9-based acetyltransferase activates genes from promoters and enhancers. *Nature biotechnology*, *33*(5), 510-517.

Hirano, S., Abudayyeh, O. O., Gootenberg, J. S., Horii, T., Ishitani, R., Hatada, I., ... & Nureki, O. (2019). Structural basis for the promiscuous PAM recognition by Corynebacterium diphtheriae Cas9. Nature Communications, 10(1), 1968.

Horvath, P., & Barrangou, R. (2010). CRISPR/Cas, the immune system of bacteria and archaea. *Science*, *327*(5962), 167-170.

Hou, N., Du, X., & Wu, S. (2022). Advances in pig models of human diseases. *Animal Models and Experimental Medicine*, *5*(2), 141-152.

Hou, Z., Zhang, Y., Propson, N. E., Howden, S. E., Chu, L. F., Sontheimer, E. J., & Thomson, J. A. (2013). Efficient genome engineering in human pluripotent stem cells using Cas9 from Neisseria meningitidis. *Proceedings of the National Academy of Sciences*, *110*(39), 15644-15649.

Hryhorowicz, M., Grześkowiak, B., Mazurkiewicz, N., Śledziński, P., Lipiński, D., & Słomski, R. (2019). Improved delivery of CRISPR/Cas9 system using magnetic nanoparticles into porcine fibroblast. *Molecular biotechnology*, *61*, 173-180.

Hu, Z., Wang, S., Zhang, C., Gao, N., Li, M., Wang, D., ... & Wang, Y. (2020). A compact Cas9 ortholog from Staphylococcus Auricularis (SauriCas9) expands the DNA targeting scope. *PLoS biology*, *18*(3), e3000686.

Huang, L., Hua, Z., Xiao, H., Cheng, Y., Xu, K., Gao, Q., ... & Li, K. (2017). CRISPR/Cas9-mediated ApoE-/-and LDLR-/-double gene knockout in pigs elevates serum LDL-C and TC levels. *Oncotarget*, *8*(23), 37751.

Huang, T. P., Newby, G. A., & Liu, D. R. (2021). Precision genome editing using cytosine and adenine base editors in mammalian cells. *Nature protocols*, *16*(2), 1089-1128.

Hunt, C., Hartford, S. A., White, D., Pefanis, E., Hanna, T., Herman, C., ... & Gong, G. (2021). Tissue-specific activation of gene expression by the Synergistic Activation Mediator (SAM) CRISPRa system in mice. *Nature Communications*, *12*(1), 2770.

Isaka, Y., & Imai, E. (2007). Electroporation-mediated gene therapy. *Expert opinion on drug delivery*, *4*(5), 561-571.

Ishii, T. (2017). The ethics of creating genetically modified children using genome editing. *Current Opinion in Endocrinology, Diabetes and Obesity*, *24*(6), 418-423.

Jabbar, A., Zulfiqar, F., Mahnoor, M., Mushtaq, N., Zaman, M. H., Din, A. S. U., ... & Ahmad, H. I. (2021). Advances and Perspectives in the Application of CRISPR-Cas9 in Livestock. *Molecular Biotechnology*, *63*(9), 757-767.

Jankowicz-Cieslak, J., Mba, C., & Till, B. J. (2017). Mutagenesis for crop breeding and functional genomics. *Biotechnologies for Plant Mutation Breeding: Protocols*, 3-18.

Ji, H., Jiang, Z., Lu, P., Ma, L., Li, C., Pan, H., ... & Zhu, H. (2016). Specific reactivation of latent HIV-1 by dCas9-SunTag-VP64-mediated guide RNA targeting the HIV-1 promoter. *Molecular Therapy*, *24*(3), 508-521.

Jiang, T., Henderson, J. M., Coote, K., Cheng, Y., Valley, H. C., Zhang, X. O., ... & Xue, W. (2020). Chemical modifications of adenine base editor mRNA and guide RNA expand its application scope. *Nature communications*, *11*(1), 1979.

Jin, L., Zeng, X., Liu, M., Deng, Y., & He, N. (2014). Current progress in gene delivery technology based on chemical methods and nano-carriers. *Theranostics*, *4*(3), 240.

Jin, Y. H., Joo, H., Lee, K., Kim, H., Didier, R., Yang, Y., ... & Lee, C. (2019). Streamlined procedure for gene knockouts using all-in-one adenoviral CRISPR-Cas9. *Scientific reports*, *9*(1), 277.

Jinek, M., Chylinski, K., Fonfara, I., Hauer, M., Doudna, J. A., & Charpentier, E. (2012). A programmable dual-RNA–guided DNA endonuclease in adaptive bacterial immunity. *science*, *337*(6096), 816-821.

Jinek, M., Chylinski, K., Fonfara, I., Hauer, M., Doudna, J. A., & Charpentier, E. (2012). A programmable dual-RNA–guided DNA endonuclease in adaptive bacterial immunity. *science*, *337*(6096), 816-821.

Joung, J. K., & Sander, J. D. (2013). TALENs: a widely applicable technology for targeted genome editing. *Nature reviews Molecular cell biology*, *14*(1), 49-55.

Joung, J., Ladha, A., Saito, M., Kim, N. G., Woolley, A. E., Segel, M., ... & Zhang, F. (2020). Detection of SARS-CoV-2 with SHERLOCK one-pot testing. *New England Journal of Medicine*, *383*(15), 1492-1494.

Kang, K., Song, Y., Kim, I., & Kim, T. J. (2022). Therapeutic applications of the CRISPR-Cas system. *Bioengineering*, *9*(9), 477.

Kantor, A., McClements, M. E., & MacLaren, R. E. (2020). CRISPR-Cas9 DNA base-editing and prime-editing. *International journal of molecular sciences*, *21*(17), 6240.

Karlson, C. K. S., Mohd-Noor, S. N., Nolte, N., & Tan, B. C. (2021). CRISPR/dCas9-based systems: Mechanisms and applications in plant sciences. *Plants*, *10*(10), 2055.

Kazi, T. A., & Biswas, S. R. (2021). CRISPR/dCas system as the modulator of gene expression. *Progress in Molecular Biology and Translational Science*, *178*, 99-122.

Kellner, M. J., Koob, J. G., Gootenberg, J. S., Abudayyeh, O. O., & Zhang, F. (2019). SHERLOCK: nucleic acid detection with CRISPR nucleases. *Nature protocols*, *14*(10), 2986-3012.

Kennedy, E. M., Kornepati, A. V., & Cullen, B. R. (2015). Targeting hepatitis B virus cccDNA using CRISPR/Cas9. *Antiviral research*, *123*, 188-192.

Khalaf, K., Janowicz, K., Dyszkiewicz-Konwińska, M., Hutchings, G., Dompe, C., Moncrieff, L., ... & Kempisty, B. (2020). CRISPR/Cas9 in cancer immunotherapy: animal models and human clinical trials. *Genes*, *11*(8), 921.

Kim, E., Koo, T., Park, S. W., Kim, D., Kim, K., Cho, H. Y., ... & Kim, J. S. (2017). In vivo genome editing with a small Cas9 orthologue derived from Campylobacter jejuni. *Nature communications*, *8*(1), 14500.

Kim, Y. B., Komor, A. C., Levy, J. M., Packer, M. S., Zhao, K. T., & Liu, D. R. (2017). Increasing the genome-targeting scope and precision of base editing with engineered Cas9-cytidine deaminase fusions. *Nature biotechnology*, *35*(4), 371-376.

Kirk, N., Kannemeyer, R., Greenaway, A., MacDonald, E., & Stronge, D. (2019). Understanding attitudes on new technologies to manage invasive species. *Pacific Conservation Biology*, *26*(1), 35-44.

Kleinstiver, B. P., Pattanayak, V., Prew, M. S., Tsai, S. Q., Nguyen, N. T., Zheng, Z., & Joung, J. K. (2016). High-fidelity CRISPR–Cas9 nucleases with no detectable genome-wide off-target effects. *Nature*, *529*(7587), 490-495.

Kleinstiver, B. P., Prew, M. S., Tsai, S. Q., Topkar, V. V., Nguyen, N. T., Zheng, Z., ... & Joung, J. K. (2015). Engineered CRISPR-Cas9 nucleases with altered PAM specificities. *Nature*, *523*(7561), 481-485.

Kochanek, S., Schiedner, G., & Volpers, C. (2001). High-capacity'gutless' adenoviral vectors. *Current opinion in molecular therapeutics*, *3*(5), 454-463.

Komor, A. C., Kim, Y. B., Packer, M. S., Zuris, J. A., & Liu, D. R. (2016). Programmable editing of a target base in genomic DNA without double-stranded DNA cleavage. *Nature*, *533*(7603), 420-424.

Konermann, S., Brigham, M. D., Trevino, A. E., Joung, J., Abudayyeh, O. O., Barcena, C., ... & Zhang, F. (2015). Genome-scale transcriptional activation by an engineered CRISPR-Cas9 complex. *Nature*, *517*(7536), 583-588.

Kovesdi, I., Brough, D. E., Bruder, J. T., & Wickham, T. J. (1997). Adenoviral vectors for gene transfer. *Current opinion in biotechnology*, *8*(5), 583-589.

Krishan, K., Kanchan, T., & Singh, B. (2016). Human genome editing and ethical considerations. *Science and engineering ethics*, *22*, 597-599.

Kulcsár, P. I., Tálas, A., Huszár, K., Ligeti, Z., Tóth, E., Weinhardt, N., ... & Welker, E. (2017). Crossing enhanced and high fidelity SpCas9 nucleases to optimize specificity and cleavage. *Genome biology*, *18*(1), 1-17.

Kuscu, C., Mammadov, R., Czikora, A., Unlu, H., Tufan, T., Fischer, N. L., ... & Adli, M. (2019). Temporal and spatial epigenome editing allows precise gene regulation in mammalian cells. *Journal of molecular biology*, *431*(1), 111-121.

Lander, E. S., Baylis, F., Zhang, F., Charpentier, E., Berg, P., Bourgain, C., ... & Winnacker, E. L. (2019). Adopt a moratorium on heritable genome editing. *Nature*, *567*(7747), 165-168.

Lanphier, E., Urnov, F., Haecker, S. E., Werner, M., & Smolenski, J. (2015). Don't edit the human germ line. *Nature*, *519*(7544), 410-411.

Lau, C. H., & Suh, Y. (2017). In vivo genome editing in animals using AAV-CRISPR system: applications to translational research of human disease. *F1000Research, 6*.

Le Rhun, A., Escalera-Maurer, A., Bratovič, M., & Charpentier, E. (2019). CRISPR-Cas in Streptococcus pyogenes. *RNA biology*, *16*(4), 380-389.

Ledford, H. (2020). CRISPR treatment inserted directly into the body for first time. *Nature*, *579*(7798), 185-186.

Lee, C. M., Cradick, T. J., & Bao, G. (2016). The Neisseria meningitidis CRISPR-Cas9 system enables specific genome editing in mammalian cells. *Molecular Therapy*, *24*(3), 645-654.

Li, L., Hu, S., & Chen, X. (2018). Non-viral delivery systems for CRISPR/Cas9-based genome editing: Challenges and opportunities. *Biomaterials*, *171*, 207-218.

Li, S. Y., Cheng, Q. X., Wang, J. M., Li, X. Y., Zhang, Z. L., Gao, S., ... & Wang, J. (2018). CRISPR-Cas12a-assisted nucleic acid detection. *Cell discovery*, *4*(1), 1-4.

Li, X., Xu, S., Fuhrmann-Aoyagi, M. B., Yuan, S., Iwama, T., Kobayashi, M., & Miura, K. (2022). CRISPR/Cas9 technique for temperature, drought, and salinity stress responses. *Current Issues in Molecular Biology*, *44*(6), 2664-2682.

Li, X., Zhou, L., Gao, B. Q., Li, G., Wang, X., Wang, Y., ... & Chen, J. (2022). Highly efficient prime editing by introducing same-sense mutations in pegRNA or stabilizing its structure. *Nature Communications*, *13*(1), 1669.

Li, Y., Li, W., & Li, J. (2021). The CRISPR/Cas9 revolution continues: from base editing to prime editing in plant science. *Journal of Genetics and Genomics*, *48*(8), 661-670.

Liebert, W. (2021). Crispr-Based Gene Drives for Combatting Malaria: Need for an Early Stage Technology Assessment. In *Area-Wide Integrated Pest Management* (pp. 795-808). CRC Press.

Lim, J. M., & Kim, H. H. (2022). Basic principles and clinical applications of CRISPR-based genome editing. *Yonsei Medical Journal*, *63*(2), 105.

Lin, Q., Zong, Y., Xue, C., Wang, S., Jin, S., Zhu, Z., ... & Gao, C. (2020). Prime genome editing in rice and wheat. *Nature biotechnology*, *38*(5), 582-585.

Lin, S., Ewen-Campen, B., Ni, X., Housden, B. E., & Perrimon, N. (2015). In vivo transcriptional activation using CRISPR/Cas9 in Drosophila. *Genetics*, *201*(2), 433-442.

Lino, C. A., Harper, J. C., Carney, J. P., & Timlin, J. A. (2018). Delivering CRISPR: a review of the challenges and approaches. *Drug delivery*, *25*(1), 1234-1257.

Liu, M. S., Gong, S., Yu, H. H., Jung, K., Johnson, K. A., & Taylor, D. W. (2020). Engineered CRISPR/Cas9 enzymes improve discrimination by slowing DNA cleavage to allow release of off-target DNA. *Nature Communications*, *11*(1), 3576.

Liu, X., Zhang, Y., Chen, Y., Li, M., Zhou, F., Li, K., ... & Xu, J. (2017). In situ capture of chromatin interactions by biotinylated dCas9. *Cell*, *170*(5), 1028-1043.

Liu, Yao, Xiangyang Li, Siting He, Shuhong Huang, Chao Li, Yulin Chen, Zhen Liu, Xingxu Huang, and Xiaolong Wang. "Efficient generation of mouse models with the prime editing system." *Cell discovery* 6, no. 1 (2020): 27.

Lo, P. A., Castillo, J., Humayun, M. S., & Yue, L. (2021). Ultrasound-directed, microbubble-assisted Crispr/Cas9 delivery in RPE cells. *Investigative Ophthalmology & Visual Science*, *62*(8), 1205-1205.

Lopatina, A., Tal, N., & Sorek, R. (2020). Abortive infection: bacterial suicide as an antiviral immune strategy. *Annual Review of Virology*, *7*, 371-384.

Lu, J., Dalbeth, N., Yin, H., Li, C., Merriman, T. R., & Wei, W. H. (2019). Mouse models for human hyperuricaemia: a critical review. *Nature Reviews Rheumatology*, *15*(7), 413-426.

Ma, Y., Yu, L., Zhang, X., Xin, C., Huang, S., Bai, L., ... & Zhang, L. (2018). Highly efficient and precise base editing by engineered dCas9-guide tRNA adenosine deaminase in rats. *Cell Discovery*, *4*(1), 39.

Makarova, K. S., & Koonin, E. V. (2015). Annotation and classification of CRISPR-Cas systems. *CRISPR*, 47-75.

Makarova, K. S., Haft, D. H., Barrangou, R., Brouns, S. J., Charpentier, E., Horvath, P., ... & Koonin, E. V. (2011). Evolution and classification of the CRISPR–Cas systems. *Nature Reviews Microbiology*, *9*(6), 467-477.

Marraffini, L. A. (2015). CRISPR-Cas immunity in prokaryotes. *Nature*, *526*(7571), 55-61.

Marshall, J. M., & Akbari, O. S. (2018). Can CRISPR-based gene drive be confined in the wild? A question for molecular and population biology. *ACS chemical biology*, *13*(2), 424-430.

Marzec, M., Brąszewska-Zalewska, A., & Hensel, G. (2020). Prime editing: a new way for genome editing. *Trends in Cell Biology*, *30*(4), 257-259.

Matharu, N., Rattanasopha, S., Tamura, S., Maliskova, L., Wang, Y., Bernard, A., ... & Ahituv, N. (2019). CRISPR-mediated activation of a promoter or enhancer rescues obesity caused by haploinsufficiency. *Science*, *363*(6424), eaau0629.

McDougald, D. S., Duong, T. T., Palozola, K. C., Marsh, A., Papp, T. E., Mills, J. A., ... & Bennett, J. (2019). CRISPR activation enhances in vitro potency of AAV vectors driven by tissue-specific promoters. *Molecular Therapy-Methods & Clinical Development*, *13*, 380-389.

Mehra, V. K., & Kumar, S. (2022). The application of CRISPR/Cas9 technology for farm animals: a review. *Agricultural Reviews*, *43*(1), 54-61.

Miller, J. C., Holmes, M. C., Wang, J., Guschin, D. Y., Lee, Y. L., Rupniewski, I., ... & Rebar, E. J. (2007). An improved zinc-finger nuclease architecture for highly specific genome editing. *Nature biotechnology*, *25*(7), 778-785.

Miller, K., Eggenberger, A. L., Lee, K., Liu, F., Kang, M., Drent, M., ... & Jiang, S. (2021). An improved biolistic delivery and analysis method for evaluation of DNA and CRISPR-Cas delivery efficacy in plant tissue. *Scientific Reports*, *11*(1), 7695.

Milone, M. C., & O'Doherty, U. (2018). Clinical use of lentiviral vectors. *Leukemia*, *32*(7), 1529-1541.

Molla, K. A., Sretenovic, S., Bansal, K. C., & Qi, Y. (2021). Precise plant genome editing using base editors and prime editors. *Nature Plants*, *7*(9), 1166-1187.

Mollanoori, H., Shahraki, H., Rahmati, Y., & Teimourian, S. (2018). CRISPR/Cas9 and CAR-T cell, collaboration of two revolutionary technologies in cancer immunotherapy, an instruction for successful cancer treatment. *Human immunology*, *79*(12), 876-882.

Monsur, M. B., Shao, G., Lv, Y., Ahmad, S., Wei, X., Hu, P., & Tang, S. (2020). Base editing: The ever expanding clustered regularly interspaced short palindromic repeats (CRISPR) tool kit for precise genome editing in plants. *Genes*, *11*(4), 466.

Montefiori, L. E., & Nobrega, M. A. (2019). Gene therapy for pathologic gene expression. *Science*, *363*(6424), 231-232.

Moreno, A. M., Fu, X., Zhu, J., Katrekar, D., Shih, Y. R. V., Marlett, J., ... & Mali, P. (2018). In situ gene therapy via AAV-CRISPR-Cas9-mediated targeted gene regulation. *Molecular Therapy*, *26*(7), 1818-1827.

Morrical, S. W. (2015). DNA-pairing and annealing processes in homologous recombination and homology-directed repair. *Cold Spring Harbor perspectives in biology*, *7*(2), a016444.

Mussolino, C., & Cathomen, T. (2012). TALE nucleases: tailored genome engineering made easy. *Current opinion in biotechnology*, *23*(5), 644-650.

Mussolino, C., Alzubi, J., Fine, E. J., Morbitzer, R., Cradick, T. J., Lahaye, T., ... & Cathomen, T. (2014). TALENs facilitate targeted genome editing in human cells with high specificity and low cytotoxicity. *Nucleic acids research*, *42*(10), 6762-6773.

Mustafa, M. I., & Makhawi, A. M. (2020). SHERLOCK and DETECTR: CRISPR-Cas Systems as Potential Rapid Diagnostic Tools for Emerging Infectious Diseases. *Journal of Clinical Microbiology*.

Nakagawa, R., Ishiguro, S., Okazaki, S., Mori, H., Tanaka, M., Aburatani, H., ... & Nureki, O. (2022). Engineered Campylobacter jejuni Cas9 variant with enhanced activity and broader targeting range. *Communications Biology*, *5*(1), 211.

Nelson, J. W., Randolph, P. B., Shen, S. P., Everette, K. A., Chen, P. J., Anzalone, A. V., ... & Liu, D. R. (2022). Engineered pegRNAs improve prime editing efficiency. *Nature biotechnology*, *40*(3), 402-410.

Nihongaki, Y., Furuhata, Y., Otabe, T., Hasegawa, S., Yoshimoto, K., & Sato, M. (2017). CRISPR–Cas9-based photoactivatable transcription systems to induce neuronal differentiation. *Nature methods*, *14*(10), 963-966.

Niidome, T., & Huang, L. (2002). Gene therapy progress and prospects: nonviral vectors. *Gene therapy*, *9*(24), 1647-1652.

Niu, D., Wei, H. J., Lin, L., George, H., Wang, T., Lee, I. H., ... & Yang, L. (2017). Inactivation of porcine endogenous retrovirus in pigs using CRISPR-Cas9. *Science*, *357*(6357), 1303-1307.

Noble, C., Adlam, B., Church, G. M., Esvelt, K. M., & Nowak, M. A. (2018). Current CRISPR gene drive systems are likely to be highly invasive in wild populations. *Elife*, *7*, e33423.

Noble, C., Olejarz, J., Esvelt, K. M., Church, G. M., & Nowak, M. A. (2017). Evolutionary dynamics of CRISPR gene drives. *Science advances*, *3*(4), e1601964.

Oladosu, Y., Rafii, M. Y., Abdullah, N., Hussin, G., Ramli, A., Rahim, H. A., ... & Usman, M. (2016). Principle and application of plant mutagenesis in crop improvement: a review. *Biotechnology & Biotechnological Equipment*, *30*(1), 1-16.

Omachi, K., & Miner, J. H. (2021). Comparative analysis and rational design of dCas9-VP64 variants for CRISPR activation. *bioRxiv*, 2021-08.

Ortinski, P. I., O'Donovan, B., Dong, X., & Kantor, B. (2017). Integrase-deficient lentiviral vector as an all-in-one platform for highly efficient CRISPR/Cas9-mediated gene editing. *Molecular Therapy-Methods & Clinical Development*, *5*, 153-164.

Ou, X., Ma, Q., Yin, W., Ma, X., & He, Z. (2021). CRISPR/Cas9 gene-editing in cancer immunotherapy: promoting the present revolution in cancer therapy and exploring more. *Frontiers in Cell and Developmental Biology*, 1179.

Palermo, G., Ricci, C. G., Fernando, A., Basak, R., Jinek, M., Rivalta, I., ... & McCammon, J. A. (2017). Protospacer adjacent motif-induced allostery activates CRISPR-Cas9. *Journal of the American Chemical Society*, *139*(45), 16028-16031

Panda, S. K., Wefers, B., Ortiz, O., Floss, T., Schmid, B., Haass, C., ... & Kühn, R. (2013). Highly efficient targeted mutagenesis in mice using TALENs. *Genetics*, *195*(3), 703-713.

Park, J. E., & Silva, A. C. (2019). Generation of genetically engineered non-human primate models of brain function and neurological disorders. *American journal of primatology*, *81*(2), e22931.

Park, S. H., & Bao, G. (2021). CRISPR/Cas9 gene editing for curing sickle cell disease. *Transfusion and Apheresis Science*, *60*(1), 103060.

Parker, A. L., Newman, C., Briggs, S., Seymour, L., & Sheridan, P. J. (2003). Nonviral gene delivery: techniques and implications for molecular medicine. *Expert reviews in molecular medicine*, *5*(22), 1-15.

Pattanayak, V., Guilinger, J. P., & Liu, D. R. (2014). Determining the specificities of TALENs, Cas9, and other genome-editing enzymes. *Methods in enzymology*, *546*, 47-78.

Perleberg, C., Kind, A., & Schnieke, A. (2018). Genetically engineered pigs as models for human disease. *Disease models & mechanisms*, *11*(1), dmm030783.

Pinilla-Redondo, R., Mayo-Muñoz, D., Russel, J., Garrett, R. A., Randau, L., Sørensen, S. J., & Shah, S. A. (2020). Type IV CRISPR–Cas systems are highly diverse and involved in competition between plasmids. *Nucleic acids research*, *48*(4), 2000-2012.

Pleška, M., Qian, L., Okura, R., Bergmiller, T., Wakamoto, Y., Kussell, E., & Guet, C. C. (2016). Bacterial autoimmunity due to a restriction-modification system. *Current Biology*, *26*(3), 404-409.

Porteus, M. H., & Carroll, D. (2005). Gene targeting using zinc finger nucleases. *Nature biotechnology*, *23*(8), 967-973.

Prins, M., Laimer, M., Noris, E., Schubert, J., Wassenegger, M., & Tepfer, M. (2008). Strategies for antiviral resistance in transgenic plants. *Molecular plant pathology*, *9*(1), 73-83.

Pyenson, N. C., & Marraffini, L. A. (2017). Type III CRISPR-Cas systems: when DNA cleavage just isn't enough. *Current opinion in microbiology*, *37*, 150-154.

Qi, L. S., Larson, M. H., Gilbert, L. A., Doudna, J. A., Weissman, J. S., Arkin, A. P., & Lim, W. A. (2013). Repurposing CRISPR as an RNA-guided platform for sequence-specific control of gene expression. *Cell*, *152*(5), 1173-1183.

Quazi, S. (2022). Elucidation of CRISPR-Cas9 application in novel cellular immunotherapy. *Molecular Biology Reports*, *49*(7), 7069-7077.

Rabinowitz, R., & Offen, D. (2021). Single-base resolution: increasing the specificity of the CRISPR-Cas system in gene editing. *Molecular Therapy*, *29*(3), 937-948.

Ramachandran, A., Huyke, D. A., Sharma, E., Sahoo, M. K., Huang, C., Banaei, N., ... & Santiago, J. G. (2020). Electric field-driven microfluidics for rapid CRISPR-based diagnostics and its application to detection of SARS-CoV-2. *Proceedings of the National Academy of Sciences*, *117*(47), 29518-29525.

Ramalingam, S., Kandavelou, K., Rajenderan, R., & Chandrasegaran, S. (2011). Creating designed zinc-finger nucleases with minimal cytotoxicity. *Journal of molecular biology*, *405*(3), 630-641.

Ramanan, V., Shlomai, A., Cox, D. B., Schwartz, R. E., Michailidis, E., Bhatta, A., ... & Bhatia, S. N. (2015). CRISPR/Cas9 cleavage of viral DNA efficiently suppresses hepatitis B virus. *Scientific reports*, *5*(1), 10833.

Rani, S. J., & Usha, R. (2013). Transgenic plants: Types, benefits, public concerns and future. *Journal of Pharmacy Research*, *6*(8), 879-883.

Rath, D., Amlinger, L., Rath, A., & Lundgren, M. (2015). The CRISPR-Cas immune system: biology, mechanisms and applications. *Biochimie*, *117*, 119-128.

Rathbone, T., Ates, I., Fernando, L., Addlestone, E., Lee, C. M., Richards, V. P., & Cottle, R. N. (2022). Electroporation-Mediated Delivery of Cas9 Ribonucleoproteins Results in High Levels of Gene Editing in Primary Hepatocytes. *The CRISPR Journal*, *5*(3), 397-409.

Rautela, I., Uniyal, P., Thapliyal, P., Chauhan, N., Sinha, V. B., & Sharma, M. D. (2021). An extensive review to facilitate understanding of CRISPR technology as a gene editing possibility for enhanced therapeutic applications. *Gene*, *785*, 145615.

Razeghian, E., Nasution, M. K., Rahman, H. S., Gardanova, Z. R., Abdelbasset, W. K., Aravindhan, S., ... & Khiavi, F. M. (2021). A deep insight into CRISPR/Cas9 application in CAR-T cell-based tumor immunotherapies. *Stem Cell Research & Therapy*, *12*(1), 1-17.

Rees, H. A., & Liu, D. R. (2018). Base editing: precision chemistry on the genome and transcriptome of living cells. *Nature reviews genetics*, *19*(12), 770-788.

Riedmayr, L. M., Hinrichsmeyer, K. S., Karguth, N., Böhm, S., Splith, V., Michalakis, S., & Becirovic, E. (2022). dCas9-VPR-mediated transcriptional activation of functionally equivalent genes for gene therapy. *Nature Protocols*, *17*(3), 781-818.

Ross, M. J., & Coates, P. T. (2018). Using CRISPR to inactivate endogenous retroviruses in pigs: an important step toward safe xenotransplantation?. *Kidney international*, *93*(1), 4-6.

Ryczek, N., Hryhorowicz, M., Zeyland, J., Lipiński, D., & Słomski, R. (2021). CRISPR/Cas technology in pig-to-human xenotransplantation research. *International Journal of Molecular Sciences*, *22*(6), 3196.

Ryø, L. B., Thomsen, E. A., & Mikkelsen, J. G. (2019). Production and validation of lentiviral vectors for CRISPR/Cas9 delivery. *CRISPR Gene Editing: Methods and Protocols*, 93-109.

Saayman, S., Ali, S. A., Morris, K. V., & Weinberg, M. S. (2015). The therapeutic application of CRISPR/Cas9 technologies for HIV. *Expert opinion on biological therapy*, *15*(6), 819-830.

Saayman, S., Ali, S. A., Morris, K. V., & Weinberg, M. S. (2015). The therapeutic application of CRISPR/Cas9 technologies for HIV. *Expert opinion on biological therapy*, *15*(6), 819-830.

Saifaldeen, M., Al-Ansari, D. E., Ramotar, D., & Aouida, M. (2020). CRISPR FokI dead Cas9 system: principles and applications in genome engineering. *Cells*, *9*(11), 2518.

Sakuma, T., Barry, M. A., & Ikeda, Y. (2012). Lentiviral vectors: basic to translational. *Biochemical Journal*, *443*(3), 603-618.

Samai, P., Pyenson, N., Jiang, W., Goldberg, G. W., Hatoum-Aslan, A., & Marraffini, L. A. (2015). Co-transcriptional DNA and RNA cleavage during type III CRISPR-Cas immunity. Cell, 161(5), 1164-1174.

Sandhya, D., Jogam, P., Allini, V. R., Abbagani, S., & Alok, A. (2020). The present and potential future methods for delivering CRISPR/Cas9 components in plants. *Journal of Genetic Engineering and Biotechnology*, *18*, 1-11.

Santosh Kumar, V. V., Verma, R. K., Yadav, S. K., Yadav, P., Watts, A., Rao, M. V., & Chinnusamy, V. (2020). CRISPR-Cas9 mediated genome editing of drought and salt tolerance (OsDST) gene in indica mega rice cultivar MTU1010. *Physiology and Molecular Biology of Plants*, *26*, 1099-1110.

Saraswat, P., & Ranjan, R. (2021). CRISPR-Cas system: a precise tool for plant genome editing. *The Nucleus*, 1-18.

Schleidgen, S., Dederer, H. G., Sgodda, S., Cravcisin, S., Lüneburg, L., Cantz, T., & Heinemann, T. (2020). Human germline editing in the era of CRISPR-Cas: risk and uncertainty, inter-generational responsibility, therapeutic legitimacy. *BMC medical ethics*, *21*, 1-12.

Schmidtmann, E., Anton, T., Rombaut, P., Herzog, F., & Leonhardt, H. (2016). Determination of local chromatin composition by CasID. *Nucleus*, *7*(5), 476-484.
Schuler, T. H., Poppy, G. M., Kerry, B. R., & Denholm, I. (1998). Insect-resistant transgenic plants. *Trends in biotechnology*, *16*(4), 168-175.

Seeger, C., & Sohn, J. A. (2014). Targeting hepatitis B virus with CRISPR/Cas9. *Molecular Therapy-Nucleic Acids*, *3*, e216.

Senís, E., Fatouros, C., Große, S., Wiedtke, E., Niopek, D., Mueller, A. K., ... & Grimm, D. (2014). CRISPR/Cas9-mediated genome engineering: an adeno-associated viral (AAV) vector toolbox. *Biotechnology journal*, *9*(11), 1402-1412.

Sharma, G., Sharma, A. R., Bhattacharya, M., Lee, S. S., & Chakraborty, C. (2021). CRISPR-Cas9: a preclinical and clinical perspective for the treatment of human diseases. *Molecular Therapy*, *29*(2), 571-586.

Shcherban, A. B. (2022). Plant genome modification: from induced mutagenesis to genome editing. *Vavilov Journal of Genetics and Breeding*, *26*(7), 684-696.

Shen, B., Zhang, W., Zhang, J., Zhou, J., Wang, J., Chen, L., ... & Skarnes, W. C. (2014). Efficient genome modification by CRISPR-Cas9 nickase with minimal off-target effects. *Nature methods*, *11*(4), 399-402.

Shinwari, Z. K., Tanveer, F., & Khalil, A. T. (2018). Ethical issues regarding CRISPR mediated genome editing. *Current Issues in Molecular Biology*, *26*(1), 103-110.

Shrock, E., & Güell, M. (2017). CRISPR in animals and animal models. *Progress in molecular biology and translational science*, *152*, 95-114.

Siksnys, V., & Gasiunas, G. (2016). Rewiring Cas9 to target new PAM sequences. *Molecular cell*, *61*(6), 793-794.

Simoni, A., Hammond, A. M., Beaghton, A. K., Galizi, R., Taxiarchi, C., Kyrou, K., ... & Crisanti, A. (2020). A male-biased sex-distorter gene drive for the human malaria vector Anopheles gambiae. *Nature biotechnology*, *38*(9), 1054-1060.

Singh, P., & Ali, S. A. (2021). Impact of CRISPR-Cas9-based genome engineering in farm animals. *Veterinary Sciences*, *8*(7), 122.

Song, F., & Stieger, K. (2017). Optimizing the DNA donor template for homology-directed repair of double-strand breaks. *Molecular Therapy-Nucleic Acids*, *7*, 53-60.

Sousa, D. A., Gaspar, R., Ferreira, C. J., Baltazar, F., Rodrigues, L. R., & Silva, B. F. (2022). In Vitro CRISPR/Cas9 Transfection and Gene-Editing

Mediated by Multivalent Cationic Liposome–DNA Complexes. *Pharmaceutics*, *14*(5), 1087.

St George, J. A. (2003). Gene therapy progress and prospects: adenoviral vectors. *Gene therapy*, *10*(14), 1135-1141.

Stoll, G. A., Pandiloski, N., Douse, C. H., & Modis, Y. (2022). Structure and functional mapping of the KRAB-KAP1 repressor complex. *The EMBO Journal*, *41*(24), e111179.

Sun, N., & Zhao, H. (2013). Transcription activator-like effector nucleases (TALENs): a highly efficient and versatile tool for genome editing. *Biotechnology and bioengineering*, *110*(7), 1811-1821.

Sung, Y. H., Baek, I. J., Kim, D. H., Jeon, J., Lee, J., Lee, K., ... & Lee, H. W. (2013). Knockout mice created by TALEN-mediated gene targeting. *Nature biotechnology*, *31*(1), 23-24.

Suvvari, T. K., Nawaz, M. D., & Mantha, M. K. (2020). FNCas9 editor-linked uniform detection assay: An innovative COVID-19 sleuth. *Biomedical and Biotechnology Research Journal (BBRJ)*, *4*(4), 302.

Svitashev, S., Schwartz, C., Lenderts, B., Young, J. K., & Mark Cigan, A. (2016). Genome editing in maize directed by CRISPR–Cas9 ribonucleoprotein complexes. *Nature communications*, *7*(1), 13274.

Tabebordbar, M., Lagerborg, K. A., Stanton, A., King, E. M., Ye, S., Tellez, L., ... & Sabeti, P. C. (2021). Directed evolution of a family of AAV capsid variants enabling potent muscle-directed gene delivery across species. *Cell*, *184*(19), 4919-4938.

Tadić, V., Josipović, G., Zoldoš, V., & Vojta, A. (2019). CRISPR/Cas9-based epigenome editing: An overview of dCas9-based tools with special emphasis on off-target activity. *Methods*, *164*, 109-119.

Tamulaitis, G., Venclovas, Č., & Siksnys, V. (2017). Type III CRISPR-Cas immunity: major differences brushed aside. Trends in microbiology, 25(1), 49-61.

Tan, Y., Chu, A. H., Bao, S., Hoang, D. A., Kebede, F. T., Xiong, W., ... & Zheng, Z. (2019). Rationally engineered Staphylococcus aureus Cas9 nucleases with high genome-wide specificity. *Proceedings of the National Academy of Sciences*, *116*(42), 20969-20976.

Tanaka, H., Stone, H. A., & Nelson, D. R. (2017). Spatial gene drives and pushed genetic waves. *Proceedings of the National Academy of Sciences*, *114*(32), 8452-8457.

Tanenbaum, M. E., Gilbert, L. A., Qi, L. S., Weissman, J. S., & Vale, R. D. (2014). A protein-tagging system for signal amplification in gene expression and fluorescence imaging. *Cell*, *159*(3), 635-646.

Tanenbaum, M. E., Gilbert, L. A., Qi, L. S., Weissman, J. S., & Vale, R. D. (2014). A protein-tagging system for signal amplification in gene expression and fluorescence imaging. *Cell*, *159*(3), 635-646.

Tang, L., Zeng, Y., Du, H., Gong, M., Peng, J., Zhang, B., ... & Liu, J. (2017). CRISPR/Cas9-mediated gene editing in human zygotes using Cas9 protein. *Molecular genetics and genomics*, *292*(3), 525-533.

Tarafdar, A. V. I. J. I. T., Kamle, M. A. D. H. U., Prakash, A. R. U. L., & Padaria, J. C. (2014). Transgenic plants: issues and future prospects. *Biotechnology*, *2*, 1-47.

Tesson, L., Usal, C., Ménoret, S., Leung, E., Niles, B. J., Remy, S., ... & Cost, G. J. (2011). Knockout rats generated by embryo microinjection of TALENs. *Nature biotechnology*, *29*(8), 695-696.

Thapa, B., & Narain, R. (2016). Mechanism, current challenges and new approaches for non viral gene delivery. In *Polymers and nanomaterials for gene therapy* (pp. 1-27). Woodhead Publishing.

Trevino, A. E., & Zhang, F. (2014). Genome editing using Cas9 nickases. In *Methods in enzymology* (Vol. 546, pp. 161-174). Academic Press.

Tripathi, L., Ntui, V. O., & Tripathi, J. N. (2020). CRISPR/Cas9-based genome editing of banana for disease resistance. Current Opinion in Plant Biology, 56, 118-126.

Tu, Z., Zhao, H., Li, B., Yan, S., Wang, L., Tang, Y., ... & Li, X. J. (2019). CRISPR/Cas9-mediated disruption of SHANK3 in monkey leads to drug-treatable autism-like symptoms. *Human molecular genetics*, *28*(4), 561-571.

Tyagi, S., Kumar, R., Kumar, V., Won, S. Y., & Shukla, P. (2021). Engineering disease resistant plants through CRISPR-Cas9 technology. *GM Crops & Food*, *12*(1), 125-144.

Tycko, J., Myer, V. E., & Hsu, P. D. (2016). Methods for optimizing CRISPR-Cas9 genome editing specificity. *Molecular cell*, *63*(3), 355-370.

Ubalua, A. O. (2009). Transgenic plants: Successes and controversies. *Biotechnol Mol Biol Rev*, *4*(6), 118-127.

Ugur, E., Bartoschek, M. D., & Leonhardt, H. (2020). Locus-specific chromatin proteome revealed by mass spectrometry-based CasID. *The Nucleus*, 109-121.

Ureña-Bailén, G., Lamsfus-Calle, A., Daniel-Moreno, A., Raju, J., Schlegel, P., Seitz, C., ... & Mezger, M. (2020). CRISPR/Cas9 technology: towards a new generation of improved CAR-T cells for anticancer therapies. Briefings in functional genomics, 19(3), 191-200.

Urnov, F. D., Miller, J. C., Lee, Y. L., Beausejour, C. M., Rock, J. M., Augustus, S., ... & Holmes, M. C. (2005). Highly efficient endogenous human gene correction using designed zinc-finger nucleases. *Nature*, *435*(7042), 646-651.

Urnov, F. D., Rebar, E. J., Holmes, M. C., Zhang, H. S., & Gregory, P. D. (2010). Genome editing with engineered zinc finger nucleases. *Nature Reviews Genetics*, *11*(9), 636-646.

Vad-Nielsen, J., Nielsen, A. L., & Luo, Y. (2018). Simple method for assembly of CRISPR synergistic activation mediator gRNA expression array. *Journal of biotechnology*, *274*, 54-57.

van Diemen, F. R., & Lebbink, R. J. (2017). CRISPR/Cas9, a powerful tool to target human herpesviruses. *Cellular microbiology*, *19*(2), e12694.

Vijayraghavan, S., & Kantor, B. (2017). A protocol for the production of integrase-deficient lentiviral vectors for CRISPR/Cas9-mediated gene knockout in dividing cells. *Journal of Visualized Experiments: JoVE*, (130).

Volpers, C., & Kochanek, S. (2004). Adenoviral vectors for gene transfer and therapy. *The Journal of Gene Medicine: A cross-disciplinary journal for research on the science of gene transfer and its clinical applications*, *6*(S1), S164-S171.

Wada, N., Ueta, R., Osakabe, Y., & Osakabe, K. (2020). Precision genome editing in plants: state-of-the-art in CRISPR/Cas9-based genome engineering. *BMC Plant Biology*, *20*, 1-12.

Wang, D., Tai, P. W., & Gao, G. (2019). Adeno-associated virus vector as a platform for gene therapy delivery. *Nature reviews Drug discovery*, *18*(5), 358-378.

Wang, G., Chow, R. D., Bai, Z., Zhu, L., Errami, Y., Dai, X., ... & Chen, S. (2019). Multiplexed activation of endogenous genes by CRISPRa elicits potent antitumor immunity. *Nature immunology*, *20*(11), 1494-1505.

Wang, R., Ruan, M., Zhang, R., Chen, L., Li, X., Fang, B., ... & Dai, Y. (2019). Antigenicity of tissues and organs from GGTA1/CMAH/β4GalNT2 triple gene knockout pigs. *Journal of Biomedical Research*, *33*(4), 235.

Whyte, J. J., Zhao, J., Wells, K. D., Samuel, M. S., Whitworth, K. M., Walters, E. M., ... & Prather, R. S. (2011). Gene targeting with zinc finger nucleases to produce cloned eGFP knockout pigs. *Molecular reproduction and development*, *78*(1), 2.

Wise, I. J., & Borry, P. (2022). An ethical overview of the CRISPR-based elimination of Anopheles gambiae to combat malaria. *Journal of Bioethical Inquiry*, *19*(3), 371-380.

Woo, J. W., Kim, J., Kwon, S. I., Corvalán, C., Cho, S. W., Kim, H., ... & Kim, J. S. (2015). DNA-free genome editing in plants with preassembled CRISPR-Cas9 ribonucleoproteins. *Nature biotechnology*, *33*(11), 1162-1164.

Wood, A. J., Lo, T. W., Zeitler, B., Pickle, C. S., Ralston, E. J., Lee, A. H., ... & Meyer, B. J. (2011). Targeted genome editing across species using ZFNs and TALENs. *Science*, *333*(6040), 307-307.

Wu, H. Y., & Cao, C. Y. (2019). The application of CRISPR-Cas9 genome editing tool in cancer immunotherapy. *Briefings in functional genomics*, *18*(2), 129-132.

Wu, X., Mao, S., Ying, Y., Krueger, C. J., & Chen, A. K. (2019). Progress and challenges for live-cell imaging of genomic loci using CRISPR-based platforms. *Genomics, proteomics & bioinformatics*, *17*(2), 119-128.

Xia, A. L., He, Q. F., Wang, J. C., Zhu, J., Sha, Y. Q., Sun, B., & Lu, X. J. (2019). Applications and advances of CRISPR-Cas9 in cancer immunotherapy. *Journal of medical genetics*, *56*(1), 4-9.

Xiao, Q., Guo, D., & Chen, S. (2019). Application of CRISPR/Cas9-based gene editing in HIV-1/AIDS therapy. *Frontiers in cellular and infection microbiology*, *9*, 69.

Xiao, Y., Ng, S., Nam, K. H., & Ke, A. (2017). How type II CRISPR–Cas establish immunity through Cas1–Cas2-mediated spacer integration. *Nature*, *550*(7674), 137-141.

Xiong, K., Zhou, Y., Hyttel, P., Bolund, L., Freude, K. K., & Luo, Y. (2016). Generation of induced pluripotent stem cells (iPSCs) stably expressing CRISPR-based synergistic activation mediator (SAM). *Stem Cell Research*, *17*(3), 665-669.

Xu, C. L., Ruan, M. Z., Mahajan, V. B., & Tsang, S. H. (2019). Viral delivery systems for CRISPR. *Viruses*, *11*(1), 28.

Yang, H., Ren, S., Yu, S., Pan, H., Li, T., Ge, S., ... & Xia, N. (2020). Methods favoring homology-directed repair choice in response to CRISPR/Cas9 induced-double strand breaks. *International journal of molecular sciences*, *21*(18), 6461.

Yao, J., Huang, J., & Zhao, J. (2016). Genome editing revolutionize the creation of genetically modified pigs for modeling human diseases. *Human genetics*, *135*, 1093-1105.

Ye, H., Rong, Z., & Lin, Y. (2017). Live cell imaging of genomic loci using dCas9-SunTag system and a bright fluorescent protein. *Protein & cell*, *8*(11), 853-855.

Yeh, P., & Perricaudet, M. (1997). Advances in adenoviral vectors: from genetic engineering to their biology. *The FASEB journal*, *11*(8), 615-623.

Young, J. L., & Dean, D. A. (2015). Electroporation-mediated gene delivery. *Advances in genetics*, *89*, 49-88.

Yu, Q. H., Wang, B., Li, N., Tang, Y., Yang, S., Yang, T., ... & Asmutola, P. (2017). CRISPR/Cas9-induced targeted mutagenesis and gene replacement to generate long-shelf life tomato lines. *Scientific reports*, *7*(1), 11874.

Yunes, M. C., Osório-Santos, Z., von Keyserlingk, M. A., & Hötzel, M. J. (2021). Gene editing for improved animal welfare and production traits in cattle: Will this technology be embraced or rejected by the public?. *Sustainability*, *13*(9), 4966.

Zarei, A., Razban, V., Hosseini, S. E., & Tabei, S. M. B. (2019). Creating cell and animal models of human disease by genome editing using CRISPR/Cas9. *The journal of gene medicine*, *21*(4), e3082.

Zhang, A., Liu, Y., Wang, F., Li, T., Chen, Z., Kong, D., ... & Luo, L. (2019). Enhanced rice salinity tolerance via CRISPR/Cas9-targeted mutagenesis of the OsRR22 gene. *Molecular breeding*, *39*, 1-10.

Zhang, D., Das, D. B., & Rielly, C. D. (2014). Potential of microneedle-assisted micro-particle delivery by gene guns: a review. *Drug delivery*, *21*(8), 571-587.

Zhang, F., Abudayyeh, O. O., & Gootenberg, J. S. (2020). A protocol for detection of COVID-19 using CRISPR diagnostics. *A protocol for detection of COVID-19 using CRISPR diagnostics*, *8*.

Zhang, J., Lei, Y., Dhaliwal, A., Ng, Q. K., Du, J., Yan, M., ... & Segura, T. (2011). Protein− polymer nanoparticles for nonviral gene delivery. *Biomacromolecules*, *12*(4), 1006-1014.

Zhang, R., Wang, Y., Chen, L., Wang, R., Li, C., Li, X., ... & Dai, Y. (2018). Reducing immunoreactivity of porcine bioprosthetic heart valves by genetically-deleting three major glycan antigens, GGTA1/β4GalNT2/CMAH. *Acta biomaterialia*, *72*, 196-205.

Zhang, X. H., Tee, L. Y., Wang, X. G., Huang, Q. S., & Yang, S. H. (2015). Off-target effects in CRISPR/Cas9-mediated genome engineering. *Molecular Therapy-Nucleic Acids*, *4*, e264.

Zhang, Y., Yin, C., Zhang, T., Li, F., Yang, W., Kaminski, R., ... & Hu, W. (2015). CRISPR/gRNA-directed synergistic activation mediator (SAM) induces specific, persistent and robust reactivation of the HIV-1 latent reservoirs. *Scientific reports*, *5*(1), 16277.

Zhao, H., Jiang, Y. H., & Zhang, Y. Q. (2018). Modeling autism in non-human primates: Opportunities and challenges. *Autism Research*, *11*(5), 686-694.

Zhao, J., Lai, L., Ji, W., & Zhou, Q. (2019). Genome editing in large animals: current status and future prospects. *National Science Review*, *6*(3), 402-420.
Zhen, S., & Li, X. (2020). Liposomal delivery of CRISPR/Cas9. *Cancer gene therapy*, *27*(7-8), 515-527.

Zu, Y., Tong, X., Wang, Z., Liu, D., Pan, R., Li, Z., ... & Lin, S. (2013). TALEN-mediated precise genome modification by homologous recombination in zebrafish. *Nature methods*, *10*(4), 329-331.

Made in the USA
Monee, IL
19 October 2026

66159900R00216